Handbook of INDOOR AIR QUALITY MANAGEMENT

Donald W. Moffat

PRENTICE HALL
Englewood Cliffs, NJ 07632

Library of Congress Cataloging-in-Publication Data
Moffat, Donald W., 1925–
Handbook of indoor air quality management / Donald W. Moffat.
p. cm.
Includes index.
ISBN 0-13-235300-8
1. Indoor air pollution. 2. Air quality management.
3. Industrial management—Environmental aspects. I. Title.
TD883.17.M64 1996
628.5′3—dc20 96-46133
CIP

Printed in the United States of America

10 9 8 7 6 5 4 3 2 1

ISBN 0-13-235300-8

PRENTICE HALL
Career & Personal Development
Englewood Cliffs, NJ 07632
A Simon & Schuster Company
On the World Wide Web at http://www.phdirect.com

Prentice-Hall International (UK) Limited, *London*
Prentice-Hall of Australia Pty. Limited, *Sydney*
Prentice-Hall Canada Inc., *Toronto*
Prentice-Hall Hispanoamericana, S.A., *Mexico*
Prentice-Hall of India Private Limited, *New Delhi*
Prentice-Hall of Japan, Inc., *Tokyo*
Simon & Schuster Asia Pte. Ltd., *Singapore*
Editora Prentice-Hall do Brasil, Ltda., *Rio de Janeiro*

Dedicated to Darcy

About the Author

Donald W. Moffat has 26 years of experience, at various levels from construction mechanic to project engineer, in-plant teacher, and program manager. He has presented nine papers to national and international professional conferences and has had over 55 articles published. He has also taught industrial management and business at Miramar College and Mesa College in San Diego, upper division and graduate courses at La Verne College, and mathematics at Oregon State University. Previously, he chaired The Bishop's Schools' Mathematics Department.

His other books include *Plant Monitoring and Inspecting Handbook, Plant Engineer's Handbook of Formulas, Charts, and Tables, 3rd edition; Plant Manager's Handbook of Model Reports & Formats; Handbook of Manufacturing and Production Management Formulas, Charts & Tables; Plant Manager's Daily Planner; Elementary Statistics for IBM PCs; Concise Desk Book of Business Finance, 2nd edition; Plant Engineer's Portable Problem Solver, and Plant Management Guide to Accounting and Finance* (coauthored with Greg A. Poage), all published by Prentice Hall; *Charts & Nomographs for Electronics Technicians and Engineers (Gernsback)*; and *Economics Dictionary, 2nd edition* (Elsevier International).

His formal education includes a B.S. degree in physics from Hofstra University, graduate studies at Syracuse University and UCLA, an M.S. degree in Management Science from United States International University, and a teaching credential from University of San Diego. He is a member of Sigma Pi Sigma, National Physics Honor Society.

Preface

You can find many books and articles that voice concern about indoor air quality—starting with the energy shortages in the 1970s—but a more accurate view is that what we saw in the 1970s was the knee of the curve. Concern has been with us, and increasing slowly, for a long time; since the energy shortages, many factors have had a cumulative effect in bringing about a rapid increase in concern.

Awareness of bad air accumulation was evident from the days when miners used to take caged canaries into the mines because the birds died at concentrations of methane and other gases that were undetectable but potentially harmful to humans. Before that, American Indians' teepees had center flaps that could be opened to allow smoke to get out. We can guess that it didn't take many episodes of burning eyes, coughing, and irritated throats (familiar complaints today) to convince them that a smoke-filled indoor environment was, at the very least, uncomfortable.

THE KNEE OF THE CURVE

Before the prices of petroleum, gas, electricity, and other sources of energy jumped in the 1970s, it was hardly cost-effective to seal leaks in buildings, insulate heavily, and recirculate conditioned air. Few realized how much indoor air was exchanged with fresh air through building leaks. In addition, the wasteful practice of bringing in outside air, heating it, distributing it to working areas, and then exhausting it was not yet a major item in the profit and loss analysis. Indoor air quality was a minor concern, partly because we knew less about it, but mostly because indoor air was quite clean.

Then the modern spiral began. As energy prices increased, building owners sealed workplaces and recirculated air. At the same time, new chemical compounds were being developed and used to improve efficiency in many areas, especially manufacturing. Illnesses and complaints increased. Investigations revealed that not only was the air in each area not being freshened, but exhaled air, allergens, manufacturing contaminants, and every other type of harmful air within the building was being combined, distributed to all areas, and recirculated. Sick building syndrome, environmental illness, and many other terms came into being. Complaints led to increased investigations, which led to more knowledge about indoor air quality, which led to increased concern.

WHAT IS HAPPENING TODAY

Improvements have been made, but indoor air problems are certainly not a thing of the past. Studies show that most of us spend well over three fourths of our lives indoors—where the air is often unhealthier than outdoor air in smoggy cities. Indoor concentrations of specific pollutants are sometimes at levels that would be illegal in outdoor air.

The main motivation for management to clean the air has been a moral desire to not cause harm to workers. In addition, there is a growing realization that poor indoor air quality affects the bottom line, albeit indirectly, through absenteeism, fewer work years (early deaths), lower productivy, lower morale, and higher insurance rates. Still, some managements with the best of intentions have put off action on satisfactory improvements because they plan to work it into "next year's" budget.

The motivation to give a higher priority to cleaning indoor air came from OSHA and other regulators. Managers now have Permissible Exposure Limits,

Short Term Exposure Limits, Ceiling Limits, and other limits they must apply to a long list of contaminants and potential contaminants. ASHRAE Standard 62-1989, by the American Society of Heating, Refrigerating and Air-Conditioning Engineers, has been adopted by building codes and is presently applied to all types of indoor environments (see Chapters 22 and 23).

INDOOR AIR QUALITY MANAGEMENT PLAN

With literally millions of chemical compounds in the environment, many hundreds of them known to be hazardous, and new compounds being developed every day, keeping the workplace up to standards is a continuing challenge. In addition to the decreased productivity and other costs mentioned, we are seeing a negative effect on the bottom line due to increases in the litigation that grows out of IAQ complaints.

One source of strength in this battle is the IAQ Management Plan (see Chapter 16). Preparing the plan forces management to look at current conditions and assess vulnerable areas; following the plan keeps the plant in a better position to meet requirements.

INDOOR AIR QUALITY EMERGENCY MANAGEMENT PLAN

The goal of the IAQ Management Plan described in Chapter 16 is to keep indoor air quality safe and under control. However, no plan can guarantee complete freedom from the possibility of an IAQ emergency—which might even occur beyond the plant's control. Having both a continuing management plan and an emergency management plan (see Chapter 24) further protects all components of the plant—workers, equipment, finances, and management.

ORGANIZATION OF THE BOOK

Chapters that follow in Part I (Chapters 1–4) emphasize differences in what clean and acceptable air means to different groups, how we classify undesirable elements, sources of impurities, and health effects of impurities. Part II (Chapters 5–9) looks in more detail at regulations and other reasons for maintaining clean air, and

Part III (Chapters 10–14) follows with chapters about monitoring air and individual employee exposure. The last two groups of chapters, Parts IV and V (Chapters 15–23 and 24–28 respectively), describe clean air maintenance procedures, and how to prepare for and handle emergencies. Appendices end the book with case studies, a list of exposure limits, medical reference material, a glossary, and addresses of applicable organizations.

Acknowledgments

Many sources in the industrial and individual sectors, and in different levels of government, helped with information for this book. Footnotes in the chapters identify contributors from the private sector, and the following is a list of government documents (in italics) and agencies from which this book drew material.

Agency for Toxic Substances and Disease Registry, Division of Toxicology

Air Quality Review—Central Office, State of Oregon

All About OSHA

Building Air Quality, prepared by the U. S. Environmental Protection Agency and the U. S. Department of Health and Human Services

Carcinogenicity of Acetaldehyde and Malonaldehyde, and Mutagenicity of Related Low-Molecular-Weight Aldehydes, NIOSH Current Intelligence Bulletin 55

Carcinogenic Effects of Exposure to Diesel Exhaust, by NIOSH

Central Office Air Quality Review, March 8, 1994, State of Oregon, Employment Department

Employee Workplace Rights, OSHA

Environmental Protection Agency

Federal Register

Memo to All Employees, November 21, 1985, State of Oregon, Employment Division

National Institute for Occupational Safety and Health

NIOSH Guide to Industrial Respirator Protection

NIOSH Health Hazard Evaluation Report, HETA 92-0374-2402

NIOSH Pocket Guide to Chemical Hazards

NIOSH Respirator Decision Logic

Occupational Safety and Health Administration

Reducing the Potential Risk of Deveoping Cancer from Exposure to Gallium Arsenide in the Microelectronics Industry, by NIOSH

Standards Cited for All SICs; 250+ Employees, OSHA

Working in Hot Environments, Centers for Disease Control

Workplace Exposure to Asbestos, NIOSH-OSHA Asbestos Work Group

World Health Organization

Contents

PART V HANDLING IAQ EMERGENCIES

PART I

AN OVERVIEW

CHAPTER 1

How Sound Indoor Air Quality Benefits Your Company

Reasons for managing the quality of indoor air have progressed from altruism to concern for health and safety to product quality to government requirements. Bottom-line improvements were seen as management cleaned the indoor air—healthy employees lasted longer on the job and absentee rates declined. Miniaturized products became less tolerant of microscopic particles in the air, and clean air reduced rejection rates.

It is unlikely that any employer needs convincing with regard to lethal and unhealthful contaminants in the workplace environment (who wouldn't have the plant evacuated immediately on learning that it had a cyanide concentration of 50 milligrams per cubic meter [mg/m^3]?). Further along the continuum are situations involving less hazardous contaminants and those for which definite statistics are unavailable because our society does not allow controlled testing with humans. There are contaminants that, although they may not be hazardous or unhealthy, are definitely not healthy or desirable. A common complaint is that the air is "stuffy" or "stale," which suggests a vague condition that might mean the oxygen-to-carbon dioxide ratio is a little lower than ideal—or it might be an early warning that a toxic chemical is leaking into the indoor atmosphere. All these conditions represent deviations from clean air; they cost the organization money, and they will cost more if they are not cleaned up.

However, one complicating factor is that expenses to clean the air are felt in the short term, whereas benefits of a cleanup are generally realized in the long term. A single employer wanting to provide a healthy working environment creates a short-term competitive disadvantage that can extend long enough to make the company fade from the scene, leaving its former employees worse off. That negative incentive emphasized the need for an overall agent to ensure that *all* competitors provide healthy working environments.

THE OCCUPATIONAL SAFETY AND HEALTH ACT OF 1970

Government's function is to step in and do what individuals (and individual organizations) cannot do separately. In 1970 Congress became concerned about the following annual health and safety statistics:

- There were an estimated 300,000 new cases of occupational diseases.
- Job-related accidents accounted for more than 14,000 worker deaths.
- Nearly 2 ½ million workers were disabled.
- Ten times as many worker-days were lost from job-related disabilities as from strikes.

Congress passed the Occupational Safety and Health Act of 1970 (William Steiger Act) ". . . to assure so far as possible every working man and woman in the nation safe and healthful working conditions and to preserve our human resources." The act created the Occupational Safety and Health Administration (OSHA) within the Department of Labor, giving it several responsibilities, including one "to develop mandatory job safety and health standards and enforce them effectively." The key words *mandatory* and *enforce* leave no doubt that employers were given another reason for concern over indoor air quality.

Regulatory Coverage

The act required that OSHA encourage states to establish their own job safety and health plans. Therefore, OSHA now functions directly and through approved state programs that are required to provide standards and enforcement programs, as well as voluntary compliance activities, that are at least as effective as the federal program. About half the states have plans in operation.

In general, between federal OSHA and state OSHAs, the act of 1970 extends to all employers and their employees in the 50 states, the District of Columbia, Puerto Rico, and all other territories under federal government jurisdiction. Coverage is quite extensive; only the following groups are not covered:

- Self-employed persons
- Farms at which only immediate members of the farm employer's family are employed
- Employees in jobs regulated by other federal agencies under other federal statutes

How Strict Is Enforcement?

An employer was recently fined over $45,000 and another was fined almost $49,000 because both allowed employees to work in confined spaces that were not cleared of a noxious airborne contaminant and did not provide adequate ventilation. Another employer received nearly $200,000 in citations for a series of violations that included not having made a formal evaluation of the hazard potential, not testing the indoor atmosphere, and not having a written entry permit policy.

OSHA imposes penalties commensurate with the severity of the violation, including the following:

Other than Serious Violation has a direct relationship to job safety and health but probably would not cause death or serious physical harm. A proposed penalty of up to $7000 for each violation is discretionary and may be adjusted downward by as much as 95 percent, depending on the employer's good faith (demonstrated by efforts to comply with the act), history of previous violations, and size of the business.

Serious Violation occurs when there is substantial probability that death or serious physical harm could result and that the employer knew, or should have known, of the hazard. A mandatory penalty of up to $7000 is proposed for each violation and may be adjusted downward.

Willful Violation is one that an employer intentionally and knowingly commits. Penalties up to $70,000 may be proposed for each willful violation, with a minimum penalty of $5000 for each violation. Although good faith is not a factor in willful violations, these penalties may be adjusted downward for certain-size businesses and if there is no history of previous violations. When there is an employee death, the penalty can be a court-imposed imprisonment or a fine of up to $250,000 for an individual or $500,000 for a corporation.

Repeat Violation is a violation of any standard, regulation, rule, or order where, upon reinspection, a substantially similar violation is found. Each such violation can bring a fine of up to $70,000.

Failure to Correct Prior Violation may bring a penalty of up to $7000 for each day the violation continues beyond the prescribed abatement date.

Falsifying Records, Reports, or Applications can bring a fine of $10,000 or up to six months in jail, or both.

Violations of Posting Requirements can bring a civil penalty of up to $7000.

Assaulting a Compliance Officer or otherwise resisting, opposing, intimidating, or interfering with a compliance officer in the performance of his or her duties is a criminal offense. These actions are subject to a fine of up to $5000 and imprisonment for not more than three years.

Statistics demonstrate that state OSHAs, as required, are at least as strict in their regulations and enforcement as federal OSHA. A report released October 3, 1990, by the Occupational Safety and Health State Plan Association (OSHSPA) compared the actions of 23 state OSHAs (21 states plus Puerto Rico and the Virgin Islands) with those of federal OSHA. It included the following highlights:

- State programs conduct nearly 70 percent of the Occupational Safety and Health Act enforcement inspections in the United States.
- Between October 1, 1988, and September 30, 1989, the 23 state-plan OSHA programs conducted more than 113,500 inspections, issuing 386,723 citations, as compared with 54,557 inspections and 184,620 citations issued by federal OSHA in the other 27 states.
- State programs have smaller jurisdictions and are flexible and responsive to local needs and working environments.

OSHA Inspections

One of OSHA's main enforcement tools is unannounced inspections. A compliance officer arrives without advance notice; in fact, alerting an employer of an inspection can bring a fine of up to $1000 and/or a six-month jail term. This rule applies to federal as well as to state inspectors.

There are special circumstances under which OSHA may give notice before arriving, but such notice is normally less than 24 hours. These circumstances include the following:

- Imminent danger situations that require correction as soon as possible
- Inspections that must take place after regular business hours or that require special preparation
- Cases in which notice is required to ensure that the employer and employee representative or other personnel will be present
- Cases in which an inspection must be delayed for more than five working days when there is good cause
- Situations in which the OSHA area director determines that giving notice would produce a more thorough or effective inspection

CHAPTER 2

Basic Issues Driving IAQ Analysis

The plant's staff will be involved to some degree in any investigation of indoor air quality (IAQ). The involvement can range from the assignment of one or two employees to choose and hire an outside consultant, to the use of the plant engineering department to perform analyses, to the participation of the entire staff in experiments and questionnaires. Because of the wide range of skills and involvement possible, it is important to understand the following assumptions and limitations that should control an investigation:

- Facilities managers are seeking economical and efficient ways to maintain their indoor air quality at or above required levels.
- The preferred, and usually most economical, method is to *prevent* problems rather than to solve them after allowing them to develop.
- Indoor air quality problems can often be resolved using skills that are available in-house. Such skills might cover the spectrum from mounting fans in outside walls to using sophisticated instruments to analyze the indoor air.

- Many IAQ problems can be prevented by educating facility management, staff, and occupants about the factors that create such problems.
- Although many basic issues and activities involved in preventing and resolving IAQ problems are independent of the area's design and use, there are other issues and activities that apply just to certain plants or groups of plants. For example, allergens brought in on workers' clothing could be a problem in any part of a plant, but hazardous conditions related to gallium arsenide are limited to production and laboratory facilities.

ISSUES TO KEEP IN MIND

Solutions to indoor air quality and other problems are usually complicated by interacting effects that can sometimes be far-reaching. A thorough understanding of all the factors that interact to create indoor air quality problems can help prevent undesirable outcomes. The following cautions should be kept in mind:

- Modifications of building functions to remedy air quality complaints may create or exacerbate other problems.
- Guidance in this book is not a substitute for appropriate emergency action in the event of a hazardous situation that may be imminently threatening to life or safety.
- Implementation of recommendations that result from an IAQ evaluation should always be done in accordance with local laws and good practice.
- Changes to the overall design and operation of a plant should be coordinated with a registered professional engineer or other registered or certified professionals.
- If medical records are used in the course of evaluating an IAQ problem, appropriate legal and thoughtful confidentiality must be maintained.

HOW PROBLEMS AND SOLUTIONS CAN VARY WITHIN ONE PLANT

Consider a factory area containing many grinding and sanding stations that put more dust into the air than workers should breathe. One way to provide breathable air would be to design a large-capacity filtering system that would change the air every few minutes, replacing it with acceptable, filtered air. It would be necessary to replace indoor air with clean air faster than the grinding and sanding operations can cause dust buildup. Another way to provide breathable air would be to have each worker wear a dust mask, a respirator, or some other device that would provide acceptable air directly to each worker's lungs.

The concept of clean air in this factory depends on which of these two cleaning methods management chooses. Both methods would have the same specifications for the air that workers breathe, but workers would breathe ambient air with the first method, and with the second method they would breathe air that was either filtered in their mask or supplied by a respirator cylinder. When workers wear dust masks, air that is clean enough for the masks to bring it to acceptable levels for breathing is acceptable ambient air.

The preceding sentence suggests another possibility: Air could be so loaded with particles that dust masks could not remove enough to make it breathable, or the particles could cause dust masks to clog so frequently that they would be costly to maintain. In this situation the plant would probably have to prefilter the air to a concentration specified by the dust mask manufacturer so that the masks could efficiently remove the rest of the particles to make the air breathable.

The following sections discuss some air quality concerns that are typically specific to certain areas of a plant. The actual conditions in each plant depend on many factors such as types of products manufactured, integration of functions within the plant, plant design, age of the plant, and automatic controls.

Factory and Manufacturing Areas

In addition to generating dust, manufacturing today usually involves chemicals that have varying degrees of toxicity. Liquid chemicals can give off vapors; solid chemicals can give off particles and vapors.

Traditional manufacturing by heavy machinery involves constant flows of lubricants, which are responsible for the characteristic factory odor because they heat in doing their job of reducing friction. Sometimes, as with many refineries, much of the process takes place outdoors or in open facilities where byproducts blow away or otherwise dissipate. However, it is common to find a factory area adjacent to the outdoor facility, where weather conditions, system faults, and other unplanned events bring the byproducts indoors.

Factory areas are often associated with loading docks, where the combination of large truck motors and large doors brings exhaust fumes into the indoor atmosphere. Temperature, humidity, and other weather-related items, also factors in indoor air quality, are difficult to control near loading docks.

Hi-tech manufacturing usually involves exotic, stronger, and more caustic chemicals, as well as closer temperature and humidity control. There is always the chance of a spill, and when exotic chemicals are used in manufacturing, the cleanup chemicals themselves might be harmful to workers who are not wearing protective equipment. Special training for those who handle and store exotic chemicals is necessary for the safety of all.

Sometimes the products being manufactured are highly vulnerable to air quality degradation. Even the standards for human protection are often not stringent enough to ensure an acceptable rejection rate for certain delicate and fragile products.

Clean Rooms

Because many hi-tech products are so sensitive to even microscopic air contaminants, they are often produced in clean rooms, where the atmosphere is carefully controlled. Incoming air is filtered to eliminate micron-sized particles, and workers' skin and hair are covered with specially designed lint-free clothing.

Confined Spaces

The subject of confined spaces reappears throughout the book because the material of several chapters relates to it. Confined spaces are especially vulnerable to problems that result from isolation and inadequate ventilation. In addition to providing sufficient breathable, nonexplosive air, it is important to provide communication between workers in the confined space and those who monitor them.

Offices

Once considered clean places that did not have the air quality problems known to exist in factories, offices are now recognized as having several serious air quality problems that cannot be ignored. High-volume paper operations such as copying, printing, cutting, stripping, and separating produce airborne paper dust, which is harmful to many workers. High voltages in video display terminals can create ozone, and many office machines give off odors and fumes that are unpleasant or noxious.

As will be demonstrated by the case studies in this book, buildings whose occupancies are strictly offices are the sources of air quality problems as often as are buildings with other functions. These problems are frequently the most difficult to diagnose and cure.

Laboratories

Many, but not all, industrial laboratories have much in common with clean rooms. They require an atmosphere that must be controlled to unusually tight standards; small amounts of contamination can invalidate their efforts. Results of work in laboratories that require a tightly controlled atmosphere might have a disproportionate effect on company operations. For example, tests in the laboratory might determine the exact mix of raw materials that will yield the desired final product, and large sums of money will be committed when the production line is given a full go-ahead.

Building Occupants

The goal of OSHA is ". . . to assure so far as possible every working man and woman in the nation safe and healthful working conditions. . . ," and it is sometimes necessary to exceed the standards in order to accommodate certain individuals. For example, air quality that passes an OSHA inspection might not be acceptable to individuals with asthma or who are especially prone to allergies. Workers whose immune systems are suppressed due to chemotherapy, radiation therapy, disease, or other causes might require cleaner air than is normally acceptable.

Depending on limitations of your current ventilation system, providing cleaner air might be as easy as changing your filters or as complex as redesigning the entire system. If your system needs to be redesigned, an expert on staff should determine whether local and national human rights standards would allow transferring vulnerable employees to other areas of the plant, with employer-paid retraining if necessary.

CHAPTER 3

Identifying Undesirable Elements in Air

A thorough analysis of indoor air quality must include an understanding of what constitutes desirable air, what undesirable elements might be present, and what effects undesirable elements might have on workers. This chapter provides a brief look at these items and tells where you can find tables, additional information, or related information.

DEFINING PROPERTIES OF IDEAL AIR

As a general rule, the safest and best air for most situations is the pristine air in which humans developed. It consists of approximately 4 parts nitrogen and 1 part oxygen, plus small percentages of carbon dioxide and other nontoxic gases. Except for special circumstances, an atmosphere that approximates this air is the goal, and additions or significant changes to this composition represent a degradation.

There are other properties of air you must consider. Allergens, bacteria, and molds are sometimes the causes of air quality complaints as well as diseases such as Legionnaires' disease and Pontiac fever. Not all workers are affected by every contaminant, nor to the same degree by any contaminant. Illnesses such as allergic rhinitis, allergic asthma, and hypersensitivity pneumonitis can be a major problem for only certain workers. Temperature and humidity of the air, odors, and changes in barometric pressure also influence workers' health, productivity, and comfort. (Later chapters include consideration of these factors, although not everyone includes them as part of indoor air quality. Some examiners place temperature and humidity under the broader category of *indoor environmental quality [IEQ].*)

The list of possible impurities is almost endless and includes byproducts of internal manufacturing and operations, results of workers' presence, and impurities from outside the plant. Following is a short list of the impurities from these sources:

Manufacturing

Dust, chemicals, and other byproducts get into the air directly from processes that include machining, etching tanks, painting, and cleaning. In addition, manufacturing areas are provided with chemicals, gases, and other materials through pipes (which sometimes leak) and other conveyances (which are not always enclosed) that pass through different parts of the plant. The following are just a few of the airborne impurities that result from manufacturing:

- Asbestos
- Silica or silicates
- Respirable quartz
- Chloroprene
- Trichloroethylene
- Lead chromate
- Sodium azide
- Nitrosamines

Operation

Concern about internally generated impurities in the air should not be limited to manufacturing areas. Even plants in the service sector and those that are strictly offices must take steps to ensure good indoor air quality. Following is a small sample of impurities that arise from operations other than manufacturing. Some items in this list can be found in manufacturing, office, and all other areas of a plant.

- Paper dust
- Toner dust
- Ammonia
- Cleaning fluids
- Overheated lubricants
- Duplicating fluids
- Conditions created by previous occupant
- Cooking vapors

Results of Workers' Presence

Workers introduce impurities into the air inside plants in two ways: They create impurities while in the plant, and they bring impurities into the plant on themselves and on their clothing. A few of the impurities introduced by workers are listed:

- Carbon dioxide
- Tobacco smoke
- Bacteria, viruses, etc.
- Pollen and other allergens
- Allergenic clothing
- Odors

Certain allergens can cause immediate and serious health problems for some people, but most items in this category range from mildly annoying to long-run killers. Body odors are usually annoying to others nearby, but they might also mask the presence of leaking natural gas. Typical of the middle range of the list is carbon dioxide, which is exhaled by all mammals. Carbon dioxide affects the balance of other components in the air, especially oxygen, but unless it rises to amounts in the range of 800 or 1000 parts per million (ppm), seldom causes problems for most people. Tobacco smoke has been proved to be the most harmful in this list for most people—smokers and nonsmokers alike.

Impurities from Outside the Plant

In addition to impurities brought in by workers and visitors, many impurities enter through windows, doors, leaks, ventilation intakes, and similar paths. The following list suggests a few typical ways that impurities enter a plant:

- Vehicle with motor running, parked near a ventilation intake
- Loading dock doors
- Elevator pits
- Uncollected garbage
- Leakage from soil
- Refuse in dumpsters
- Underground fuel tanks

HOW IAQ AFFECTS WORKERS

People enter into the indoor air quality equation in two ways: (1) People affect the indoor air, and (2) protection of people is one of the reasons for concern over indoor air quality. In both situations there is a range to consider: Not everyone affects indoor air quality to the same degree, and people are affected by airborne contaminants in different ways. Following are just a few examples of groups that can be unusually susceptible to indoor air contaminants:

- Individuals with asthma and other respiratory diseases
- People with allergies
- People whose immune systems are suppressed due to chemotherapy, radiation therapy, organ transplant, disease, or other causes
- Contact lens wearers

Some workers are susceptible to several contaminants individually, and others react only to combinations of pollutants but not to any of them individually. People with heart disease may be more affected by exposure to lower levels of carbon monoxide than healthy individuals. Children exposed to environmental tobacco smoke have been shown to be at higher risk of respiratory illnesses, and those exposed to nitrogen dioxide have been shown to be at higher risk of respiratory infections.

The diagnosis of indoor air quality problems is complicated by the fact that air pollutants affect people in different ways. A certain contaminant at a certain concentration might cause headaches in one person, dizziness in another, skin irritation in another, and have no effect on still another person. Changing the concentration of the same contaminant could change its effects among the same individuals. The effects of some pollutants or combinations of pollutants can range all the way from no effect to coma and death.

"Health" and "comfort" are used to describe a spectrum of physical sensations. For example, when the air in a room is slightly too warm for a person's activity level, that person may experience mild discomfort. However, at higher temperatures the person might feel not only increased discomfort but also symptoms such as headaches, fatigue, and disorientation.

Some complaints by building occupants are clearly related to the discomfort end of the spectrum. A common IAQ complaint is, "There's a funny smell in here." Odors are often associated with a perception of poor air quality, whether or not they cause symptoms. Too often we find an innocuous odor mysteriously triggering symptoms. On-the-job stressors such as improper lighting, noise, vibration, overcrowding, ergonomic stressors, and psychosocial problems can produce symptoms similar to those associated with poor air quality. The following symptoms most commonly result from actual poor air quality:

- Headache
- Fatigue
- Eye, nose, and throat irritation
- Skin irritation

- Shortness of breath
- Sinus congestion
- Cough
- Dizziness
- Nausea
- Sneezing

Although these are the symptoms most commonly reported because of poor air quality, not one symptom in this list is *unique* to poor air quality. They can also result from diseases, allergies, stresses, hysteria, and other causes. Some of them can be physically as well as mentally contagious. Investigators have found marginally poor air causing no problems in one plant but triggering many symptoms in another plant. The difference was that some other stress in the second plant, such as long periods of overtime work, lowered the trigger level for air quality complaints.

HOW TOXIC SUBSTANCES AFFECT WORKERS

Many contaminants are outright poisonous. This chapter gives some facts about selected toxic chemicals typically found in workplaces and other locations. (Appendix B lists exposure limits of hazardous chemicals in industrial plants.) There are many reasons for serious concern about toxic substances:

- Toxins enter the body in various ways.
- Many toxins are poisonous in very small concentrations.
- They attack various parts of the body, including the central nervous system, respiratory system, kidneys, liver, and muscles.
- Often they cannot be detected by smell, taste, sight, or other human senses.
- Effects on workers range from temporary sickness (such as pneumonia) to permanent damage (such as brain damage); from temporary annoyances (such as rashes) to cancer and death.
- In situations where health damage could be reversed if detected in time, workers are often unaware that their health is being damaged.

Regulatory bodies such as OSHA, the National Institute for Occupational Safety and Health (NIOSH), and the Environmental Protection Agency (EPA) specify exposure limits for hundreds of toxic chemicals. They establish these limits from experience records, experiments, and other sources of information; they update the limits as new information becomes available. Some of the limits established are: recommended exposure limit (REL); threshold limit value (TLV);

permissible exposure limit (PEL); flash point (Fl.P); upper explosive limit (UEL); lower explosive limit (LEL); and minimum explosive concentration (MEC).

The following section describes some toxic chemicals that occur in workplaces. They are a random selection and are not meant to point out the most or least hazardous, most or least prominent, most or least difficult, or any other extreme.

Profiles of Toxic Chemicals

Arsenic. Arsenic is a powerful poison that can cause death or illness at high levels. It has been found at 781 of 1300 National Priorities List sites identified by the EPA. Arsenic can be found in nature at low levels, mostly in compounds with oxygen, chlorine, and sulfur—called inorganic arsenic compounds—which are poisonous to humans. When combined with carbon and hydrogen in plants and animals, it is called organic arsenic and is less harmful to humans than inorganic arsenic.

One use of inorganic arsenic compounds is in preserving wood; therefore, they are a potential hazard where wood is prepared or even where it is used, cut, or sanded. The compounds are also used to make insecticides and weed killers; many plants contract with pest control firms to spray inside and outside their buildings.

Arsenic compounds are considered to be at a high level when the concentration is 60 ppm or more. Most compounds have no smell or special taste. Although breathing high levels can irritate the lungs, the main hazardous condition exists when arsenic compounds settle out of the air and can dissolve in water, contaminate food, or otherwise be transferred to the mouth and ingested. High levels of ingested inorganic arsenic damage tissues including nerves, stomach and intestines, and skin, and may even cause death. Lower levels of exposure to inorganic arsenic may cause nausea, vomiting, and diarrhea; decreased production of red and white blood cells; abnormal heart rhythm; blood vessel damage; or a "pins and needles" sensation in hands and feet.

Long-term exposure to inorganic arsenic may lead to a darkening of the skin and the appearance of small "corns" or "warts" on the palms, soles, and torso. Direct skin contact may cause redness and swelling. The Department of Health and Human Services (DHHS) has determined that arsenic is a known carcinogen that involves the following cancer risks:

- Breathing inorganic arsenic increases the risk of lung cancer.
- Ingesting inorganic arsenic increases the risk of skin cancer and tumors of the bladder, kidney, liver, and lungs.

Tests conducted on workers fall into two categories: tests to determine exposure to arsenic and tests to determine damage from exposure. Damage from exposure generally does not use tests specific to arsenic. That is, a test for skin cancer or bladder tumor would be the same whether the disease arose from arsenic exposure or from some other source.

The most reliable test for arsenic exposure is a urine analysis. Arsenic stays in the body only a short time, so urine must be analyzed soon after a worker is exposed. Tests on hair or fingernails can measure exposure to high levels of arsenic over the past 6 to 12 months. None of these tests are very useful for evaluating low levels of exposure, nor do they predict whether a worker has actually suffered any harmful health effects.

OSHA has established a maximum permissible exposure limit for workplace airborne arsenic of 10 micrograms per cubic meter ($\mu g/m^3$). (It might be of interest to note that the EPA has set a limit of 0.05 ppm for arsenic in drinking water, and it is considering lowering this limit.)

Benzene. Exposure to benzene occurs mostly from breathing contaminated air from industry, automobile exhaust, tobacco smoke, or gasoline fumes. Breathing very high levels of benzene can be fatal. Breathing low levels over a long period of time can harm blood cells and cause cancer. This chemical has been found in at least 743 of 1300 National Priorities List sites identified by the EPA.

Benzene is a colorless liquid whose vapor has a sweet odor. It is also called benzol. The main source of benzene in the environment is its use as a major industrial chemical and from the use of oil and gasoline. Industry uses benzene to make chemicals for styrofoam, plastics, resins, nylon, and synthetic fibers. It is also used to make some types of rubber, lubricants, dyes, detergents, drugs, and pesticides. Glues, paints, and furniture wax are common sources, and vapors from contaminated water or soil can get into the indoor atmosphere. This partial list of applications indicates that benzene is a potential problem throughout a large portion of industry.

When benzene enters the environment, the following can happen individually or in sequences:

- As a vapor, benzene mixes very quickly with air.
- It reacts with other chemicals in the air, and breaks down within a few days.
- In liquid form, benzene mixes easily in water, then vaporizes quickly.
- It breaks down more slowly in water than in air.

The following points are important with regard to worker exposure to benzene:

- The most common exposure is from breathing benzene in the air.
- Tobacco smoke is the source of about 50 percent of most people's total exposure.
- Vehicle exhaust and industrial emissions, which are easily rerouted into indoor air, are the source of about 20 percent of most people's exposure.
- Highest levels of benzene in air may be found in the workplace.

Health effects vary and depend largely on the concentration of benzene in the air workers breathe. It is generally harmful to tissues that form blood cells. At very high levels (500,000 times the average levels), brief exposures of 5 to 10 minutes can cause death. Down one order of magnitude to high levels (50,000 times the average levels), exposure can cause unconsciousness, rapid heart rate, drowsiness, dizziness, headaches, tremors, and confusion. Most of the time these symptoms will stop soon after exposure ends and the worker breathes fresh air.

Breathing lower levels for longer periods may damage blood cells and bone marrow, causing anemia or excessive bleeding or cancer of the white blood cells (leukemia). DHHS has determined that benzene is a known human carcinogen and is associated with leukemia. Benzene may also harm the immune system and increase the chance of infection. As noted, benzene mixes easily in water; eating or drinking high levels of benzene can cause rapid heart rate, coma, death, dizziness, sleepiness, convulsions, vomiting, or irritation of the stomach.

Direct contact with the skin may cause redness and sores, and benzene may irritate and damage eyes. The health effects of eating or drinking foods containing low levels of benzene have not been established, nor have genetic effects on humans. However, studies with animals indicate that benzene may damage genes and may affect the ability to have healthy children.

Tests can measure the amount of benzene in breath and blood but cannot tell how much benzene a person has been exposed to. The tests do not determine health effects of exposure, but other medical tests give those answers. For example, laboratory blood analysis reveals the condition of white blood cells.

Most tests must be completed soon after the exposure because benzene does not stay in the body for a long time. The blood test is accurate for recent exposures only, and the breath test is not useful for very low levels of exposure. A urine test gives results indirectly by measuring phenol, into which the body converts benzene.

OSHA sets an occupational exposure limit of 1 ppm in air for an 8-hour workday, 40-hour workweek. EPA sets a maximum permissible level of benzene in drinking water at 5 parts per billion (ppb) per day for a lifetime of exposure. The maximum permissible level of benzene in water for short-term exposures (10 days) for children is 235 ppb.

Vinyl Chloride. Exposure to vinyl chloride occurs mostly from breathing contaminated air in the workplace or near plastics industries, hazardous waste sites, and landfills. Vinyl chloride may damage the liver, nerves, and immune system. It has been found in at least 458 of 1300 National Priorities List sites identified by the EPA.

Vinyl chloride, also known as chloroethene, chloroethylene, ethylene monochloride, or monochloroethylene, is a colorless vapor with a mild, sweet odor. The majority of people can easily smell vinyl chloride at high levels. Most of the vinyl chloride produced in the United States is used to make polyvinyl chloride (PVC), which is used in a variety of products including pipes, wire and cable coatings, and packaging materials.

The highest exposures are from breathing workplace air in or near plastics industries. In addition, tobacco smoke contains some vinyl chloride. Most of the information available about vinyl chloride comes from studies on male workers in the plastics industry and from animal studies. Studies on workers show that at sufficient doses vinyl chloride can damage the liver, nerves, and immune system. Fortunately, most work exposures are considerably lower than the levels that cause these harmful effects. However, when vinyl chloride concentrations become extremely high, breathing the fumes can cause death. A person who breathes high levels for up to five minutes can become dizzy or sleepy and may pass out. Recovery from short exposures is usually rapid if exposure is stopped and the person breathes fresh air.

Some people who have breathed vinyl chloride over several years suffer damaged livers. This damage is most often associated with high levels of vinyl chloride. Nerve damage has been reported by some, and others have developed an immune reaction (a reaction of the immune system to the chemical insult, manifesting as acroosteolysis, joint and muscle pain, enhanced collagen deposition, stiffness of the hands, and scleroderma-like skin changes). The low threshold exposure limit that causes liver changes, nerve damage, and the immune reaction in humans has not been determined.

Men who work with vinyl chloride complain of a diminution of sex drive, and women report irregular menstrual periods. Some women have developed high blood pressure during pregnancy. Workers who spilled liquid vinyl chloride on their skin have reported numbness, redness, and blisters.

DHHS has determined that vinyl chloride is a known carcinogen. This conclusion is based on studies of workers who breathed vinyl chloride for many years and had higher rates of liver cancer. Animal studies support these findings.

Vinyl chloride can be measured in a worker's breath, and its major breakdown product, thiodiglycolic acid, can be measured in urine. Neither test indicates the level of exposure, whether there was compound exposure, or the health effects of exposure. The tests must be performed shortly after a suspected exposure but are not routinely available in doctors' offices.

OSHA sets the maximum allowable amount of vinyl chloride in workroom air at 1 ppm during an 8-hour workday, 40-hour workweek. The maximum amount allowed in any 15-minute period is 5 ppm. NIOSH recommends that workers exposed to any measurable amount of vinyl chloride must wear special breathing equipment.

CHAPTER 4

Determining the Sources and Health Effects of Impurities

Cleaning the air when you know the origin of the impurities you must remove is an easy task compared with the puzzle you face when you do not know the sources of the impurities. You will see that some contaminants were there all the time, just waiting to be activated. Other problems that you might think you solved return as regularly as night and day, often *because* of the diurnal cycle.

A major part of this chapter looks at information we have about the way contaminants affect workers' health. That section covers toxic substances, risk, dose-response, routes of exposure, frequency of exposure, organs most often involved, and several related topics.

IDENTIFYING SOURCES OF CONTAMINANTS

For both identifying the source of a contaminant and for planning a procedure to correct the problem, it is helpful to place pollutant sources into categories. But there is a danger in overreliance on categories because there can be a complex of contaminants, coming from a variety of sources. When you identify and eliminate a source, don't write off the problem as solved; the same contaminant might continue to pollute your air from an entirely different source. The contaminant concentration might even have been high enough to mask the second source and to have made it impossible for you to detect it until you cleaned up the first source. The pollutants discussed in the following sections range from serious and deadly toxins, to less serious health hazards, to mere annoyances.

Factory Operations

Indoor air quality is a concern in almost all factory operations. As areas susceptible to unclean air, conventional machinery factories often carry an image of dirt everywhere, but many assemblies, even greasy ones, are clean in their own way. To prevent abrasive damage by keeping airborne particles out of the assemblies of precision parts, workers have clean areas, away from the grinding operations. Control goes a step further as products shift increasingly to miniature and hi-tech; they must be manufactured in highly controlled clean rooms that filter out particles down to micron size. As generators of air contaminants, hi-tech as well as conventional factories feature operations such as the following that contribute to indoor air degradation:

- Grinding, welding, soldering, painting, and related activities. Almost every activity that abrades, heats, cuts, and shapes will affect the air to some extent.
- Cleaning, etching, degreasing the product, lubricating the equipment, and related activities. Almost every activity that uses chemicals, especially when there are fumes or vapors, will affect the air.
- Maintenance activities often involve chemicals, machinery, heat, and other items that affect the air
- Many factory activities use compressed air, which can stir up dust and spread fumes, vapors, and other pollutants. The compressors might be located in an area where their air intakes pick up additional pollutants to add to those generated by the factory operations.

The Heating, Ventilating, and Air Conditioning System (HVAC)

The HVAC system is a major player in controlling indoor air quality because whether the air is clean or contaminated, HVAC circulates it throughout the plant. HVAC also controls the temperature component of air quality, and it sometimes is itself the source of undesirable items in the indoor air. Some of the considerations in this area are:

- Dust or dirt in ducts, motors, fans, or any place in the system
- Microbiological growth in drip pans, humidifiers, ductwork, and coils
- Improper use of biocides, insecticides, sealants, and cleaning compounds
- Improper venting of combustion products
- Refrigerant leaks
- Filters that retain contaminants after their sources have ceased to inject the contaminant into the plant's air

Other Internal Sources

Factories are not the only worksites that can have poor air quality. The following are also sources of poor air quality in offices, laboratories, duplicating facilities, and other worksites:

- Emissions such as volatile organic compounds, heat, and ozone from office machinery
- Fumes from supplies such as solvents, toners, ammonia
- Paper dust
- Emissions from shops, laboratories, storage areas, and cleaning and maintenance processes
- Electromagnetic and other types of radiation—not always considered in evaluation of indoor air quality
- Dust and dirt that settles and then is sent back into the air by vacuum cleaning, sweeping, blowing, or other activities
- Microorganisms in mist from improperly maintained cooling towers
- Volatile organic compounds from paint, caulk, adhesives, and other products
- Fumes and dust from special-use areas such as lounges, print shops, exercise rooms, and food preparation areas

Building Components and Furnishings

Several of the locations in the following list are favorable for the growth of dust mites, a problem in residences as well as industrial plants. This mite is to blame for growing numbers of allergic reactions, especially those related to breathing disorders. The problem has become so widespread that most countries now consider the dust mite to be the most frequent source of allergens, and the World Health Organization (WHO) has described dust mite allergens as "a worldwide problem." Included in the building components and furnishings category are such items as:

- Textured surfaces—carpeting, curtains, upholstering, and other textiles
- Old or deteriorated furnishings
- Materials containing damaged asbestos
- Microbiological growth on or in soiled or water-damaged furnishings
- Microbiological growth in areas of surface condensation
- Standing water from clogged or poorly designed drains
- Ever-present water because of leaks
- Gas that emanates from drains because their traps are dry

Human Presence

The production of carbon dioxide in the process of breathing and exchanging oxygen is an example of an unavoidable result of human presence. Other activities, such as smoking, are matters of personal choice. Whether inherently or by choice, humans directly influence the quality of air in the workplace, and their influence ranges from seriously hazardous to health, to unpleasant, to not noticeable. The following are just a few of the human factors that affect the workplace:

- Germs, viruses, and other contagion from sicknesses
- Body and breath odors
- Deodorizers, fragrances, and cosmetics
- Medication odors
- Smoking

Sources Outside the Building

Employees and visitors entering your plant can inadvertently bring in contaminants. Other undesirable products can enter your plant's air through open windows and doors, by way of the HVAC air intake, and by seepage through the foundation from the ground. Some items in the following list are general health hazards, and others might affect just a few sensitive individuals:

- Pollen, dust, fungal spores
- Vehicle exhaust
- Decay products, fumes, vapors from dumpsters, and items on the ground

- Exhaust from nearby buildings
- Reentrained air (exhaust that is drawn back in)
- Radon leaking from the ground
- Leakage from underground fuel tanks
- Contaminants from previous users of the site

FINDING POLLUTANT PATHWAYS

After you detect air pollution in an area of the plant, a knowledge of possible pollution pathways can shorten the task of locating the source. Airflow patterns in buildings result from the combined action of mechanical ventilation systems, human activity, and natural forces. Pressure differentials created by any or all these actions move airborne contaminants from areas of relatively higher pressure to areas of relatively lower pressure through any available openings. Finding those openings can be the key to the problem; they may be as small as cracks in walls or they may be larger than open doors. A common opening is an oversize uncaulked hole through which conduits, pipes, ducts, cables, and other items pass. They are especially likely with modifications, as when holes are made in existing walls to run cables for networking computers.

Mechanical Ventilation Systems

Air is moved primarily by the HVAC system, but other equipment can actually move a significant amount of air. For example, an elevator moving in its shaft pushes a considerable amount of air ahead of it, and elevator doors are deliberately made to allow that air to escape at each floor. Actually, all of a buildings components—walls, ceilings, floors, penetrations, HVAC equipment, other equipment, and occupants—interact to distribute contaminants. Chases, cocklofts, crawl spaces, utility tunnels, and other hidden spaces can be both sources of and pathways for pollutants.

For example, as air moves from supply registers or diffusers to return air grilles, it is diverted or obstructed by partitions, walls, and furnishings, and redirected by openings that provide pathways for air movement. Within local areas, movements of people and equipment have a major effect on pollutant movement, and the overall paths can depend on the time of day. Some pathways change as workers open doors and windows or turn desk fans on or off, and actions in one area can affect pathways in areas a long distance away.

Natural Forces

Natural forces can exert an important influence on air movement between zones and between the building's interior and exterior. Both the stack effect and wind can overpower a building's mechanical system and disrupt air circulation and ventilation, especially if the building envelope is leaky. *Stack effect* is the pressure-driven flow produced by convection (the tendency of warm air to rise). It exists whenever there is a temperature difference between vertical areas. As heated air escapes from upper levels, indoor air moves from lower to upper floors, leaving a partial vacuum that draws replacement outdoor air through openings at lower levels. Places where the stack effect transports contaminants between floors are stairwells, elevator shafts, utility chases, chimneys, and unused ducts.

Wind effects are transient, creating local areas of high pressure on the windward side and low pressure on the leeward side of buildings. When there are leakage openings in the building exterior, these pressure changes can bring about changing pressure relationships within and between rooms. Depending on the angle, wind can also drive air down into rooftop openings (creating an area of momentary positive pressure someplace in the building and causing air to move from that area) or pull air up from them (creating an area of momentary negative pressure someplace in the building and causing air to move into that area).

The basic principle of air movement from areas of relatively higher pressure to areas of relatively lower pressure can produce many patterns of contaminant distribution. *Relatively* is an important word here; both areas can be at higher (or lower) pressure than the outdoors or than other nearby areas, but air will flow from high pressure to low pressure between the two areas in question. Following are a few examples of air movement caused by natural forces:

- Local circulation in the room containing the pollutant source
- Air movement into adjacent spaces that are under lower pressure
- Recirculation of air within the zone containing the pollutant source or in adjacent zones where return systems overlap
- Movement from lower to upper levels of the building
- Air movement into the building through either infiltration of outdoor air or reentry of exhaust air

The following example was contrived to give only the basic facts that are needed for determining the pollutant's source. A real situation would be complicated with irrelevant information and other data that might take the search to dead ends.

The transitory characteristic of natural forces such as wind is worth emphasizing because it can complicate the search for a pollutant's source. For a situation that is simpler than most in the real world, suppose you have identified a contaminant in an area that has no outside walls (call it Area 1) but is adjacent to a room with pervious outside walls (call it Area 2). The wind and its direction are intermittent, and as the wind changes direction, it sometimes picks up a pollutant from its source. Having determined the range of contaminant level in Area 1, you move to Area 2 and determine that the concentration is sometimes higher and sometimes lower than the peaks in Area 1. Trying to correlate the levels with wind gusts leaves you with a jumble of data unless you can include wind direction and strength with each reading and allow the right time delay after each gust so the pollutant can penetrate to where you are taking your readings.

When an entire building is kept under a positive pressure relative to outside pressure, air movement at all leaks will be from inside to outside. However, unless the building is part of an experiment in self-sufficiency and is recycling 100 percent of its internal air, there is always some location (for example, the outdoor air intake) that is under negative pressure relative to the outdoors. Control becomes difficult because outdoor air taken in this way usually goes directly to the HVAC system for distribution throughout the plant.

Human Activity

All people in the building, whether employees, contractors, visitors, clients, or in any other capacity, are considered here. The following are several ways in which the activities of humans affect the way pollutants move within a plant:

- Humans consume oxygen and raise the carbon dioxide level.
- People move about, often via electric carts.
- People move equipment, often with the use of mechanical carts and other devices.
- People open doors to enter and leave the building.
- Workers adjust fans, thermostats, windows, and other items that affect air circulation.
- People bring many things, of varying degrees of concern, to the indoor air, such as germs, viruses, odors, cosmetics, and pollen.
- Peoples' activities such as smoking change the air's makeup.

VARIABLES IN EVALUATING HEALTH EFFECTS

As long ago as the sixteenth century, people recognized that there is no such thing as an absolutely safe chemical. The Swiss physician Paracelsus (1493–1541) said:

> All substances are poisonous; there is none which is not a poison. The right dose differentiates a poison and a remedy.

On the other hand, limiting the dose and/or controlling the exposure permit any chemical to be handled safely, no matter how poisonous. However, before the necessary degree of control can be determined for a particular exposure or situation, the toxicity of the substance in question must be known. Methods of choosing exposure limits must, because of the lack or inadequacy of dose-response information for many chemicals, rely on experience in the use of these substances and on scientific and professional judgment.

Toxicologists rank chemicals by categories that range from practically nontoxic (an adult human would have to consume a quart) to supertoxic (fewer than seven drops would be lethal for most people). In the occupational setting, it is the risk associated with a particular use of a chemical rather than its inherent toxicity that is important. *Risk* can be defined as the probability that a substance will produce harm under certain conditions of use. The converse of risk is *safety*, which is the probability that no harm will occur under specific circumstances.

A major factor affecting the ability of a chemical to elicit a toxic response is the susceptibility of the biological system or the individual. To identify the relative degree of hazard in a particular instance requires knowledge about the chemical, the exposure situation, and the exposed subject. In addition, the route of administration and the frequency of exposure must be known.

Routes of Exposure

Investigators recognize four principal routes by which toxic substances can invade humans and laboratory animals:

- *Inhalation*. This exposure route is of primary concern in most examinations of indoor air quality.
- *Ingestion*. Many air contaminants, especially particulates, are primarily harmful when breathed and secondarily harmful when they settle out of

the air and onto surfaces. Workers ingest them when the contaminants settle directly onto food. When contaminants settle onto other surfaces that workers touch, they ingest the contaminants after putting pencils, cigarettes, food, and other items into their mouths.

- *Dermal absorption.* Some air contaminants, such as cyclohexane, can be absorbed through the skin. First aid recommendations include prompt water flush. Cyclohexane is basically a colorless liquid, but its sweet, chloroform-like odor is an airborne product.
- *Parenteral administration.* Medically this term refers to administration through routes other than the intestinal canal, such as by subcutaneous, intramuscular, intravenous, or intratesticular injection. This route of exposure to air contaminants is not applicable to this book.

The route of exposure of a toxin also affects the relative toxicity of the agent. For example, a chemical that can be detoxified in the liver will be less toxic if it is received orally than if it is received systemically (i.e., inhaled). Studies that provide information about the relative toxicity of an agent via different routes of exposure can provide a considerable amount of information about the absorbability of the agent. If exposure to a certain dose of a chemical via all routes causes death within the same time period, it can be assumed that the substance in question is easily and rapidly absorbed. On the other hand, if the dermal dose of a chemical that is required to kill a subject (laboratory animal) is much higher than the dose required to produce the same effect when the chemical is ingested, one can deduce that the skin provides, to some degree, a barrier against the agent's toxicity.

Duration and Frequency of Exposure

The frequency of exposure also has an important influence on the magnitude of the toxic effect. A large single dose of an acute toxin will usually have more than three times the effect of one-third the dose given at three different times, and the same dose administered in 10 or 15 applications might have no effect whatsoever. The pattern of exposure is important because it is possible for some of the substance to be excreted between successive exposures or because the damage caused by the toxin has a chance to be partially or completely repaired between exposures. Thus a *chronic effect* is said to occur in the following ways:

1. If a toxic substance accumulates in the system of an exposed person because the dose absorbed is larger than the body's ability to transform or eliminate the substance

2. If the toxin produces irreversible adverse effects
3. If the exposure is in a manner that permits inadequate time for repair or recovery

Variation in Response

Responses to toxic insults vary in a number of ways. Some toxins have immediate effects, whereas others are associated with delayed symptom onset. The latency period for carcinogenic agents may be as long as 40 years for some types of cancer. Even some acute agents, such as some chemicals that have adverse ocular effects, may not cause overt symptoms until hours after exposure.

Another difference in type of response concerns the reversibility or irreversibility of the effect. Reversibility depends on the site of action as well as the magnitude of the insult. That is, some tissues of the body, such as the liver, have considerable ability to regenerate; others, like the kidney or central nervous system, do not.

The site of action associated with toxic substances also varies widely. Local effects are those lesions caused at the site of first contact between the agent and the organism. Examples of localized effects are skin burns caused by contact with a harmful substance and site-of-contact tumors that develop at the locus of exposure to the carcinogen.

In contrast with localized effects, systemic effects involve the absorption and distribution of the toxic agent from the point of entry to a distant site; the toxic response is manifested at this distant point. An example of a systemic poison is mercury, which produces its toxic effect on the central nervous system (CNS). Often, the site of deposition for a chemical is not the organ system most affected by the toxin. For example, although lead is deposited and concentrated in the bone, it affects the CNS. Any sites that are adversely affected by the toxic effects of exposure to a substance, whether they are sites of contact or distal sites, are called the *target organs of toxicity*.

In cases of systemic poisoning, the system most often affected is the CNS; it is common for the CNS to be involved even when another organ, such as the liver, is the primary target organ. In descending order of frequency, the following systems or organs are most often involved in cases of systemic poisoning:

- Central nervous system
- Circulatory system
- Blood and hematopoietic system
- Visceral organs (liver, kidneys, lungs)
- Skin

Dose-Response

The association of the size of a chemical exposure with the effects it causes is called the *dose-response relationship*. A single data point relating a dose to a response is sufficient to establish a dose-response relationship. As additional data become available, it is possible to expand the understanding of the relationship to cover a range of exposures. Dose-response is an important principle in toxicology, and an understanding of the association is important in establishing occupational exposure limits. Knowing how toxic substances act makes it easier to predict the potential effects of exposure.

It is generally true that lowering the dose reduces the response. However, although data are often available to demonstrate this generality, data showing that more subtle responses (e.g., those at the subcellular level) have been reduced are rarely available, so it is risky to make these assumptions.

An important principle underlying the concept of dose-response is that there must be a quantifiable means of measuring the toxicity of a substance and a method of expressing this measured toxicity. Although lethality in test animals is often used to measure toxicity, the best form of measurement would involve quantification of the sequence of molecular events occurring during the toxic response. In the absence of such endpoints, other good methods are available. For example, it is common to measure an effect believed to be related to the substance in question. The level of activity of an enzyme in the blood is often used as a measure of the effect; serum glutamic oxaloacetic transaminase (SGOT) levels are used to measure liver damage. Many different endpoints can be used to measure toxic effects, such as changes in muscle tone, heart rate, blood pressure, electrical activity of the brain, motor functioning, and behavior.

Figure 4-1 shows the classic dose-response relationship. Increases in the exposure in the zero and very low level region and in the very high region (where almost everyone is already affected) are slow to affect increasing numbers of workers. The most rapid changes are in the middle region, where almost any incremental increase in exposure will affect more workers.

SICK BUILDING SYNDROME

Sick building syndrome (SBS) is a term often used to describe situations in which building occupants report acute health and comfort effects that are apparently linked to the time they spend in the building, but in which no specific illness or

Figure 4-1
Typical Dose-Response

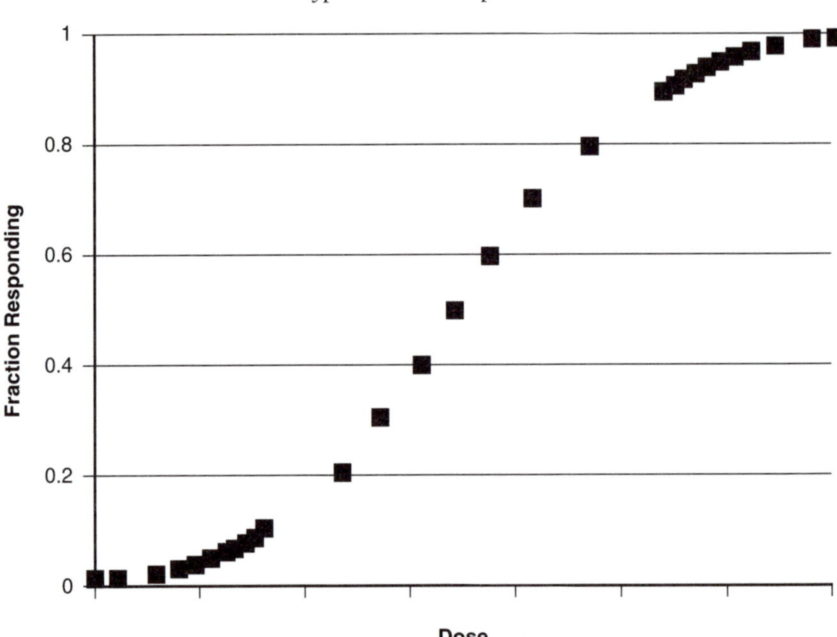

cause can be identified. The complaints may be localized in a particular room or zone or may be widespread throughout the building. Many different symptoms have been associated with SBS, including respiratory complaints, irritation, and fatigue. Analysis of air samples often fails to detect high concentrations of specific contaminants. The problem may be caused by any or all of the following, or sometimes by a combination of them:

- Combined effects of multiple pollutants at low concentrations
- Other environmental stressors such as overheating, poor lighting, noise
- Ergonomic stressors—equipment and conditions that were not designed properly for working with humans
- Job-related psychosocial stressors such as overcrowding or labor-management problems

Common Sources of SBS Problems

The World Health Organization estimates that up to 30 percent of office buildings worldwide may have significant problems. Ten to 30 percent of the occupants of the buildings experience health effects that are, or are perceived to be, related to poor indoor air quality. In the United States, NIOSH recently had 3000 telephone calls relating to poor indoor air quality in a single week, with 90 percent of them requesting investigations of the buildings they occupy or manage.

Contaminants that precipitate complaints about working in a sick building may originate from a variety of sources inside or outside the building. As with other indoor air quality problems, SBS problems can include airborne chemicals, bacteria, fungi, pollen, dust, and other items. Other factors, not always considered within the realm of air quality include lighting, noise, personal and work-related stress, preexisting health conditions, temperature, and humidity. These complaints often transfer, mentally and/or physically, to complaints about working in a sick building.

Potential sources of contaminants within the building include dust, poor maintenance of HVAC systems, tobacco smoke, cleaning supplies, pesticides, building materials, furnishings, occupant metabolic wastes (respiration and perspiration), and cosmetics. Of course, these items are typical of all occupied buildings. They lead to complaints when their concentrations become excessive, often as a result of poor air circulation and replacement.

Dusty surfaces, stagnant water, and damp materials provide an environment ripe for microbial growth. When mold spores and other microbial particles become airborne, some building occupants may experience allergic reactions. One potentially fatal but, fortunately, rare infection is caused by *Legionella* bacteria.

Cigarette smoke presents a serious health risk to those exposed because it contains carbon monoxide, formaldehyde, and thousands of other chemicals. Smoke in the air (second-hand smoke) comes directly off the tip of the cigarette and therefore has higher concentrations of dangerous products than does the smoke taken in by the smoker, which is filtered through the tube of tobacco and the filter. In addition to EPA's well-publicized listing of tobacco as a confirmed cancer-causing agent, studies have shown that exposure to second-hand smoke may result in asthma, lung cancer, and problems as unexpected as inner ear infections.

The Importance of Ventilation

The HVAC system must not only provide a comfortable environment but must also be one of the main tools used to control contaminants. Poor indoor air quality occurs when ventilation is inadequate to the task of keeping contaminant concentrations below levels that produce occupant health problems. The perception of still or stale air, odor, draftiness, or errant temperature and humidity leads to discomfort. Most IAQ complaints originate from the HVAC system's inability to meet occupant needs. Of the more than 1200 IAQ investigations performed in recent years by NIOSH, over half were attributed to inadequate ventilation.

Common Complaints

Records show that the most frequent complaint about indoor air quality is that it is either too hot or too cold. However, it seldom stops there; workers take it from there and develop a variety of symptoms that confuse the investigation. Sometimes, because they know the air temperature is not in the comfort range, investigators neglect tests that might uncover more serious air problems. Sometimes investigators test and test and test and find no problem except with air temperature.

The second most common complaint is about air movement—the air is too drafty or too still. Ideally, the HVAC system should change the air but workers should not feel the movement. Other common comfort-related complaints involve humidity—the air is too dry or too muggy. Of course, temperature and humidity together affect human comfort, especially when the temperature is too high. When the humidity is high, air temperatures that might otherwise have been in the upper range of comfort become highly uncomfortable. Later chapters will look at *relative* humidity and its effect on equipment and material as well as on workers.

Sometimes discomfort arising from improper room temperature is confused with discomfort from more serious problems of air quality. Many health-related complaints associated with poor air quality are the same as those caused by a flu or a cold—headaches, sinus problems, congestion, dizziness, nausea, fatigue, and irritation of the eyes, nose, or throat. It is easy to deny that indoor air quality is the source of the problem unless many of the occupants share the symptoms, they are unreasonably persistent, there is a distinct and suspect quality to the air, or the symptoms arise only inside the building.

Health-related complaints may be due to allergic reactions. A general rule is that in the presence of an allergen 10 percent or more of the workers with the same exposure may exhibit symptoms including sneezing, swollen airways, and asthma-like attacks. However, this is a *general* rule—some individuals are extremely sensitive to certain allergens that have little or no effect on an average mix of workers. Some workers with a building-related allergy will experience similar symptoms in other environments if the particular allergen (dust mites, mold spores, etc.) is present.

Most Likely Times of Complaints

One time to clearly expect complaints is when a new or changed procedure affects the air quality and causes symptoms such as eye irritation or any of those previously mentioned. Another common complaint-trigger is new or changed procedures that cause symptoms but do not deteriorate the air quality with dangerous or toxic chemicals. In general, any odor that employees do not know about can be expected to start a series of complaints about the air quality; odors from obvious sources, as when an area is being painted, are less likely to cause complaints.

Other changes that increase the likelihood of complaints include new furnishings, carpet, renovation activities, and opening of new or previously unused work areas. In fact, many items that involve a characteristic odor cause complaints. Changes in air circulation and persistent moisture often trigger complaints. Poor air quality is often claimed when there is a stressful work environment such as impending layoffs, extensive overtime, or an ongoing employee/employer conflict.

OTHER GROUPS OF AIR PROBLEMS

In addition to sick building syndrome, various names have been attached to groups of problems to facilitate investigations and remedies. There is considerable overlap among the groupings, and some are full or partial subsets of others. This section briefly discusses some of them.

Building-Related Illness

Building-related illness (BRI) is a term that refers to illness brought on by exposure to the building air in which symptoms of diagnosable illness are identified (e.g., certain allergies or infections) and can be directly attributed to environmental agents in the air. Legionnaires' disease and hypersensitivity pneumonitis are examples of BRI that can have serious, and even life-threatening, consequences.

Multiple Chemical Sensitivity

A small percentage of the population may be sensitive to a number of chemicals in indoor air, each of which may occur at very low concentrations. The existence of this condition, known as *multiple chemical sensitivity* (MCS), is a matter of considerable controversy. MCS is not currently recognized by the major medical organizations, but medical opinion is divided, and further research is needed.

Applicability of access for the disabled and workers compensation regulations to people who believe they are chemically sensitive may become concerns for facility managers.

Environmental Illness[1]

Environmental illness (EI) is defined as a multisystemic and multisymptomatic illness caused by multiple allergens or reactants. Little is known about the causes of EI; there are theories suggesting that people with this illness had a virus when they were children that affected their immune system drastically.

With EI we enter an area that overlaps many other areas of human health and transcends all boundaries. There are no defined symptoms because the illness can cause all the symptoms of every disease or disorder known, including both psychological as well as medical disorders. Lupus, Lyme arthritis, multiple sclerosis, and many other disorders are related in some fashion. Allergies and EI or MCS seem to follow each other: A person with allergies may also have MCS.

A difference that helps investigators separate environmental illness from sick building syndrome is that SBS occurs when a building circulates chemicals or pollutants without the ability to filter them. If a reactant (ozone, photocopier fluid, paper dust, paint, glue, perfume, mold, or anything else that can float freely in the air) floats throughout a building, and a person reacts to it no matter where he or she goes in the building, then the building is sick.

Long-term health effects of EI are said to depend on a worker's exposure. One average rule, that has not been proved with a scientifically controlled test, is that every year of exposure and every very severe reaction take two years off a person's maximum life span. Other health effects can include constantly feeling sick or tired, losing control, and severe mood swings.

Chronic Fatigue Syndrome

Chronic fatigue syndrome (CFS) is characterized by debilitating fatigue, manifested as severe exhaustion and extremely poor stamina, neurological problems, and a variety of flu-like symptoms. This condition is also known as chronic fatigue immune dysfunction syndrome (CFIDS). Outside the United States it is usually

1. Thanks to Julie A. May, 51 Melrose Cres., Sherwood Park, AB, Canada, T8A 3V2, for providing this information. Her E-mail address is jmay@orbital.net.

known as myalgic encephalomyelitis (ME). In the past the syndrome has been known as chronic Epstein-Barr virus (CEBV).

The core symptoms include excessive fatigue, general pain, mental fogginess, and often gastrointestinal problems. Many other symptoms are also present, however they are typically different among different individuals. These symptoms include fatigue following stressful activities, headaches, sore throat, sleep disorder, abnormal temperature, and others. As with so many other illnesses that are hard to define, CFS is hard to diagnose because its symptoms are also the symptoms of other health problems.

Although researchers have not uniquely identified the causes of CFS, the theories include viruses, environmental toxins, and other impurities that are often found in air that circulates through industrial plants. Investigations are ongoing, and those responsible for indoor air quality should be aware of the possibilities that CFS usually results from a combination of conditions, and physical or emotional stress seems to make it worse.

PART II

REGULATIONS AND OTHER CONSIDERATIONS

CHAPTER 5

Regulatory, Advisory, and Standards Organizations

One of the main sets of responsibilities given OSHA when it was created in 1970 was to "develop mandatory job safety and health standards and enforce them effectively." A few other agencies, with various degrees of advisory, regulatory, and enforcement authority, were created to support OSHA. Major agencies this chapter examines are OSHA and NIOSH.

Although OSHA and NIOSH were created by the same act of Congress, they are two distinct agencies with separate responsibilities. OSHA is part of the Department of Labor and is responsible for creating and enforcing workplace safety and health regulations. NIOSH is part of the Department of Health and Human Services and is a research agency that is part of the Centers for Disease Control and Prevention (CDC). NIOSH identifies the causes of work-related diseases and injuries and the potential hazards of new work technologies and practices. With this information, NIOSH determines new and effective ways to protect workers from chemicals, machinery, and hazardous working conditions. It makes recommendations for prevention of work-related illnesses and injuries. In general, creating new ways to *prevent* workplace hazards is NIOSH's job.

OCCUPATIONAL SAFETY AND HEALTH ADMINISTRATION (OSHA)

Almost any discussion about air quality in the workplace will include OSHA, the standards it sets, and its enforcement policies. A few of this agency's functions that apply are:

- OSHA sets standards for health and safety. In the field of air quality, these standards set maximum worker exposure levels for contaminants that could be in industrial air.
- OSHA provides for research to update information on air contaminants and to develop ways of dealing with air contaminants.
- OSHA establishes reporting and recording systems, and cites employers who fail to meet the systems' requirements.
- OSHA requires employers to train employees in proper use of safety equipment.

Employers Affected by OSHA

In general, OSHA's coverage includes all employers and their employees in the 50 states, the District of Columbia, Puerto Rico, and all other territories under federal government jurisdiction.

Employees of the United States or of any state or political subdivision of a state are not covered by Federal OSHA, but OSHA-approved state plans apply to public employees in their states. In addition, the Occupational Safety and Health Act of 1970 makes federal agency heads responsible for providing safe and healthful working conditions for their employees. The act requires agencies to comply with standards consistent with those OSHA issues for private sector employers. OSHA conducts federal workplace inspections in response to employees' reports of hazards and as part of a special program that identifies federal workplaces with higher than average rates of injuries and illnesses.

Federal agency heads are required to operate comprehensive occupational safety and health programs, and OSHA conducts extensive evaluations of these programs to assess their effectiveness. These programs include:

- Recording and analyzing illness and injury data
- Providing training to all personnel
- Conducting self-inspections to ensure compliance with OSHA standards

Employment rules of state and local governments are not affected by OSHA. However, one requirement of any state that seeks approval of a state-OSHA plan is that it must include coverage of state and local government workers. In fact, a state plan may cover the public sector only but may not cover the private sector only.

Standards

OSHA is responsible for promulgating legally enforceable standards, and employers are responsible for becoming familiar with whatever standards apply to their establishments.

General duty clause. If no specific standard applies in a certain area of a company's work, the employer is responsible for following the General Duty Clause. It includes the general statement that each employer ". . . shall furnish . . . a place of employment which is free from recognized hazards that are causing or are likely to cause death or serious physical harm to his employees." Toxins, allergens, particulates, and other contaminants in the workplace air are recognized as hazards as much as slippery stairs, falling debris, and unprotected moving machinery.

Where to get copies of standards. The *Federal Register* is one of the best sources of the latest information on standards, because all OSHA standards, as well as amendments, corrections, insertions, and deletions, are published there when adopted. Many public libraries have the *Federal Register*, and subscriptions are available from the Superintendent of Documents, U.S. Government Printing Office, Washington, D.C. 20402.

A quick and convenient way to check and search specific issues of the *Federal Register* is through the Internet. Starting with Gopher://gabby.osha-slc.gov:70/11/.fed brings up a screen of hyperlinked years as follows:

Federal Register for 1995

Federal Register for 1994

Federal Register for 1993

and so on through 1971. From there you can click on the appropriate year, month, and issue in successive screens.

Each year the Office of the Federal Register publishes all current regulations and standards in the *Code of Federal Regulations* (CFR), available at many libraries and from the Government Printing Office. OSHA's regulations are collected in Title 29 of the CFR, Part 1900–1999, but they would fill a bookshelf and be difficult for a firm's staff member to search through to learn about applicable regulations. More convenient ways to learn of OSHA regulations include the following:

- Call a local OSHA office with specific questions.
- Use the services of a consultant who is familiar with the regulations.
- Purchase commercial computer disks that contain the regulations. One advantage of these disks is that computers can search almost instantly for given words or phrases.
- Several sites on the Internet provide free OSHA information, and it is usually easy to hyperlink among them. For example,

  http://www.osha.slc.gov/OshStd toc/OSHA Std toc.html

  brings up a table of contents, and

  http://www.osha-slc.gov/wais_search.html

  provides for a search of "OSHA Regulations, Documents, Technical Info., & Training."

Numerous OSHA publications and informational materials are available from OSHA Publications Office, 200 Constitution Avenue, NW, Room N-3101, Washington, D.C. 20210, (202) 523-9667. The list includes pamphlets, booklets, books, newsletters, and other forms—some are free and some have a charge.

Emergency temporary standards. Under certain limited conditions, as when OSHA determines that workers are in grave danger due to exposure to toxic substances or agents determined to be toxic, the agency is authorized to set emergency temporary standards. They take effect immediately and are in effect until superseded by a permanent standard. Each temporary standard is published in the *Federal Register,* where it also serves as a proposed permanent standard. It is then subject to the same approval procedure as a new standard that does not go through a temporary status, except that a final ruling should be made within six months.

Recordkeeping and Reporting

OSHA continually reviews its standards to keep pace with developing and changing industrial technology, it tracks the effectiveness of its standards and regulations, and it monitors for developing problems. All of these activities require a consistent, nationwide procedure for keeping records and reporting results.

Employers of 11 or more workers must maintain records of occupational injuries and illnesses as they occur. Those with 10 or fewer employees are exempt from keeping such records unless they are selected by the Bureau of Labor Statistics (BLS) to participate in the Annual Survey of Occupational Injuries and Illnesses. The survey uses sampling techniques on data from small businesses to ensure that published statistics represent businesses of all sizes.

Employers in some nonmanufacturing industries, such as retail trade, finance, insurance, real estate, and service industries, are generally exempt from OSHA recordkeeping. However, a few of the regularly exempt employers will have to maintain records if they are selected to participate in the Annual Survey of Occupational Injuries and Illnesses. Selected employers are notified in advance and supplied with the necessary forms and instructions. Whether exempt or not, all employers (with the exceptions mentioned in the *Employers Affected by OSHA* section) must comply with OSHA standards, display the OSHA poster, and report to OSHA within 48 hours any accident that results in one or more fatalities or the hospitalization of five or more employees.

Workplace Inspections

Every establishment covered by the Occupational Safety and Health Act of 1970 is subject to inspection by OSHA safety and health officers, as a means of enforcing its standards. The act provides that, after presenting appropriate credentials, compliance officers are authorized to:

- Enter without delay and at reasonable times any . . . workplace or environment where work is performed by an employee.
- Inspect and investigate during regular working hours, and at other reasonable times, and within reasonable limits and in a reasonable manner, any such place of employment and all pertinent conditions, structures, machines, apparatus, devices, equipment and materials.
- Question privately any such employer, owner, operator, agent, or employee.

If an employer refuses to admit an OSHA compliance officer, or attempts to interfere with the inspection, the act permits appropriate legal action.

Inspection Process

Details of an actual inspection depend on so many factors that there is no such thing as a "typical inspection." However, there are usually four phases to a complete inspection: inspector's credentials, opening conference, inspection tour, and closing conference.

Inspector's credentials. An OSHA compliance officer carries U.S. Department of Labor credentials bearing his or her photograph and a serial number. Employers may telephone the nearest OSHA office to verify the officer's identity. Anyone who tries to collect a penalty at the time of inspection or promotes the sale of a product or service at any time is not an OSHA compliance officer and should be reported promptly to local law enforcement agencies.

Opening conference. The compliance officer explains the purpose of the visit, the scope of the inspection, and the standards that apply. He or she gives the employer copies of applicable safety and health standards plus a copy of any employee complaint (with name removed if the employee requests) that may be involved.

Employer and employee representatives to accompany the compliance officer during the inspection can be selected at the opening conference. If there is no authorized employee representative, the compliance officer must consult privately with a reasonable number of employees concerning safety and health matters in the workplace.

Inspection tour. The compliance officer determines the route and duration of the inspection but makes every effort to minimize work interruptions. At various places the officer might take photographs, note instrument readings, and examine records. When the officer observes trade secrets, he or she must keep them confidential or be subject to a $1000 fine and/or a year in jail.

Closing conference. After completing the inspection, the compliance officer meets with the employer for a closing conference. That is a time for free discussion of problems and needs, a time for frank questions and answers.

The compliance officer describes all unsafe or unhealthful conditions observed and explains all apparent violations for which a citation may be issued or recommended. However, the officer does not assign penalties; only the OSHA area director has that authority, and only after having received a full report. During the closing conference the employer may wish to produce records to show compliance efforts and to provide information that can help OSHA determine how much time may be needed to abate an alleged violation. When appropriate, such as when health hazards are being evaluated or when laboratory reports are required, more than one closing conference may be held.

TABLE 5-1
IAQ Citations Among the 20 Most Frequently Cited General Industry Standards of Fiscal 1994

Standard Section	*Standard*	*Topic*	*Alleged Violations*
1910.1200(e)(1)	HazCom	Written program	4728
1910.1200(h)	HazCom	Employee training	3833
1903.2(a)(1)	Posting of notices	OSH Act protections	2901
1910.1200(f)(5)(i)	HazCom	Container labels	1729
1910.1200(g)(1)	HazCom	MSDSs	1627
1910.1200(f)(5)(ii)	HazCom	Container labels	1571
1910.1200(g)(8)	HazCom	MSDSs	1002

The employees' representative may be present at the closing conference. There may also be a closing discussion with the employees or with their representative to discuss matters of direct interest to employees.

Citations and Penalties

The Occupational Safety and Health Act gave OSHA responsibility for enforcing its regulations and standards; it also gave OSHA the necessary tools for enforcement, in the form of citations and penalties. Typically, fewer than half the inspections result in citations, with state-plan OSHAs issuing more than a proportionate number. In 1994 OSHA's Hazard Communication Standard headed the list of the most frequently cited general industry standard. Violators of the same standard were also the most heavily fined ($2.43 million for 15,586 violations of 1910.1200—Hazard Communications). Table 5-1 gives some figures for indoor air quality violations that were among the 20 most frequent categories cited. (MSDS in the Topic column refers to Material Safety Data Sheet, a form that must identify and accompany hazardous materials being shipped. MSDSs must also be available in the workplace wherever such materials are used.)

Citations inform employers and employees of the regulations and standards alleged to have been violated and of the proposed length of time set for their abatement. Employers receive citations and notices of proposed penalties by certified mail and must post a copy of each citation at or near the place a violation occurred. The posting must remain for three days or until the violation is abated, whichever is longer.

Chapter 1 summarized eight categories of violations that may be cited and the penalties that may be proposed.

NATIONAL INSTITUTE FOR OCCUPATIONAL SAFETY AND HEALTH (NIOSH)

NIOSH develops and periodically revises recommended exposure limits (RELs) for hazardous substances or conditions in the workplace. The agency also recommends appropriate preventive measures to reduce or eliminate the adverse health and safety effects of these hazards. To formulate recommendations, NIOSH evaluates available medical, biological, engineering, chemical, trade, and other information relevant to the hazard. These recommendations are then published and transmitted to OSHA and the Mine Safety and Health Administration (MSHA) for use in promulgating legal standards.

NIOSH recommendations and other publications, many of which are free, are generally available to businesses and the public. When seeking publications, it is helpful to know what to expect in each of the categories.

- *Criteria documents* recommend workplace exposure limits and appropriate preventive measures to reduce or eliminate adverse health effects and accidental injuries.
- *Current Intelligence Bulletins* (CIBs) are issued to disseminate new scientific information about occupational hazards. A CIB may draw attention to a formerly unrecognized hazard, report new data on a known hazard, or present information on hazard control.
- *Alerts*, *Special Hazard Reviews*, *Occupational Hazard Assessments*, and *Technical Guidelines* support and complement the other standards development activities of the agency. Their purpose is to assess the safety and health problems associated with a given agent or hazard (e.g., the potential for injury or for carcinogenic, mutagenic, or teratogenic effects) and to recommend appropriate control and surveillance methods. Although these documents are not intended to supplant the more comprehensive criteria documents, they are prepared to assist OSHA and MSHA in formulating regulations.

Examples of NIOSH Work

As mentioned, NIOSH performs a considerable amount of research. Much of it uses actual records for long-term occupational (versus animal) health statistical studies, allowing direct data application to workers without interpolation or other adjustments that might raise questions. One of the most important studies ever conducted on the health effects of dioxin was carried out by NIOSH on a highly exposed group of U.S. chemical workers. This study has been credited throughout the world for providing data needed to improve our assessment of the human risks of dioxin.

NIOSH has data on its work in each of the 50 states. It evaluates workplace hazards and recommends solutions when requested by employers, workers, or state or federal agencies. The next few paragraphs discuss information from two states: Oregon, because it is on the west coast and has a large wood products industry (logging, mills, etc.), and Maryland, because it is on the east coast and has a low workplace fatality rate.

Oregon. Between 1980 and 1993 NIOSH responded to 55 requests involving a variety of Oregon workplaces, including seafood processing facilities, paper mills, ceramics products manufacturing, gypsy moth control, and small precision parts casting.

For example, the U.S. Department of Agriculture Federal Grain Inspection Service in Portland, Oregon, requested that NIOSH evaluate the exposure to chemical fumigants by grain inspectors working in ports and railcars. The inspectors were required to rely on their own sensory perception to detect fumigants. NIOSH found the inspectors to be at risk of serious health effects, including cancer. The agency then recommended that inspectors wear personal protective equipment, that the fumigated grain be aerated, and that the "sniff test" be modified or dropped as a routine procedure.

In another example from Portland, a foundry requested that NIOSH researchers study the reproductive health problems of men working with the solvent 2-ethoxyethanol. The potential fertility problems identified in these workers contributed significantly to subsequent efforts that will protect workers nationwide from health effects associated with this solvent.

Maryland. The average annual rate of workplace fatalities in Maryland was 5.3 per 100,000 workers—lower than the national average of 7.0 per 100,000. Since 1980 NIOSH has responded to 145 requests from Maryland workplaces ranging from Federal office buildings to silk screening establishments. NIOSH has investigated respiratory problems among sewage treatment plant workers, lead exposures associated with repainting outdoor steel structures, and eye problems among video display terminal users.

The first example is one for which the result was confirmation that existing practices satisfactorily protected workers. This evaluation stemmed from a request for assistance from the Anne Arundel County Department of Utilities. County officials were concerned about potential lead exposures of workers and the surrounding community residents from abrasive blasting removal of lead-based paint. Lead can cause irreversible brain and nervous system damage, mental retardation, infertility, anemia, kidney damage, birth defects, behavioral disorders, and death. NIOSH found that the lead hazards for construction workers and the community residents were well controlled through a comprehensive worker protection program that included innovative engineering controls. The results filled an important research gap by demonstrating a feasible approach to protecting workers during abrasive blasting of lead paint, an increasingly common construction activity accompanying state and national programs to renovate deteriorating bridges and elevated highways.

The last example is a NIOSH evaluation that involved 18 workers at a compressor manufacturing plant in Walkersville, Maryland, who suddenly became ill and were rushed to a hospital. Because the cause of the illness was not known, the plant was evacuated and NIOSH was contacted for emergency assistance. NIOSH researchers identified the cause of the problem as exposure to hazardous concentrations of carbon monoxide from poorly maintained propane space heaters. The problem was complicated by exposure to high concentrations of tetrachloroethylene from manufacturing equipment. These problems were corrected and the plant was back in full production within three days of the evaluation.

AMERICAN SOCIETY OF HEATING, REFRIGERATING, AND AIR-CONDITIONING ENGINEERS (ASHRAE)

ASHRAE establishes minimum ventilation rate standards for the industry, based on a consensus of the HVAC design community. Occupant tolerance, industry knowledge of indoor air contaminants, HVAC system operating costs, building envelope permeability, and the growing number of contaminants generated within buildings have been the basis for frequent changes in the standard over the years.

As concern about pollutants grew, especially in the 1980s, the building industry identified, and took steps to correct, many of the causes of poor indoor air quality. ASHRAE responded by revising a key standard: ASHRAE Standard 62-1989, *Ventilation for Acceptable Indoor Air Quality*. This revision increased the minimum per-person outdoor air requirement to generally between 15 and 20 cubic feet per minute (cfm) for occupied spaces, and 60 cfm for smoking areas. Specific ventilation rates for occupied spaces depend on space type and usage. For example operating rooms require a higher outdoor airflow per person than do auditoriums—30 cfm versus 15 cfm.

The standard quickly became the basis for building codes. In 1992 the Southern Building Code adopted it; in 1993 it became the basis for ventilation requirements in the Uniform Building Code and in the Mechanical Code of the Building Officials and Code Administrators. Today it is the guiding document for most indoor environments.

ASHRAE Standard 62-1989 establishes good source control practices, sets minimum ventilation rates, and defines acceptable indoor air quality that will "avoid adverse health effects." It is particularly important because poor ventilation or lack of outdoor air inflow is generally considered the predominant IAQ problem. Studies indicate that 20 to 50 percent of today's buildings exhibit ventilation-related problems.

Legally, ASHRAE Standard 62-1989 is viewed as the standard of care for designers and building owners to ensure adequate ventilation. It describes general system and equipment requirements for outdoor air inflow, air distribution, design documentation, and microbial growth control. In addition, Section 6.0 of the Standard offers two alternative methods to help designers satisfy this requirement: The Ventilation Rate Procedure specifies minimum outdoor airflow rates for adequate dilution, whereas the IAQ Procedure specifies contaminant levels and subjective evaluation for acceptable indoor air quality.

Chapters 22 and 23 provide a commentary on Standard 62-1989.

CHAPTER 6

Examining Specific Health Concerns

Would the perfect indoor air, if it could be found or produced, be air that simply contains no dangerous or hazardous components?

- Dangerous or hazardous to whom? Some people are immune to certain air contaminants that are deadly to others. Some people *appear* to be immune to certain hazards but are accumulating the effects until a threshold time exposure sensitizes them to that hazard. Some people are not bothered by a certain toxin, but it sensitizes them to other toxins.
- Dangerous or hazardous when? There are elements in the air that are innocuous except at certain temperatures, except at certain times, or except when a person is stressed.
- Dangerous or hazardous when a person is exposed in what way? A given person might be harmed when exposed to a certain chemical for 1 eight-hour session but might not be bothered by 100 four-hour exposures with 20 hours between exposures.

These are some of the realities you must face when examining indoor air quality. This chapter looks at health concerns, including examples of research and investigation by NIOSH. A list of chemical hazards arranged by target organs and by symptoms is also included.

EXAMPLES OF NIOSH INVESTIGATIONS

Chapter 5 discussed different types of NIOSH documents—Criteria Documents, Current Intelligence Bulletins, Alerts, Special Hazard Reviews, Occupational Hazard Assessments, and Technical Guidelines. These are generally conservative, giving information and test results, and making recommendations. Almost no one would condone controlled experiments with humans, even volunteers, so the main sources of data are animal experiments and incidental statistics. NIOSH investigators use experience and judgment in basing recommendations on such data.

Gallium Arsenide in the Electronics Industry

In an Alert NIOSH describes animal studies on gallium arsenide, determines that the paths and organs involved function similarly in humans, and makes exposure limit recommendations.

This Alert gives background information, cites three research reports it uses, draws exposure limit conclusions, and makes recommendations for work practices, personal protective equipment, and waste disposal.

Background. Gallium arsenide is used in the microelectronics industry in manufacturing certain semiconductor devices, and the potential exists for worker exposure to particulates. Results from three animal studies demonstrate that gallium arsenide dissociates in mammals into gallium and arsenic in biological tissues. Further, inorganic arsenic has been determined to be a carcinogen, and since gallium arsenide can dissociate to gallium and arsenic, NIOSH recommends that gallium arsenide be regarded as a potential occupational carcinogen.

Conclusions. NIOSH's evaluation of the three published animal studies concurs with their conclusions that gallium arsenide dissociates into gallium and arsenic and that inorganic arsenic is biologically available and distributed throughout the bodies of exposed animals. There is no evidence that gallium arsenide reacts differently in humans; this compound therefore presents the potential for worker exposure to arsenic, a known carcinogen. NIOSH recommends that worker exposure to gallium arsenide be controlled by observing the NIOSH REL (recommended exposure limit) for inorganic arsenic (2 $\mu g/m^3$ of air as a 15-minute ceiling). NIOSH also recommends that the concentration of gallium arsenide in air be estimated by determination of arsenic, which can be done by using analytical method 7900 in the *NIOSH Manual of Analytical Methods* (NIOSH 1984).

Recommendations. NIOSH makes the following recommendations for minimizing the risk of exposure to gallium arsenide:

- Workers should be made aware of and trained to recognize the hazards of gallium arsenide exposure.
- Engineering controls and work practices should be implemented to reduce gallium arsenide and arsenic exposures in production areas of gallium arsenide semiconductor manufacturing.
- Workers should be provided with and required to use personal protective clothing and equipment.
- Procedures for decontamination, waste removal, transport, and disposal should be established for removing gallium arsenide or arsenic from contaminated materials.

Air and wipe samples should be collected regularly from work areas that have the potential for worker exposure to gallium arsenide particulates or that have the potential for surface contamination. These areas include crystal growing, crystal puller cleaning, crystal sawing, and water polishing and dicing. In some areas of the gallium arsenide process, both arsenic and gallium arsenide particulates may be present.

Engineering Controls and Work Practices. The following recommendations for engineering controls and work practices are intended to reduce gallium arsenide and arsenic exposures in production areas of semiconductor manufacturing.

Crystal growth. Quartz ampoules used in horizontal Bridgeman and gradient freeze processes should be enclosed during crystal growth to prevent emissions of gallium arsenide or arsenic in the event of ampoule failure. If such failures occur, contaminated surfaces should be vacuum cleaned with a system that uses a high-efficiency particulate air (HEPA) filter, or it should be wet-wiped, or both. Use of

HEPA-filtered units requires particular care—personal protective equipment and proper filter disposal—when changing filters.

Crystal puller cleaning. The crystal puller area should be physically isolated from other process areas and maintained at negative pressure in relation to surrounding areas to prevent gallium arsenide and arsenic contamination. Access to the puller area should be limited to necessary and trained personnel. The puller should be allowed to cool before being opened for cleaning, to minimize gallium arsenide and arsenic exposures. During cleaning of crystal pullers for the liquid-encapsulated Czochralski process, local exhaust ventilation should be used to minimize potential exposure to gallium arsenide and arsenic particulates. A HEPA-filtered vacuum cleaner should be used for the initial cleaning. Any additional wiping or scrubbing should be done wet. Floors and other exposed surfaces should be wet mopped after completion of puller cleaning.

Crystal surface grinding and sawing. The crystal grinding and sawing area should be physically isolated from other process areas and maintained at a negative pressure in relation to surrounding areas to prevent gallium arsenide and arsenic contamination from migrating to other areas. Access to this area should be limited to necessary and trained personnel. Wet grinding or sawing can reduce particulate emissions from this process, but aerosolization of the gallium arsenide–contaminated coolant liquid may occur. Thus the coolant catch basin should be enclosed as much as possible. The gallium arsenide–laden coolant liquid should be properly disposed of. If air and wipe sampling indicate incomplete control of emissions, local exhaust ventilation should be applied to the saw. Floors and exposed surfaces should be wet mopped on a regular basis unless wipe sampling indicates no surface contamination with arsenic.

Wafer polishing, backlapping, and dicing. Surfaces that may be contaminated should be wet wiped on a regular basis unless wipe sampling indicates that there is no surface contamination with arsenic. If laser dicing is used, local exhaust ventilation should be provided for this operation.

Personal protective clothing and equipment. Workers should be provided with and required to use disposable suits, gloves, foot coverings, and other appropriate protective clothing necessary to prevent skin contact with gallium arsenide particulates. If respiratory protection is required, employers should institute a respiratory protection program that meets OSHA requirements as specified in 29 CFR 1910.134. In addition to the selection of respirators approved by MSHA and

NIOSH, a complete respiratory protection program should include, at a minimum, an evaluation of workers' ability to perform the work while wearing a respirator, regular training, fit testing, periodic environmental monitoring, maintenance, inspection, and cleaning of respirators.

Decontamination and waste disposal. Decontamination procedures should be established and implemented where feasible to remove gallium arsenide or arsenic from materials and equipment that have become contaminated. When decontamination is not possible or feasible, removal, transport, and disposal of contaminated materials should be in accordance with regulations of the EPA, the U.S. Department of Transportation, and/or state and local authorities.

Exposure to Diesel Exhaust

It is widely known that byproducts of internal combustion engine operation, especially carbon monoxide, are poisonous. Taking the knowledge of the dangers of diesel exhaust further, NIOSH has studied the added danger of developing cancer because of breathing exhausts. Diesel engines were studied separately from gasoline engines because they are the predominant source of industrial power throughout the world for units up to about 5000 horsepower. They can operate with less highly refined fuel, and they consume less fuel per horsepower per hour.

In addition to diesel engines that power vehicles outdoors, many of the engines operate in enclosed or semienclosed areas where NIOSH estimates that 1.35 million workers are occupationally exposed to combustion products in approximately 80,000 workplaces in the United States. Also, exhausts of outdoor vehicles often get into indoor air distribution systems. Workers who are most likely to be exposed include auto, truck, and bus garage workers, farmworkers, forklift drivers, truck drivers, loading dock workers, railroad workers, bridge and tunnel workers, and many people who work in confined spaces.

Diesel engine emissions. Emissions from diesel engines consist of both gaseous and particulate fractions. Gaseous components include carbon dioxide, carbon monoxide, nitric oxide, nitrogen dioxide, oxides of sulfur, and hydrocarbons (e.g., ethylene, formaldehyde, methane, benzene, phenol, 1,3-butadiene, acrolein, and polynuclear aromatic hydrocarbons [PAHs]). Particulates, generally recognized as soot, include solid carbon cores that are produced during the combustion process and that tend to form chain or cluster aggregates. More than 95 percent of these particulates are less than 1 micron (μm) in size. Estimates show that as many as 18,000 different substances from the combustion process can be absorbed onto

TABLE 6-1

Comparison of Limits for Occupational Exposure to Selected Components of the Gaseous Fraction of Diesel Exhaust

Component	OSHA PEL	NIOSH REL
Carbon dioxide	5000 ppm (9000 mg/m^3), 8-hr TWA; 30,000 ppm (54,000 mg/m^3), STEL	10,000 ppm (18,000 mg/m^3), 8-hr TWA; 30,000 ppm (54,000 mg/m^3), 10-min ceiling
Carbon monoxide	50 ppm (55 mg/m^3), 8-hr TWA	35 ppm (40 mg/m^3), 8-hr TWA; 200 ppm (230 mg/m^3), ceiling (no minimum time)
Formaldehyde	1 ppm, 8-hr TWA; 2 ppm, 26-minute STEL	0.016 ppm (0.020 mg/m^3), 8-hr TWA; 0.1 ppm (0.12 mg/m^3), 15-min ceiling
Nitrogen dioxide	5 ppm (9 mg/m^3), ceiling	1 ppm (1.8 mg/m^3), 15-min ceiling
Nitric oxide	25 ppm (30 mg/m^3), 8-hr TWA	25 ppm (30 mg/m^3), 10-hr TWA
Sulfur dioxide	5 ppm (13 mg/m^3), 8-hr TWA	0.5 ppm (1.3 mg/m^3), 10-hr TWA

diesel exhaust particulates. Absorbed material constitutes 15 to 65 percent of total particulate mass and includes such compounds as PAHs, several of which are carcinogens.

Table 6-1 lists OSHA permissible exposure limits (PELs) and NIOSH recommended exposure limits (RELs) for some gases typically found in diesel exhaust. Because diesel emission particulates are of respirable size, the presence of diesel equipment contributes to the total burden of respirable dust present in an occupational environment. Existing limits for occupational exposures to other respirable dusts also limit exposures to the particulate fraction of diesel emissions. Table 6-2 lists OSHA and NIOSH limits relevant to the particulate fraction of diesel engine emissions. In both tables, TWA means time-weighted average, and STEL means short-term exposure limit.

The limits in the two respirable dust columns in Table 6-2 were not intended directly for diesel exhaust particulates, but they would inadvertently limit airborne

TABLE 6-2
Comparison of Limits for Occupational Exposure to the Particulate Fraction of Diesel Exhaust

Component	OSHA PEL	NIOSH REL
Respirable dust	5 mg/m^3	No REL
Respirable dust when quartz content is more than 5% of total	$\frac{10\text{mg/m}^3}{\%\ SiO_2 + 2}$	REL is specific to quartz
Coal tar pitch volatiles (CTPV)	Not applicable to diesel emissions	0.1 mg/m^3 10-hr TWA (cyclohexane-extractables)
Polynuclear aromatic hydrocarbons	No PEL	No REL

concentrations because diesel particulates would be included in respirable dust samples taken where diesel engines are operating.

Conclusions and recommendations. Recent animal studies confirm an association between the induction of cancer and exposure to whole diesel exhaust. The lung is the primary site identified with carcinogenic or tumorigenic responses following inhalation exposures. Limited epidemiological evidence suggests an association between occupational exposure to diesel engine emissions and lung cancer. The consistency of these toxicological and epidemiological findings indicates that a potential occupational carcinogenic hazard exists in human exposure to diesel exhaust.

It is the particulates in diesel exhaust that are primarily associated with tumor induction. Limited evidence indicates that the gaseous fraction of diesel exhaust may be carcinogenic as well.

NIOSH and OSHA consider that whole diesel exhaust be regarded as a potential occupational carcinogen. OSHA has formalized the definition of "Potential occupational carcinogen" in the Code of Federal Regulations (29 CFR 1990) as follows:

> "Potential occupational carcinogen" means any substance, or combination or mixture of substances, which causes an increased incidence of benign and/or malignant neoplasms, or a substantial decrease in the latency period between

> exposure and onset of neoplasms in humans or in one or more experimental mammalian species as the result of any oral, respiratory, or dermal exposure, or any other exposure which results in the induction of tumors at a site other than the site of administration. This definition also includes any substance which is metabolized into one or more potential occupational carcinogens by mammals.

The excess cancer risk for workers exposed to diesel exhaust has not yet been quantified, but minimizing worker exposure should decrease the probability of developing cancer. As prudent public health policy, employers should assess the conditions under which workers may be exposed to diesel exhaust and reduce exposures to the lowest feasible limits.

SYMPTOMS CAUSED BY CHEMICAL HAZARDS[1]

This section presents information about chemicals that harm workers via inhalation. Some of the chemicals listed can enter a worker's system via other routes as well, such as through the skin or via ingestion.

The *NIOSH Pocket Guide to Chemical Hazards* contains 340 pages of hazardous chemicals and over 150 symptoms and target organs. It is not practical to make a separate listing of all the chemicals that could cause abdominal pain, then all those that could cause anorexia, and so forth through all the symptoms. In addition to not being practical, the value of the listing would be questionable because searching the list would not give you a qualified medical diagnosis, and it could lead you to suspect several chemicals that are not present in your work environment.

This section provides two lists of chemicals that often cause some common groups of symptoms. The following precautions for using these lists apply:

- Not everyone with the indicated exposure will exhibit the indicated symptoms.
- The symptoms of a person with the indicated exposure will not necessarily be limited to the symptoms shown.
- Exhibiting the indicated symptoms does not necessarily mean the person has been exposed to the indicated chemical.

1. Thanks to Health Evaluation Programs, Park Ridge, Illinois, for permission to take this information from the chart they prepared.

- Reactions to exposure to multiple chemicals can have unpredictable results.

These lists give you a starting point for investigating problems. Obtain a qualified medical diagnosis before taking action.

Headaches and Irritated Eyes

The following chemicals have been found to cause both headaches and irritated eyes in workers exposed to them:

Acetone
Acetone cyanohydrin
Acetylene tetrabromide
Acrylonitrile
2-Aminopyridine
Antimony
Benzene
Benzenethiol
Benzyl chloride
2-Butanone
2-Butoxyethanol
2-Butoxyethanol acetate
n-Butyl acetate
sec-Butyl acetate
tert-Butyl acetate
n-Butyl alcohol
Butylamine
n-Butyl mercaptan
Calcium cyanamide
Caprolactam
Chloroform
Cyclohexanone
Cyclonite
Cyhexatin
DDT
1-Decanethiol
Demeton
Diazomethane
Dibutyl phosphate
Dichlorvos
Dicyclopentadiene
Diethyl phthalate
Diisobutyl ketone
Diisopropylamine
Dioxane
Dioxathion
Diphenyl
Disulfiram
Disulfoton
Endosulfan
Ethion
Ethyl alcohol
Ethyl benzene
Ethyl butyl ketone
Ethyleneimine
Ethylene oxide
Ethyl ether
Furfural
Gasoline
Glycerin (mist)
Glycolonitrile
1-Heptanethiol
n-Hexane
Hexane isomers
Hexone
sec-Hexyl acetate
Hexylene glycol
Iodine
Isobutyl acetate
Isobutyl alcohol
Isobutyronitrile
Isophorone
Isopropyl alcohol
Kerosene
Lindane
Malononitrile
Methyl acetate
Methyl alcohol
Methyl Cellosolve®
Methyl chloroform
Methylcyclohexanol
Methylisoamyl ketone
Methyl isobutyl carbinol
Methyl parathion
Naphthalene
1-Nitropropane
2-Nitropropane
Nonane
Osmium tetroxide (as Os)
Oxygen difluoride

Parathion
Pentaborane
Petroleum distillates (naphtha)
Phorate
Phosdrin®
Propylene glycol dinitrate
Pyridine
Selenium
Sodium azide
Sodium cyanide (as CN)
Styrene
Subtilisins
Succinonitrile
TEDP
p-Terphenyl
Tetrahydrofuran
Tetranitromethane
Toluene
Tributyl phosphate
Trichloroethylene
Turpentine

Nausea and Dizziness

The following chemicals have been found to cause both nausea and dizziness in workers exposed to them:

Aldrin
2-Aminopyridine
Antimony
Arsine
Benzenethiol
n-Butyronitrile
Camphor (synthetic)
Caprolactam
Chloroform
Chloropentafluoroethane
Cyanogen chloride
Cyclohexanethiol
Cyclohexylamine
Cyclonite
Cyclopentane
Decaborane
1-Decanethiol
Diazinon®
Dicrotophos
Dieldrin
Diethyl phthalate
1-Dodecanethiol
Endrin
Ethylene glycol
Ethylene glycol dinitrate
Ethyleneimine
Ethyl ether
Glycolonitrile
Halothane
1-Hexadecanethiol
Hexane isomers (excluding *n*-hexane)
Hexylene glycol
Hydroquinone
Iodoform
Iron pentacarbonyl (as Fe)
Isoamyl alcohol (primary)
Isoamyl alcohol (secondary)
Isobutyronitrile
Isophorone
Malononitrile
Methyl alcohol
Methyl demeton
Nicotine
Nitroglycerine
o-Nitrotoluene
m-Nitrotoluene
p-Nitrotoluene
Nonane
1-Octadecanethiol
Pentachlorophenol
1-Pentanethiol
Petroleum distillates (naphtha)
1-Propanethiol
Pyridine
Succinonitrile
Temephos
Tetrachloroethylene
Tetrahydrofuran
1,2,4-Trimethylbenzene
1,3,5-Trimethylbenzene
1-Undecanethiol
Vinylbromide
Vinylfluoride
Vinylidene chloride
Vinylidene fluoride
o-Xylene
p-Xylene

CHEMICAL HAZARDS LISTED BY TARGET ORGANS

As expected in a book about indoor air quality, all the hazards mentioned in the preceding lists include inhalation as a route of exposure. Most of the chemicals can enter the body via other routes such as skin absorption or ingestion—a fact that complicates your overall task, because ensuring that workers cannot breathe a certain chemical (for instance, to arbitrarily pick one of hundreds of chemicals, arsine) does not necessarily mean that workers are fully protected from it.

As with the lists organized by symptoms, the following lists give you a starting point for investigation. A word of caution: Not everyone whose indicated target organs are affected has been exposed to the indicated chemical, nor will everyone exposed to the indicated chemical suffer damage to the indicated target organs. Obtain medical confirmation before taking action.

Chemical Hazards that Target Kidneys and Liver

Common targets of toxic chemicals are kidneys and liver. The damage inflicted ranges from temporary to permanent, and often to cancer. The following list identifies chemicals that enter the body via inhalation and that target both liver and kidneys. Most of them target other organs as well.

Acetone cyanohydrin
Acetonitrile
2-Acetylaminofluorene
Acetylsalicylic acid
Aldrin
Allyl chloride
Aniline (and homologs)
o-Anisidine
p-Anisidine
Arsenic
Arsine
Benzenethiol
Benzidine
Bromine pentafluoride
Bromoform
2-Butoxyethanol
2-Butoxyethanol acetate
n-Butyl mercaptan
p-*tert*-Butyltoluene
Caprolactam
Captafol
Captan
Carbon disulfide
Carbon tetrachloride
Chlordane
Chlorobromomethane
Chlorodifluoromethane
Chloroform
o-Chlorostyrene
o-Chlorotoluene
Chromic acid
Chromates
Copper (dusts and mists)
Crag® herbicide
o-Cresol
m-Cresol
p-Cresol
Cyclohexanone
Cyhexatin
2,4-D
DDT
Decaborane
o-Dianisidine
Diborane
o-Dichlorobenzene
p-Dichlorobenzene
1,1-Dichloroethane
1,3-Dichloropropene

Dieldrin
Diisobutyl ketone
Dimethylformamide
Dimethyl sulfate
Dioxane
Diquat (Diquat dibromide)
Endosulfan
Epichlorohydrin
2-Ethoxyethanol
Ethyl bromide
Ethyl chloride
Ethylene chlorohydrin
Ethylenediamine
Ethylene dibromide
Ethylene dichloride
Ethylene glycol dinitrate
Ethyleneimine
Ethylene oxide
Ethylidene norbomene
Ethyl mercaptan
Ethyl silicate
Flourine
Gasoline
Halothane
Hexone
Hydrazine
Indene
Indium
Iodoform
Isophorone
N-Isopropylaniline
Kepone
Lindane
Mesityl oxide
Methomyl
Methoxychlor
Methoxyflurane
Methyl chloride
Methylcyclohexanol
o-Methylcyclohexanone
Methylisoamylketone
Molybdenum
Monomethyl aniline
Morpholine
Naphtha (coal tar)
Naphthalene
Nitrobenzene
p-Nitrochlorobenzene
Nitrogen trifluoride
1-Nitropropane
2-Nitropropane
N-Nitrosodimethylamine
Paraquat
Pentachloroethane
Pentachlorophenol
Perchloromethyl
Phenol
Phenothiazine
Phenylhydrazine
Phosphorus (yellow)
Phthalic anhydride
Pictoram
Picric acid
1-Propanethiol
Propargyl alcohol
Propionitrile
Propoxur
Propylene dichloride
Propylene glycol dinitrate
Pyridine
Resorcinol
Selenium
Sodium fluoroacetate
Stibine
o-Terphenyl
m-Terphenyl
p-Terphenyl
Tetrachloroethylene
Tetramethyl succinonitrile
Tetryl
Tin (organic compounds as Sn)
o-Tolidine
Toluene
o-Toluidine
1,1,2-Trichloroethane
1,2,3-Trichloropropane
Triethylamine
2,4,6-Trinitrotoluene
Vinylidene chloride
m-Xylene
o-Xylene
p-Xylene
Uranium (soluble compounds as U)
Xylidine

Chemical Hazards that Attack Heart or Blood

Chemicals in the following list enter via inhalation, and target either the heart or the blood—items selected because the result is almost always serious. As with the preceding list, many of the chemicals listed here attack other organs as well.

Acetylsalicylic acid
Acrolein
Arsine
Allyl glycidyl ether
Aniline (and homologs)
p-Anisidine
ANTU
Barium chloride (as Ba)
Barium nitrate (as Ba)
Benzene
Benzidine
2-Butoxyethanol
2-Butoxyethanol acetate
n-Butyl glycidyl ether
Cadmium fume (as Cd)
Carbon monoxide
Chloroform
Chromic acid
Chromates
Dicyclopentadienyl iron
N,*N*-Dimethylaniline
1,1-Dimethylhydrazine
Dinitolmide
o-Dinitrobenzene
m-Dinitrobenzene
Dinitrotoluene
Diphenylamine
1-Dodecanethiol
2-Ethoxyethanol
Ethylene glycol dinitrate
Ethylene oxide
Ethyl mercaptan
Ethyl silicate
1-Heptanethiol
1-Hexadecanethiol
n-Hexanethiol
Hydrogen cyanide
Indium
Iodoform
2-Isopropylamine
N-Isopropylaniline
Isopropyl glycidyl ether
Lead
Lindane
Malathion
Manganese tetroxide (as Mn)
Methyl Cellosolve®
Methyl hydrazine
Methyl mercaptan
Molybdenum (soluble compounds as Mo)
Monomethylaniline
Naphthalene
Nitric oxide
p-Nitroaniline
Nitrobenzene
4-Nitrobiphenyl
p-Nitrochlorobenzene
Nitrogen trifluoride
Nitroglycerine
o-Nitrotoluene
m-Nitrotoluene
p-Nitrotoluene
1-Octadecanethiol
Paraquat
Perchloryl fluoride
Phenylhydrazine
Phenylphosphine
Phosphorus (yellow)
Picric acid
Pindone
1-Propenethiol
Propylene glycol dinitrate
n-Propyl nitrate
Resorcinol
Selenium
Sodium cyanide (as CN)
Tellurium
Tetranitromethane
Toluenediamine
o-Toluidine
p-Toluidine
1,2,3-Trimethylbenzene
1,2,4-Trimethylbenzene
1,3,5-Trimethylbenzene
2,4,6-Trinitrotoluene
Triphenyl phosphate
Tungsten
Tungsten carbide (cemented)
Uranium (soluble compounds as U)
Vinyl chloride
Vinyl cyclohexene dioxide
Warfarin
o-Xylene
m-Xylene
p-Xylene
Xylidine

CHAPTER 7

Selecting the Most Effective Respirators for Your Company

Because respirators facilitate breathing, they are a prominent consideration in the field of indoor air quality. This chapter discusses various types of respirators and their purposes and provides a detailed decision tree for selecting the proper type for your given conditions.

This chapter draws heavily from what is probably the ultimate source of expertise on this subject: the *Joint Respirator Committee*, established in 1973 by NIOSH and OSHA. Their purpose was to develop standard respirator selection criteria and tables for the approximately 400 hazardous materials regulated by OSHA.

Today NIOSH has established an evaluation and certification program and issues certifications jointly with MSHA. The certification program's goal is to help increase worker protection from airborne contaminants by certifying respirators that meet the minimum performance requirements that appear in regulation 30 CFR 11. A NIOSH certification evaluation includes a laboratory evaluation of the respirator, an evaluation of the manufacturer's quality control (QC) plan, audit testing of certified respirators, and investigations of problems.

NIOSH also monitors respirators over the lifetime of their certification. Samples of "off the shelf" respirators are evaluated in NIOSH laboratories to see if they continue to meet applicable minimum performance requirements. In addition, NIOSH performs in-plant QC audits to determine if manufacturers are complying with the QC plans submitted in their approval applications. Reports of problems received from regulatory agencies, labor organizations, respirator users, and respirator manufacturers are investigated and resolved.

When are respirators used?

- They are used routinely by workers such as grinders and sandblasters who are constantly in a suspect atmosphere.
- They are used by workers who handle toxic materials.
- They are used by workers such as firefighters who work in an oxygen-deficient atmosphere.
- They are used in emergency situations to allow people extra time to escape from an area in which the atmosphere has become toxic.

EXAMINING TYPES OF RESPIRATORS

The basic purpose of any respirator is, simply, to protect the respiratory system from inhalation of hazardous atmospheres. Respirators provide protection either by removing contaminants from the air before it is inhaled or by supplying an independent source of respirable air. Those that remove contaminants from the ambient air are called *air-purifying respirators*; those that provide air from a source other than the surrounding atmosphere are *atmosphere-supplying respirators*. Both types can be further subclassified by the mode of operation and by the facepiece, hood, suit, or other device—called the *inlet covering*—that serves as a barrier separating contaminated air from the worker.

Tight-Fitting Coverings

Tight-fitting coverings, usually called *facepieces*, are made of flexible molded rubber, silicone, neoprene, or other materials. Present designs incorporate rubber or woven elastic headstraps that are attached at two to six points. They either buckle at the back of the head or may form a continuous loop of material.

Facepieces are available in three basic configurations. The first, a *quarter-mask*, covers the mouth and nose, with the lower sealing surface resting between chin and mouth. Quarter-masks, found most commonly on dust and mist respirators, provide good protection but are more easily dislodged than other types. A second type, the *half-mask*, fits over the nose and under the chin. Half-masks are designed to seal more reliably than quarter-masks and are therefore preferred for use against more toxic materials. The *full-facepiece* inlet covering provides cover from roughly the hairline to below the chin. It generally gives the greatest protection, usually seals most reliably, and provides some eye protection as well. Full-facepiece respirators, both air-purifying and atmosphere-supplying, are designed for use in higher concentrations of toxic materials than are quarter- or half-mask respirators.

Mouthpiece respirators have a clamp to close the nostrils and a mouthpiece that a person holds in the teeth, allowing the lips to seal around it. This type provides a good seal but precludes vocal communication, causes fatigue, and provides no eye protection. It is certified for use as an escape-only respirator.

Loose-Fitting Coverings

Loose-fitting coverings come in a wide variety of designs that include hoods, helmets, suits, and blouses. A light, flexible device covering only the head and neck, or head, neck, and shoulders is called a *hood*. If rigid headgear is incorporated into the design, it is called a *helmet*. *Blouses* extend down to the waist, and some have wrist-length sleeves. The enclosure includes a system through which clean compressed air is distributed around the breathing zone.

A special type of loose-fitting covering in common use is the abrasive-blasting hood. The hood material is designed to withstand rebounding particles, and there is usually an impact-resistant glass or plastic viewing lens with additional plastic, glass, or woven wire shielding that deflects rebounding particles.

EXAMINING AIR-PURIFYING RESPIRATORS

We now look at respirators that filter or clean ambient air to provide safe air for the wearer to breathe. These respirators are subdivided into those that filter out particulates and those that remove vapors and gases.

Particulate-Filtering Respirators

Particulate-filtering respirators provide protection against dusts, fumes, and/or mists. A *dust* is a solid, mechanically produced particle; a *fume* is a solid condensation particulate, usually of a vaporized metal; a *mist* is a liquid condensation particulate.

All current particulate-filtering respirators use fibrous material to filter out the contaminant. Efficiency is evaluated as the probability that a single particle will be trapped by the fibers; filters are not designed to be 100 percent efficient because they would be unacceptably difficult to breathe through. The efficiency depends on such factors as particle size relative to the fiber size, its velocity, and to some extent, the composition, shape, and electrical charge of both particle and fiber. As particulates collect on the filter they completely or partially clog openings, which usually results in more efficient trapping of other particles, along with an increase in breathing resistance.

These considerations lead to two ways of identifying particulate filters: absolute and nonabsolute. *Absolute* filters use screening to remove particles from the air. That is, they exclude particles that are larger than the pores. However, most respirator filters are *nonabsolute*, which means they contain pores that are larger than the particles to be removed. They use combinations of interception capture, sedimentation capture, inertial impaction capture, diffusion capture, and electrostatic capture to remove the particles.

The exact combination of filtration mechanisms that come into play depends on the flowrate through the filter and the size of a particle. In general:

- Large heavy particles are usually removed by inertial impaction and interception.
- Large light particles are removed by diffusion and interception.
- Very small particles are removed by diffusion.

Brief descriptions of these filtration mechanisms follow.

Interception Capture. As the air streams approach a fiber lying perpendicular to their path, they split and compress in order to flow around the fiber and rejoin on the other side. If the center of a particle in these airstreams comes within one particle radius of the fiber, it encounters the fiber surface and is captured. As particle size increases, the probability of interception capture increases. Particles do not deviate from their original streamline in this mechanism.

Diffusion Capture. The motion of smaller particles is affected by air molecules colliding with them, and particles can randomly cross the airstream and encounter a fiber as they pass. This random motion depends on particle size and air temperature. As particle size decreases and air temperature increases, the particle's diffusive activity increases, increasing the probability of capture. Lower flowrate through the filter also increases the probability of capture because the particle spends more time in the area of a fiber.

Sedimentation Capture. Only large particles (2 microns and larger) are captured by sedimentation. This type of capture relies on gravity to pull particles from the airstream, so flowrate through the filter must be low.

Inertial Impaction Capture. As airstreams split and change direction suddenly to go around a fiber, particles with sufficient inertia cannot change direction sufficiently fast to avoid the fiber, and they impact on the fiber's surface. A particle's size, density, speed, and shape determine its inertia.

Electrostatic Capture. In electrostatic capture, particles are charged with one polarity and filter fibers are charged with the opposite polarity, causing particles to be attracted to filter fibers. Usually the electrostatic capture mechanism is designed to aid other capture mechanisms, especially interception and diffusion.

Three Types of Particulate Filters

There are several types of particulate filters, but three types predominate. The most common type presently available is a machine-made flat disk of random-laid nonwoven fiber material, carefully controlled to produce maximum filter efficiency and minimum breathing resistance.

A second type is a flat disk of compressed natural wool or synthetic fiber felt, or a blend that is given an electrostatic charge during filter manufacture. The charge results from impregnating the material with a resin and mechanically beating or "needling" it. Increased filter efficiency protects adequately against most industrial dusts. However, certain agents, such as oil mists, can remove the electrostatic charge, as can storage in very humid air. Therefore this type of filter should be stored in its original package, kept out of oil mists and high humidity (over 80 percent), and used as soon as possible after purchase. You can identify a resin-impregnated felt filter by rubbing it between two fingers and then rubbing the fingers together. The fingers will feel slightly sticky.

In the third type of dust filter, the filtering medium is only loosely packed in the filter container, making it much thicker than the compressed type. Such filters are usually made of fibrous glass, but nonfelted, resin-impregnated natural wool fibers have been used. They are not as common as the felted type.

MSHA/NIOSH-Certified Particulate Respirators

For the 30 CFR 11 Subpart K certification tests, particulate respirators are classified as designed for protection against a variety of dusts, fumes, and mists. The following types are presently certified by MSHA/NIOSH:

Replaceable or Reusable Dust and Mist. Respirators, with either replaceable or reusable filters, designed as respiratory protection against (1) dusts and mists having an exposure limit of not less than 0.05 mg/m^3 of air or (2) dusts and mists having an exposure limit of not less than 2 million particles per cubic foot of air.

Replaceable Fume. Respirators, with replaceable filters, designed as respiratory protection against fumes of various metals having an exposure limit of not less than 0.05 mg/m^3 of air.

Replaceable Dust, Fume, and Mist. Respirators, with replaceable filters, designed as respiratory protection against dusts, fumes, and mists of materials having an exposure limit of less than 0.05 mg/m^3 or 2 million particles per cubic foot of air.

Single Use. Respirators designed as respiratory protection against pneumoconiosis- and fibrosis-producing dusts, or dusts and mist. The filter is either an integral part of the facepiece or it is the entire facepiece itself.

High-Efficiency and Lower-Efficiency Respirators

The classification schedules allow for identifying two levels of efficiency—high efficiency and lower efficiency. As shown in the following definitions, efficiency is mainly a function of the smallest particle expected to be removed from breathable air.

High Efficiency. The highest efficiency filters (99.97 percent against a 0.3 micron dioctyl phthlate particle) are used in high-efficiency respirators certified for protection against dusts, fumes, and mists having an exposure limit of less than 0.05 mg/m^3 or 2 million particles per cubic foot of air.

Lower Efficiency. Respirators for dust, fumes, and mists having an exposure limit of not less than 0.05 mg/m^3 of air have lower-efficiency filters as classified in 30 CFR 11. Their efficiency is approximately 99 percent against a silica dust particle with a geometric mean diameter of 0.4 to 0.6 micron and a standard geometric mean deviation not greater than 2.

Vapor- and Gas-Removing Respirators

The other major class of respirators for removal of airborne contaminants provides protection against both specific gases and vapors, such as ammonia gas and mercury vapor, and classes of gases and vapors, such as acid gases and organic vapors. In contrast with filters, which are effective to some degree no matter what the particulate, cartridges and canisters used for vapor and gas removal are designed for protection against specific contaminants.

Vapor- and gas-removing respirators normally remove the contaminant by interaction of its molecules with a granular, porous material, commonly called the *sorbent*. The process is called *sorption*. In addition to sorption, some respirators use catalysts that react with the contaminant to produce a less toxic gas or vapor.

The sorbent and catalyst processes are essentially 100 percent efficient until the sorbent's capacity to absorb gas and vapor or to catalyze their reaction is exhausted. After that the contaminant will pass through the sorbent and into the facepiece. This action is in contrast with mechanical particulate removal, which becomes more efficient as matter collects on and plugs the spaces between fibers. Three removal mechanisms are used in vapor- and gas-removing respirators: adsorption, absorption, and catalysis, and the sorbents may be housed in either cartridges or canisters.

Adsorption. Adsorption retains the contaminant molecule on the surface of the sorbent granule by physical attraction. The intensity of the attraction varies with the type of sorbent and contaminant. Adsorption by physical attraction holds the adsorbed molecules only weakly, but chemisorption (involving chemical forces) is usually also involved. With this process, the bonds holding the molecules to the sorbent granules are much stronger.

A characteristic common to all adsorbents is a large specific surface area, up to 1500 m^2/g of sorbent. Activated charcoal is the most common adsorbent. It is used primarily to remove organic vapors, although it does have some capacity for adsorbing acid gases. Activated charcoal also can be impregnated with other substances to make it more selective against specific gases and vapors. Examples are activated charcoal impregnated with iodine to remove mercury vapor, with metallic oxides to remove acid gases, and with salts of metals to remove ammonia gas. Other sorbents that could be used in vapor- and gas-removing respirators include molecular sieves, activated alumina, and silica gel.

Absorption. Absorbents may also be used to remove gases and vapors. Absorbents differ from adsorbents in that although absorbents are porous, they do not have as large a specific surface area. Absorption is also different from adsorption in that gas or vapor molecules usually penetrate deep into the molecular spaces throughout the sorbent and are held there chemically. Absorption probably cannot occur without prior adsorption on the surface of the particles. Furthermore, adsorption occurs instantaneously, whereas absorption is slower. Most absorbents are used for protection against acid gases. They include mixtures of sodium hydroxide or potassium hydroxide with lime and/or caustic silicates.

Catalysis. A catalyst is a substance that influences the rate of chemical reaction between other substances. In respirator cartridges and canisters, one catalyst often used is hopcalite, a mixture of porous granules of manganese and copper oxides. This mixture speeds the reaction between toxic carbon monoxide and oxygen to form carbon dioxide.

Cartridges and canisters. Vapor- and gas-removing sorbents may be contained in either cartridges or canisters. The basic difference is in the volume of sorbent contained, not in its function. Cartridges are vapor- and gas-removing elements that may be used singly or in pairs on quarter- and half-masks and on full facepieces. The sorbent volume is small, about 50 to 200 cubic centimeters (cc), so the useful lifetime is short, particularly in high gas or vapor concentrations. Therefore, the use of respirators with cartridges generally is restricted to low concentrations of vapors and gases. Users should refer to NIOSH recommendations, certification labels, or specific standards set forth by regulatory agencies for specific maximum use concentrations.

Canisters have a larger sorbent volume and may be chin-, front-, or back-mounted. Respirators with canisters can be used in higher vapor and gas concentrations (up to the immediately dangerous to life or health level) than those with cartridges. Chin-style canisters have a volume of about 250 to 500 cc and are used on full-facepiece respirators. Front- or back-mounted canisters are held in place by a harness and are connected to the facepiece by a corrugated, flexible breathing tube. They have a sorbent volume of 1000 to 2000 cc. Front- or back-mounted and chin-style canisters are used with full facepieces as part of "gas masks." The gas mask is certified for single or specific classes of gases and vapors. It differs from the chemical cartridge respirator only in its larger sorbent volume and the higher concentrations of vapors and gases against which it provides protection.

Because vapor- and gas-removing cartridges and canisters are designed for protection against specific contaminants or classes of contaminants, it is important to check the printed certification label for the list of applicable contaminants. In addition American National Standard ANSI K13.1-1973 has established a color code for the various types of sorbent cartridges and canisters to identify the contaminants they protect against. However, users should not rely on memorizing the color code; they should always *read the label!* (The color code of the ANSI K13.1 standard has been included verbatim in OSHA regulation 29 CFR 1910.134[g].)

Powered Air-Purifying Respirators

A powered air-purifying respirator (PAPR) uses a blower to pass contaminated air through an element that removes the contaminants and supplies the purified air to a respirator inlet covering. The purifying element may be a filter to remove particulates, a cartridge to remove vapors and gases, or a combination filter and cartridge, canister, or canister and filter. The covering may be a facepiece, helmet, or hood. These respirators are certified under 30 CFR 11, subparts I, K, L, and M.

PAPRs come in several configurations. In one configuration the air-purifying element(s) is attached to a small blower on the wearer's belt and is connected to the respiratory inlet covering by a flexible tube. This type of device is usually powered by a small battery, either mounted on the belt separately or as part of the blower. Some units are powered by an external DC or AC source.

In another configuration the air-purifying element is attached to a stationary blower, usually mounted on a vehicle and powered by a battery or an external power source. It is connected to the respiratory inlet covering by a long flexible tube.

In a third configuration the powered air-purifying respirator consists of a helmet or facepiece to which the air-purifying element and blower are attached. Only the battery is carried on the belt.

The respiratory inlet covering for a powered air-purifying respirator may be a tight-fitting half-mask or full facepiece, or a loose-fitting hood or helmet. With a tight-fitting facepiece, the PAPR must deliver at least 4 cfm (115 liters per minute [lpm]). When used with a loose-fitting facepiece, the PAPR must deliver at least 6 cfm (170 lpm) at all times.

One potential disadvantage of PAPRs is that there is a constant flow through the air-purifying element instead of flow only during inhalation. The result is that the useful service lifetimes of the elements could be shorter than the service lifetimes of comparable elements attached to a negative-pressure respirator. To overcome this problem, some PAPRs have a spring-loaded exhalation valve assembly that causes the blower assembly to slow down when the wearer exhales.

ATMOSPHERE-SUPPLYING RESPIRATORS

This section discusses respirators that provide breathing gas from a source independent of the surrounding atmosphere instead of purifying gas drawn from the ambient atmosphere. Different types are classified according to the method by which the breathing gas is supplied and used and the method used to regulate the gas supply.

The following types are covered in this section:

- *Self-contained breathing apparatus (SCBA)*. With these respirators the wearer need not be connected to a stationary breathing gas source, such as an air compressor. Instead, enough air or oxygen for up to four hours, depending on the design, is carried by the wearer. SCBAs are classified as *closed circuit* or *open circuit*.
- *Supplied-air respirators (SARs)*. Variations of these respirators are *airline respirators* and *hose masks*.
- *Combination respirators*. MSHA/NIOSH may certify respirators assembled from two or more types of respirators in combination, as prescribed in 30 CFR 11.63(b). Several types of air-purifying units or SCBA in combination with the Type C supplied-air respirator have been certified.

SCBA Closed-Circuit Respirators

These respirators are also known as *rebreather* devices, a self-descriptive name. The wearer rebreathes exhaled gas after the device removes carbon dioxide and restores the oxygen content from a compressed or liquid oxygen source or an oxygen-generating solid. Descriptions and certification tests are given in subpart H of 30 CFR 11.

Rebreathers are designed primarily for one to four hours' use in oxygen-deficient and/or atmospheres immediately dangerous to life or health (IDLH) such as might be encountered during mine rescues or in confined spaces. Because negative pressure is created in the facepiece of non-positive-pressure apparatus during inhalation, there is increased leakage potential. Therefore, negative-pressure closed-circuit SCBA should be used in IDLH atmospheres only where their long-term-use capability is necessary. No such limitation applies to use in oxygen-deficient atmospheres over long periods.

Two basic types of closed-circuit SCBA are presently available: One uses a cylinder of compressed oxygen and the other uses a solid oxygen-generating substance. With the cylinder version, breathable air is supplied from an inflatable bag. Exhaled air passes through a granular solid adsorbent that removes carbon dioxide, thereby reducing the flow back into the breathing bag. The bag collapses, so that a pressure plate bears against the admission valve and admits more pure oxygen, which reinflates the bag.

The advantage of the rebreathing system is extended protection time (one to four hours), because most of the air components are recirculated, and only the oxygen supply has to be provided. Disadvantages include the bulk of the SCBA and the negative pressure created in the facepiece during inhalation from some closed-circuit SCBAs.

A second type of closed-circuit SCBA provides oxygen from an oxygen-generating solid, usually potassium superoxide (KO_2). Water and carbon dioxide in the exhaled breath react with the KO_2 to release oxygen (O_2) according to the following equations:

$$2\ KO_2 + CO_2 + H_2O \rightarrow K_2CO_3 + 1.5\ O_2 + H_2O$$
$$2\ KO_2 + 2\ CO_2 + H_2O \rightarrow 2\ KHCO_3 + 1.5\ O_2$$

There is a short time lag between canister initiation and the start of oxygen flow because oxygen is not released until the wearer's exhaled breath reaches the canister. Some devices provide a quick-start feature known as a *chlorate candle*, which is a canister section filled with mixed sodium chlorate and iron. The wearer starts oxygen flow by striking the device, somewhat like lighting a match. This design provides oxygen until the potassium superoxide begins to function.

A closed-circuit apparatus with an oxygen-generating solid is lighter and simpler than the cylinder type. However, it is useful for only about one hour and, once initiated, cannot be turned off. The precautions are the same as for the compressed oxygen unit.

SCBA Open-Circuit Respirators

An open-circuit SCBA exhausts exhaled air to the atmosphere instead of amending and recirculating it. Regulation 30 CFR 11 does not specify which breathing gas must be used, but it is almost always compressed air. Compressed oxygen cannot be used in a device designed for compressed air because minute amounts of oil or other foreign matter in the device components can cause an explosion. In fact, 30 CFR 11 prohibits certification of any device designed to permit interchangeable use of oxygen and air. The accepted safety rule is:

NEVER USE OXYGEN IN A DEVICE UNLESS IT IS SPECIFICALLY DESIGNED FOR THAT PURPOSE!

In a typical open-circuit SCBA a cylinder of compressed air (2000 to 4500 pounds per square inch [psi]) supplies air to a regulator that has two main functions: It reduces the pressure for delivery to the facepiece, and it regulates air flow by passing air to the facepiece on demand. The regulator is either mounted directly to the facepiece or connected via a flexible corrugated hose to the respiratory inlet covering, usually a full facepiece.

The service life of an open-circuit SCBA is usually shorter than for the closed-circuit version because the former has to provide the total breathing volume requirements without recirculation. Thirty to sixty minutes is typical. SCBA with less than thirty minutes of service time are certified generally for escape use only.

Supplied-Air Airline Respirators (Types C and CE)

Airline respirators as described in 30 CFR 11 subpart J use compressed air from a stationary source delivered through a hose under pressure, typically under 125 psi at the point where the hose attaches to the air supply. Certified systems deliver at least 170 Lpm to a helmet or hood at the lowest pressure and longest hose length. At the highest pressure and shortest hose length, the flowrate must not exceed 425 Lpm to a helmet or hood. Equivalent airflows to a tight-fitting facepiece are 115 Lpm and 425 Lpm, respectively.

A demand or pressure-demand airline respirator is similar in basic operation to a demand or pressure-demand open-circuit SCBA, except that the air is supplied through a small-diameter hose from a stationary source of compressed air rather than from a portable air source. Because the air pressure is limited to 125 psi, regulators for airline respirators need only single-stage reduction.

Supplied-Air Hose Masks

Hose masks supply air from an uncontaminated source through a strong, large-diameter hose to a respiratory inlet covering. Two types are available; one has a blower and the other does not.

The blower on the first type can be hand or motor operated to push low-pressure air through the hose to the respiratory inlet covering. The blower is designed so that air flows freely through it when it is not blowing, allowing the wearer to inhale respirable air by breathing normally in the event of a blower failure. Because the second type has no blower, the wearer always inhales through the hose.

The hose mask with a blower is categorized by 30 CFR 11 subpart J as a Type A supplied-air respirator and is certified for use in atmospheres not immediately dangerous to life or health. Hose masks without a blower are categorized as Type B and are certified for use only in atmospheres not immediately dangerous to life or health.

A certified hose mask with blower may have up to 300 ft of air supply hose in multiples of 25 ft; masks without blower are limited to 75 ft of hose in multiples of 25 ft. A hand- or motor-operated blower must deliver air through the maximum length of hose at not less than 50 Lpm. The motor-operated blower of a device with 50 ft of hose must deliver no more than 150 Lpm. No maximum air flow rate is specified for a hand-operated blower.

Currently there are only three hose masks certified, and they are not widely used. Characteristics such as being heavy and cumbersome, and offering a very low protection factor, make them unpopular.

Combination Respirators

Regulation 30 CFR 11.63(b) provides for MSHA/NIOSH to certify respirators assembled as combinations of two or more types of respirators. Both agencies have certified several types of air-purifying units or SCBA in combinations with the Type C supplied-air respirator.

Advantages and Disadvantages of Atmosphere-Supplying Respirators

As can be expected, no one type of respirator is superior for all situations. Characteristics to consider include weight, cost, efficiency, safety, dependability, work that must be done, and mobility. Following is a short summary of some of the features previously mentioned and a comparison of advantages and disadvantages of airline respirators, hose masks, and self-contained breathing apparatus.

Airline respirators. A significant advantage is that they may be used for long continuous periods. Other advantages are minimal breathing resistance and discomfort, light weight, low bulk, moderate initial cost, and relatively low operating cost.

The biggest disadvantage of supplied-air respirators is that losing the source of respirable air supplied to the respiratory inlet covering eliminates all protection to the wearer. Such loss may be caused by cutting, burning, kinking, or crushing the supply air hose, by compressor failure, or by depletion of the respirable air in a storage tank. The possibility of such loss supports the NIOSH recommendations against airline respirator use in IDLH atmospheres. However, an airline respirator with an auxiliary self-contained air supply could be used in such atmospheres because the auxiliary self-contained air supply can be used for escape.

The trailing air supply hose may make the airline respirator unsuitable for those who move frequently between widely separated work stations. A combination airline and self-contained breathing apparatus may be suitable if the supply of self-contained breathing air is adequate for the times required to move from place to place.

Airline respirators that operate in the demand mode have negative air pressure inside the respiratory inlet covering during inhalation, permitting contaminated atmosphere to leak in if the covering fits poorly. However, airline respirators that operate in the pressure-demand mode are designed to have positive air pressure inside the respiratory inlet covering, which helps keep contaminated air from leaking in.

Hose Masks. Advantages of the hose mask without blower are its theoretically long use periods and its simple construction, low bulk, easy maintenance, and minimal operating cost. An advantage of the hose mask with blower is its minimal resistance to breathing.

Air pressure inside the respiratory inlet covering of the hose mask with no blower is negative during inhalation, so contaminated air may leak in if the covering fits poorly. Therefore, hose masks, with or without blower, are certified only for use in non-IDLH atmospheres.

The trailing air supply hose of the hose mask limits mobility. The maximum hose length of 75 ft for a hose mask without blower is a severe restriction. It also requires the wearer to inhale against the resistance to air flow offered by the air hose, which can become significant during heavy work. Inhaling against this resistance strains the wearer and leads to fatigue.

Self-contained breathing apparatus. Because SCBA wearers carry their own supply of respirable air, they are independent of surrounding atmosphere. Comparatively free movement over an unlimited area is a major advantage of such an apparatus.

The bulk and weight of most SCBAs make them unsuitable for strenuous work or use in a constricted space, and limited service life makes them unsuitable for routine use over long continuous periods. The short service life of open-circuit-type devices generally limits them to use where the wearer can conveniently and quickly go from the hazardous atmosphere to a safe atmosphere to change the supply air tank.

Open-circuit SCBAs are normally less expensive to purchase and use than closed-circuit SCBAs. In addition, open-circuit SCBAs require less maintenance and fewer inspections.

Demand-type open-circuit SCBAs and most closed-circuit SCBAs have negative air pressure inside the respiratory inlet covering during inhalation, so contaminated air can leak in if they fit poorly. Pressure-demand-type open-circuit SCBAs and those closed-circuit SCBAs that are positive-pressure devices provide very good protection because the air inside the respiratory inlet covering is normally at positive pressure, helping to keep contaminated atmosphere from leaking in.

GUIDELINES FOR MAKING YOUR RESPIRATOR SELECTION

After identifying, examining, and evaluating all criteria, and after meeting the requirements and restrictions of the respiratory protection program, you can use this section to help narrow the choice of respirator. This sequence of questions leads toward identifying the class of respirators that should provide adequate respiratory protection.

1. Is the respirator intended for use during firefighting?
   a. If yes, only a self-contained breathing apparatus (SCBA) with a full facepiece operated in pressure-demand or other positive-pressure mode is recommended.
   b. If no, proceed to Question 2.
2. Is the respirator intended for use in an oxygen-deficient atmosphere, i.e., less than 19.5 percent oxygen at sea level? (See the section *Oxygen-Deficient Atmosphere* following the decision logic questions for more information on this subject.)
   a. If yes, any type of SCBA or supplied-air respirator (SAR) with an auxiliary SCBA is recommended. The auxiliary SCBA must be of sufficient duration to permit escape to safety if the air supply is interrupted. If additional contaminants are present, proceed to Question 3.
   b. If no, proceed to Question 3.
3. Is the respirator intended for use during emergency situations?
   a. If yes, two types of respirators are recommended: an SCBA with a full facepiece operated in pressure-demand or other positive-pressure mode, or an SAR with full facepiece operated in pressure-demand or other positive-pressure mode in combination with an auxiliary SCBA operated in

pressure-demand or other positive-pressure mode. The auxiliary SCBA must be of sufficient duration to permit escape to safety if the air supply is interrupted.

b. If no, proceed to Question 4.

4. Is the contaminant regulated by the Department of Labor as a potential occupational carcinogen or identified by NIOSH as a potential human carcinogen in the workplace, and is the contaminant detectable in the atmosphere?
   a. If yes, two types of respirators are recommended: an SCBA with a full facepiece operated in pressure-demand or other positive-pressure mode, or an SAR with full facepiece operated in pressure demand or other positive-pressure mode in combination with an auxiliary SCBA operated in pressure-demand or other positive-pressure mode. The auxiliary SCBA must be of sufficient duration to permit escape to safety if the air supply is interrupted.
   b. If no, proceed to Question 5.
5. Is the exposure concentration of the contaminant, as determined by acceptable industrial hygiene methods, less than the NIOSH REL or other applicable exposure limit? (Whenever a worker is given a respirator to use on a voluntary basis when ambient levels are below applicable limits, OSHA requires the implementation of a complete respiratory protection program, which includes medical evaluation, training, fit testing, periodic environmental monitoring, and all other requirements in 29 CFR 1910.134.)
   a. If yes, a respirator would not be required except for an escape situation. Proceed to Question 7.
   b. If no, proceed to Question 6.
6. Are conditions such that a worker who is required to wear a respirator can escape from the work area and not suffer loss of life or immediate or delayed irreversible health effects if the respirator fails, i.e., are the conditions *not* immediately dangerous to life or health (IDLH)? (See the section *Immediately Dangerous to Life or Health* following the decision logic questions for more information on IDLH.)
   a. If yes, conditions are not considered to be IDLH. Proceed to Question 7.
   b. If no, conditions are considered to be IDLH. Two types of respirators are recommended: an SCBA with a full facepiece operated in pressure-demand or other positive-pressure mode or an SAR with a full face-piece operated in pressure-demand or other positive-pressure mode in combination with an auxiliary SCBA operated in pressure-demand or

other positive-pressure mode. The auxiliary SCBA must be of sufficient duration to permit escape to safety if the air supply is interrupted.

7. Is the contaminant an eye irritant, or can the contaminant cause eye damage at the exposure concentration? (See the section *Eye Irritation* at the end of the decision logic questions for more information on this subject.)
   a. If yes, a respirator equipped with a full facepiece, helmet, or hood is recommended. Proceed to Question 8.
   b. If no, a quarter- or half-mask respirator may still be an option, depending on the exposure concentration. Proceed to Question 8.
8. Divide the eight-hour time-weighted average (TWA) exposure concentration for the contaminant (or maximum exposure concentration for a contaminant with a ceiling limit) determined in Question 5 by the NIOSH REL or other applicable exposure limit to determine the minimum protection factor required. For escape respirators, determine the potential for generation of a hazardous condition caused by an accident or equipment failure. If a potential hazardous condition could occur, or if a minimum protection factor has been calculated, proceed to Question 9.
9. If the physical state of the contaminant is a particulate (solid or liquid) during periods of respirator use, proceed to Question 10; if it is a gas or vapor, proceed to Question 11; if it is a combination of gas or vapor and particulate, proceed to Question 12.
10. Particulate Respirators

10.1. Is the particulate respirator intended for escape purposes only?
   a. If yes, refer to the section *Escape Apparatus* following the decision logic questions for more information on the subject.
   b. If no, the particulate respirator is intended for use during normal work activities. Proceed to Question 10.2.

10.2. A filter medium that will provide protection against exposure to the particulate in question is recommended. Proceed to Question 10.3.

10.3. Respirators that are listed in Table 7-1 and that have assigned protection factors (APFs) equal to or greater than the minimum protection factor determined in Question 8 are recommended. (See the section *Several Variations of Protection Factors* following the decision logic questions for more information on the subject.) Maximum airborne concentrations for each level of respiratory protection can be calculated by multiplying the NIOSH REL or other applicable exposure limit by the APF for that class of respirator. Workers wearing respirators should meet NIOSH recommended medical guidelines.

11. Gas/Vapor Respirators

TABLE 7-1

Assigned Protection Factor Classifications of Respirators for Protection against Particulate Exposures[a]

Assigned Protection Factor	***Type of Respirator***
5	Single-use or quarter-mask respirator[b]
10	Any air-purifying half-mask respirator including disposable[c] equipped with any type of particulate filter except single-use[b]
	Any air-purifying full-facepiece respirator equipped with any type of particulate filter[d]
	Any supplied-air respirator equipped with a half-mask and operated in a demand (negative-pressure) mode[b]
25	Any powered air-purifying respirator equipped with a hood or helmet and any type of particulate filter
	Any supplied-air respirator equipped with a hood or helmet and operated in a continuous-flow mode
50	Any air-purifying full-facepiece respirator equipped with a high-efficiency filter[b]
	Any powered air-purifying respirator equipped with a tight-fitting facepiece and a high-efficiency filter
	Any supplied-air respirator equipped with a full facepiece and operated in a demand (negative-pressure) mode[b]
	Any supplied-air respirator equipped with a tight-fitting facepiece and operated in a continuous-flow mode
	Any self-contained respirator equipped with a full facepiece and operated in a demand (negative-pressure) mode[b]
1,000	Any supplied-air respirator equipped with a half-mask and operated in a pressure-demand or other positive-pressure mode[b]
2,000	Any supplied-air respirator equipped with a full facepiece and operated in a pressure-demand or other positive-pressure mode[b]
10,000	Any self-contained respirator equipped with a full facepiece and operated in a pressure-demand or other positive-pressure mode[b]

TABLE 7-1
Continued

Assigned Protection Factor	*Type of Respirator*
	Any supplied-air respirator equipped with a full facepiece operated in a pressure-demand or other positive-pressure mode in combination with an auxiliary self-contained breathing apparatus operated in a pressure-demand or other positive-pressure mode[b]

Notes
[a] Only high efficiency filters are permitted for protection against particulates having exposure limits of less than 0.05 mg/m^3.

[b]Assigned protection factors were determined by Los Alamos National Laboratories by conducting fit testing on a panel of human volunteers.

[c] An APF of 10 can be assigned to disposable particulate respirators if they have been properly fitted using a quantitative fit test.

[d] The APF was based on consideration of efficiency of dust, fume, and/or mist filters.

11.1. Is the gas/vapor respirator intended for escape purposes only?
   a. If yes, refer to the section *Escape Apparatus* for a discussion on selection.
   b. If no, the gas/vapor respirator is intended for use during normal work activities. Proceed to Question 11.2.

11.2. Are the warning properties for the gas/vapor contaminant adequate at or below the NIOSH REL or other applicable exposure limit?
   a. If yes, Proceed to Question 11.3.
   b. If no, an air-purifying respirator equipped with an effective end-of-service-life indicator (ESLI), a supplied-air respirator, or a self-contained breathing apparatus is recommended. Proceed to Question 11.4.

11.3. An air-purifying chemical cartridge/canister respirator is recommended that has a sorbent suitable for the chemical properties of the anticipated gas/vapor contaminant(s) and for the anticipated exposure levels. (See the section *Limitations of Respirators for Gases and Vapors* following the decision logic questions for recommended maximum-use concentrations of air-purifying chemical cartridge/canister respirators.)

11.4. Respirators that are listed in Table 7-2 and that have APFs equal to or greater than the minimum protection factor determined in Question 8 are recommended. Maximum airborne concentrations for each class of

TABLE 7-2
Assigned Protection Factor Classifications of Respirators for Protection against Gas/Vapor Exposures

APF[a]	Type of Respirator
10	Any air-purifying half-mask respirator (including disposable) equipped with appropriate gas/vapor cartridges[b]
	Any supplied-air respirator equipped with a half-mask and operated in a demand (negative-pressure) mode[b]
25	Any powered air-purifying respirator with a loose-fitting hood or helmet[c]
	Any supplied-air respirator equipped with a hood or helmet and operated in a continuous-flow mode[c]
50	Any air-purifying full-facepiece respirator equipped with appropriate gas/vapor cartridges or gas mask (canister respirator)[b]
	Any powered air-purifying respirator equipped with a tight-fitting facepiece and appropriate gas/vapor cartridges or canisters[c]
	Any supplied-air respirator equipped with a full facepiece and operated in a demand (negative-pressure) mode[b]
	Any supplied-air respirator equipped with a tight-fitting facepiece operated in a continuous-flow mode[c]
	Any self-contained respirator equipped with a full facepiece and operated in a demand (negative-pressure) mode[b]
1,000	Any supplied-air respirator equipped with a half-mask and operated in a pressure-demand or other positive-pressure mode[b]
2,000	Any supplied-air respirator equipped with a full facepiece in a pressure-demand or other positive-pressure mode[b]
10,000	Any self-contained respirator equipped with a full facepiece and operated in a pressure-demand or other positive-pressure mode[b]
	Any supplied-air respirator equipped with a full facepiece operated in a pressure-demand or other positive-pressure mode in combination with an auxiliary self-contained breathing apparatus operated in a pressure-demand or other positive-pressure mode[b]

[a] The assigned protection factor for a given class of air-purifying respirators may be further reduced by considering the maximum use concentrations for each type of gas and vapor air-purifying element.

[b] APFs were determined by Los Alamos National Laboratories (LANL) by conducting quantitative fit testing on a panel of human volunteers.

[c] APFs were based on workplace protection factor data or laboratory data more recently reported than the LANL data.

respiratory protection can be calculated by multiplying the NIOSH REL or other applicable exposure limit by the APF for that class of respirators. The calculated maximum use concentration limits should not exceed the limitations noted in the section Limitations of Respirators for Gases and Vapors. Workers wearing respirators should meet NIOSH medical guidelines.

12. Combination Particulate and Gas/Vapor Respirators

12.1. Is the combination respirator intended for escape purposes only?
   a. If yes, refer to the section *Escape Apparatus* following the decision logic questions.
   b. If no, the combination respirator is intended for use during normal work activities. Proceed to Question 12.2.

12.2. Does the gas/vapor contaminant have adequate warning properties at or below the NIOSH REL or other applicable exposure limit?
   a. If yes, proceed to Question 12.3.
   b. If no, either an air-purifying respirator equipped with an effective ESLI, a supplied-air respirator, or a self-contained respirator is recommended. Proceed to Question 12.4.

12.3. An air-purifying chemical cartridge/canister is recommended that has a particulate prefilter suitable for the specific type(s) of gas/vapor and particulate contaminant(s) and for the exposure concentrations. (Refer to the next section for more information on limitations of respirators for gases and vapors.)

12.4. Respirators that are listed in Table 7-3 and that have APFs equal to or greater than the minimum protection factor determined in Question 8 are recommended. Maximum airborne concentrations for each level of respiratory protection can be calculated by multiplying the NIOSH REL or other applicable exposure limit by the APF for that class of respirator. The calculated maximum use concentrations limits should not exceed the limitations noted in the next section.

Limations of Respirators for Gases and Vapors

Air-purifying respirators cannot be used in IDLH atmospheres or in atmospheres containing less than 19.5 percent oxygen by volume. Gas masks (canister respirators) may be used for escape if the atmosphere is not oxygen-deficient. The IDLH value will be the maximum use concentration if the product APF times REL, or times another applicable exposure limit, is larger than the IDLH value. (See Tables 7-1, 7-2, and 7-3.) In addition, Table 7-5 gives maximum use concentrations for all gas and vapor air-purifying elements.

Table 7-3

Assigned Protection Factor Classifications of Respirators for Protection against Combination Gas/Vapor and Particulate Exposures[a]

APF[b]	Type of Respirator
10	Any air-purifying half-mask respirator equipped with appropriate gas/vapor cartridges in combination with any type of particulate filter
	Any full-facepiece respirator with appropriate gas/vapor cartridges in combination with a dust or mist or fume; dust and mist; or dust, mist, and fume filter[c]
	Any supplied-air respirator equipped with a half-mask and operated in a demand (negative-pressure) mode
25	Any powered air-purifying respirator equipped with a loose-fitting hood or helmet
	Any supplied-air respirator equipped with a hood or helmet and operated in a continuous-flow mode
50	Any air-purifying full-facepiece respirator equipped with appropriate gas/vapor cartridges in combination with a high-efficiency filter or an appropriate canister incorporating a high-efficiency filter
	Any powered air-purifying respirator with a tight-fitting facepiece equipped with appropriate gas/vapor cartridges in combination with a high efficiency filter or an appropriate canister incorporating a high-efficiency filter
	Any supplied-air respirator equipped with a full facepiece and operated in demand (negative-pressure) mode
	Any supplied-air respirator equipped with a tight-fitting facepiece and operated in a continuous-flow mode
	Any self-contained respirator equipped with a full facepiece and operated in a demand (negative-pressure) mode
1,000	Any supplied-air respirator equipped with a half-mask and operated in a pressure-demand or other positive-pressure mode
2,000	Any supplied-air respirator equipped with full facepiece and operated in a pressure demand or other positive-pressure mode
10,000	Any self-contained respirator equipped with a full facepiece and operated in a pressure-demand or other positive-pressure mode

TABLE 7-3
Continued

APF[b]	*Type of Respirator*
	Any supplied-air respirator equipped with a full facepiece operated in a pressure-demand or other positive-pressure mode in combination with an auxiliary self-contained breathing apparatus operated in a pressure-demand or other positive-pressure mode

[a] Only high-efficiency filters are permitted for protection against particulates having exposure limits of less than 0.05 mg/m^3.

[b] The assigned protection factor for a given class of air-purifying respirators may be further reduced by considering the maximum use concentrations for each type of gas and vapor air-purifying element.

[c] The APF was based on consideration of efficiency of dust, fume, and/or mist filters.

Air-purifying devices should not be allowed for either entry into or escape from hazardous environments when supporting evidence exists to demonstrate that unreasonbly short service life would occur at the maximum use concentration. Where there is reason to suspect that a sorbent has a high heat of reaction with a substance, avoid use of that sorbent in favor of a nonoxidizable sorbent.

Air-purifying respirators cannot be used for protection against gases and vapors with poor warning properties unless the respirator is approved with an effective ESLI. When respirators approved for a given class of contaminants (such as organic vapors) cannot be used due to sorbent deficiencies, there are specific air-purifying respirators that are approved by MSHA/NIOSH for protection against such gases and vapors.

Oxygen-Deficient Atmosphere

NIOSH defines an *oxygen-deficient atmosphere* as one with less than 19.5 percent oxygen at sea level. The agency certification of air-line or air-purifying respirators is limited to those respirators used in atmospheres containing at least 19.5 percent oxygen, except for those airline respirators equipped with an auxiliary self-contained breathing apparatus.

The minimum requirement of 19.5 percent oxygen at sea level provides an adequate amount of oxygen for most work assignments and includes a safety factor. The safety factor is important because oxygen-deficient atmospheres offer little warning of the danger, and it is difficult to continuously measure for oxygen deficiency.

Oxygen concentrations of less than 16 percent at sea level can cause decreased mental effectiveness, visual acuity, and muscular coordination. Concentrations below 10 percent can cause loss of consciousness, and death can occur at concentrations below 6 percent. Individuals often note only subjective changes when exposed to low concentrations of oxygen, and they can collapse without warning.

Because oxygen-deficient atmospheres are life-threatening, only the most reliable respirators are recommended. The most reliable are the self-contained breathing apparatuses or the supplied-air respirators with auxiliary self-contained units. A high protection factor is not necessary to ensure an adequate supply of oxygen, even in an atmosphere containing no oxygen, so any certified self-contained unit is adequate. All aspects of a respiratory protection program must be instituted for these recommendations to be valid.

Immediately Dangerous to Life and Health (IDLH)

An IDLH condition is defined as one that poses a threat of exposure to airborne contaminants when that exposure is likely to cause death or immediate or delayed permanent adverse health effects or prevent escape from such an environment. The purpose of establishing an IDLH exposure level is to ensure that workers can escape from a given contaminated environment if their respiratory equipment fails. The IDLH is a maximum level above which only a highly reliable breathing apparatus providing maximum worker protection is permitted. Any appropriate approved respirator may be used to its maximum use concentration up to the IDLH concentration.

In establishing the IDLH concentration, the following conditions must be assured:

- The ability to escape without loss of life or immediate or delayed irreversible health effects. Thirty minutes is considered the maximum time for escape so as to provide some margin of safety in calculating the IDLH.
- Prevention of severe eye or respiratory irritation or other reactions that would hinder escape.

The following are sources of information for determining whether the exposure limit for a contaminant represents an IDLH condition:

- Specific IDLH guidelines provided in the literature, such as the American Industrial Hygiene Association (AIHA) *Hygienic Guides* and the *NIOSH Pocket Guide to Hazardous Chemical Substances*.

- Human exposure and effects data
- Animal exposure and effects data

Where such data specific to the contaminant are lacking, toxicological data from analogous substances and chronic animal exposure data may be considered.

Eye Irritation

Eye protection in the form of respirators with full facepieces, helmets, or hoods is required for routine exposures to airborne contaminants that cause eye irritation to the mucous membranes of the conjunctivae or the cornea or that cause any reflex tearing. Eye protection is required for contaminants that cause minor subjective effects as well as for those that cause any damage, including disintegration and sloughing of conjunctival or corneal epithelium, edema, or ulceration. NIOSH is not aware of any standards for gas-tight goggles that would permit NIOSH to recommend such goggles as providing adequate eye protection.

For escape, some eye irritation is permissible if the severity of irritation does not inhibit the escape and if no irreversible scarring or ulceration of the eyes or conjunctivae is likely.

When data on threshold levels for eye irritation are insufficient, quarter- or half-mask respirators can be used, provided that the worker experiences no eye discomfort and no pathological eye effects develop. Workers should be told that if they experience any eye discomfort, they will be provided with respirators that have full facepieces, helmets, or hoods and that provide protection equivalent to the quarter- or half-mask respirators.

Escape Apparatus

Escape devices have a single function: to allow a person working in a normally safe environment sufficient time to escape from suddenly occurring respiratory hazards.

Escape devices can be separated into two categories: air-purifying respirators and self-contained breathing apparatus. Air-purifying respirators remove contaminants from the air by sorbent and/or filter media, but because they do not provide air, cannot be used in an oxygen-deficient atmosphere. Escape respirators in the air-purifying category include the escape gas mask (canister) respirator, the gas mask (canister) respirator, and the filter self-rescuer. The escape gas mask consists of a half-mask or a mouthpiece respirator. A mouthpiece respirator can be used for short periods of time to escape from low concentrations of organic vapor or acid gas. The escape gas mask, which utilizes a half-mask, filters contaminants from the air, and can also be used to escape from low concentrations of organic vapor or acid gas. Escape gas mask respirators equipped with full facepieces can be used for escape from IDLH conditions but not from oxygen-deficient atmospheres. No air-purifying device is suitable for escape from a potentially oxygen-deficient atmosphere. The filter self-rescue unit is a mouthpiece device that is designed to protect specifically against less than 1 percent carbon monoxide.

A self-contained breathing apparatus provides air to the user for escape from oxygen-deficient environments. Escape SCBA devices are commonly used with full facepieces or hoods and, depending on the supply of air, are usually rated as 3- to 60-minute units. Self-contained self-rescuer (SCSR) devices have been approved by MSHA and NIOSH for escape from mines, but these devices may also have applications in other environments. SCSRs are mouthpiece respirators that provide a source of oxygen-enriched air for up to 60 minutes. All SCBA devices can be used in oxygen-deficient atmospheres.

When selecting an escape apparatus, careful consideration must be given to potential eye irritation. This consideration is important for determining whether a gas mask or an SCBA equipped with a full facepiece should be selected rather than a device equipped with a half-mask or mouthpiece.

Most gas masks or escape gas masks can be used in situations involving gas(es), vapor(s), or particulates. For escape from environments contaminated with particulates, an air-purifying element must be selected that will provide protection against the given type of particulate. Tables 7-4 and 7-5 summarize some of the information that will help in selecting appropriate escape apparatuses.

Several Variations of Protection Factors

A potential for confusion exists because several protection factor terms have previously been used and modified, and today there some similar-sounding but different factors. For completeness, and to ensure that readers who later hear of an old expression will realize it is not a new one, this section includes some historical information.

Respirator protection factor. In 1965 the Bureau of Mines referred to the term "decontamination factor," in their Approval Schedule 21B and defined it as the ratio of the concentration of dust, fume, or mist present in the ambient atmosphere to the concentration within the facepiece while the respirator is being worn. The original definition and application have been generalized over the years, and the terms *respirator protection factor* and *protection factor* have now replaced the term *decontamination factor*.

TABLE 7-4
Selection Options for Escape Respirators

Escape Conditions	*Type of Respirator*
Short distance to exit, no obstacles, no oxygen deficiency	Any escape gas mask[a] (canister respirator) or gas mask[b] (canister respirator)
	Any escape self-contained breathing apparatus having a suitable service life[c]
	Any acceptable device for entry into emergency situations
Long distance to exit or obstacles along the way, no oxygen deficiency	Any gas mask[b]
	Any escape self-contained breathing apparatus having a suitable service life[c]
	Any self-contained self-rescuer having a suitable service life
Potential oxygen deficiency	Any escape self-contained breathing apparatus having a suitable service life[c]
	Any self-contained self-rescuer having a suitable service life

[a]An escape gas mask is a respirator designed for use only during escape from IDLH or non-IDLH atmospheres. It may consist of a half-mask facepiece or mouthpiece, appropriate air-purifying element for the contaminant, and associated connections. Maximum use concentrations for these types of respirators are designated by the manufacturer.

[b]A gas mask consists of a full facepiece and either chin-style or front- or back-mounted canisters with associated connections. Maximum use concentrations for canister air-purifying elements are listed in Table 7-5.

[c]Escape self-contained breathing apparatuses can have rated service lives of 3 to 60 minutes. All acceptable devices for entry into emergency situations can also be used.

TABLE 7-5
NIOSH-Recommended Maximum Use Concentrations, in ppm, for Gas and Vapor Air-Purifying Elements

Type of Gas or Vapor	Classification of Gas and Vapor Air-Purifying Elements		
	Cartridge(s)	Chin-style canister	Front- or back-mounted canister
Organic vapors	1,000[a]	5,000[b]	20,000[b]
Acid gases			
Sulfur dioxide (SO_2)	50	100	100
Chlorine (Cl_2)	10	25	25
Hydrogen chloride (HCl)	50	100	100
Ammonia (NH_3)	300	500	500
Methyl amine (CH_3NH_2)	100	—	—
Carbon monoxide (CO)	NA	NA	1,500

[a]Maximum use concentration will be 1000 ppm or the immediately dangerous to life or health value for the specific organic vapor, whichever is lower.

[b]Maximum use concentration for "entry into" will be limited to the value listed or to the immediately dangerous to life and health value for the specific organic vapor, whichever is lower.

Three formulas use two variables, C_I (concentration inside the facepiece) and C_O (concentration outside the facepiece), to describe the degree of protection as follows:

$$\text{Protection factor} = \text{PF} = \frac{C_O}{C_I}$$

$$\text{Penetration} = \frac{C_I}{C_O}$$

$$\text{Efficiency} = \text{E} = \frac{C_O - C_I}{C_O}$$

Given that air inside the facepiece is always cleaner than ambient air, PF will always be a number larger than 1, and efficiency will always be a positive number. Sometimes the fractional efficiency is multiplied by 100 and expressed as a percentage. Protection factor evaluations are almost always made on worker/respirator systems; penetration and efficiency evaluations are made only on component parts of a respirator system. It is important to recognize that with a worker/respirator system, C_I is a complicated function of many individual sources of penetration such as air-purifying element penetration, exhalation valve penetration, face seal penetration, other inboard penetrations, and environmental conditions that would affect penetration. To deal with multiple methods for determining and applying protection factors, other definitions described in the rest of this section have been developed.

Assigned protection factor (APF). A special application of the general protection factor concept, APF is defined as a measure of the minimum anticipated workplace level of respiratory protection that would be provided by a properly functioning respirator or class of respirators to a percentage of properly fitted and trained users. The maximum specified use concentration for a respirator is generally determined by multiplying the exposure limit for the contaminant by the protection factor assigned to a specific class of respirators.

Workplace protection factor (WPF). The ratio of C_O to C_I is defined as WPF when it is measured under actual conditions of a specific workplace and with a properly functioning respirator that is correctly worn and used. TWA samples are taken simultaneously. In practice, WPF is determined by measuring the inside and outside concentrations during the activities of a normal workday.

Simulated workplace protection factor (SWPF). This is a surrogate measure of WPF and differs only in that it is measured in a laboratory simulation of a workplace setting. Although readers will see this factor in the literature, experts consider it of questionable value for determining WPF until a reliable relationship can be demonstrated between WPF and SWPF.

CHAPTER 8

Gathering Chemical Information

This chapter looks at different types of chemical compounds to demonstrate the quantity and types of information that employers can and should gather as the first step in designing a safe working environment. Even if a book of this size could present the same amount of information for each of the hundreds of hazardous chemicals on the NIOSH and OSHA lists, the agencies update their lists and the information in them frequently. The last section in this chapter gives guidelines for obtaining information such as presented here.

FIFTEEN COMMON SOLVENTS[1]

Organic solvents are absorbed chiefly through the respiratory organs, but some can also be absorbed through the skin and digestive tract. In an average person,

1. From *How the Activated Carbon Filter in the Nederman Extractor Kit 2000 Adsorbs Solvents in Gas Form*, courtesy of Nederman, Inc., Westland, Michigan.

respiratory organs present approximately 861 to 1075 ft^2 of absorbing surface, the digestive tract approximately 108 ft^2, and the skin approximately 22 ft^2.

Another important fact to consider is that the easier a solvent dissolves fat, the easier it is absorbed by the body and the greater is its capacity for affecting the nervous system. When solvents are used, there is considerable risk that fumes will escape into the premises and be spread by the ventilation system. Wiping products and surfaces with a solvent accelerates its evaporation and creates the risk of releasing fumes over a large area. Table 8-1 presents the properties and airborne-related risks of 15 common solvents.

TABLE 8-1
Risks Associated with 15 Common Solvents

Product	*Properties*	*Uses*	*Risks*
ACETONE 2-propanone dimethylketone	Colorless liquid with sweet or sharp smell; highly volatile; miscible with water	Usually a solvent and dissolvent; used in manufacture of other chemicals and drugs	Inhalation or ingestion of large amounts can induce headache, dizziness, fatigue and, in serious cases, fainting.
INDUSTRIAL GASOLINE solvent naptha	Colorless liquid; volatile to highly volatile, depending on composition	Used as a solvent, dissolvent, and degreasing agent	Fumes irritate eyes and respiratory passages. Inhalation or ingestion can induce headache, nausea, dizziness and, in severe cases, fainting. Inhalation over a long period can damage nerves, especially if the gasoline is of the hexane type.

TABLE 8-1
Continued

Product	*Properties*	*Uses*	*Risks*
BUTANOL 1-butanol *n*-butyl alcohol	Colorless liquid with sharp smell; gives off vapor, especially when heated; soluble in water	Used as a solvent and in the production of other chemicals	Fumes irritate eyes and respiratory passages. Inhalation or contact with skin can induce headache, dizziness, fatigue and, in serious cases, fainting.
ETHANOL ethyl alcohol	Colorless liquid with characteristic smell; volatile; miscible with water	Usually used as a solvent, dissolvent, and antifreeze agent; used in manufacture of other chemicals such as acetaldehyde and acetic acid and as a fuel	Inhalation or ingestion of large amounts can induce dizziness, headache, nausea and, in severe cases, fainting.
ETHYL ACETATE acetic acid ester acetic ester	Colorless liquid with fruity smell; highly volatile; miscible with water	Usually used as a solvent and dissolvent; used in the manufacture of drugs and other chemicals; used in paint and dyestuffs, synthetic fibers, and in plastics industries	Fumes irritate eyes and respiratory passages. Inhalation or ingestion of large amounts can induce headache, nausea, dizziness and, in severe cases, fainting. Absorption over a long period of time can damage liver and kidneys.

Continued

Table 8-1
Continued

Product	*Properties*	*Uses*	*Risks*
ETHYLENE GLYCOL 1,2-ethylene diol glycol	Colorless and odorless, viscous liquid; gives off vapor when heated; miscible with water	Used as a solvent for dyes and plastics such as cellophane, and as an antifreeze agent and hydraulic fluid; used in manufacture of polyester fibers and other plastics, printing inks, explosives, cosmetics, and other chemicals	Inhalation or ingestion can induce dizziness, vomiting, stomach pains, weakness; can affect the central nervous system; can cause fainting in serious cases; is thought to cause kidney damage.
MINERAL TURPENTINE thinner with 17–22% aromatics white spirit dilutine	Colorless liquid; gives off vapor, especially when heated; sparingly soluble in water	Used as a solvent, degreasing, and cleaning agent; used in the manufacture of formaldehyde and other chemicals	Inhalation or ingestion can induce headache, nausea, dizziness, fatigue and, in severe cases, fainting. Inhalation over long periods can produce nerve damage. If both inhaled and ingested, pneumonia can result.
METHANOL methyl alcohol wood alcohol spirit	Colorless liquid with a faint smell of alcohol; volatile; miscible with water	Used as a solvent, dissolvent, antifreeze agent, industrial alcohol, and additive in motor fuel	Inhalation or ingestion can induce dizziness, headache, nausea, disturbed vision and, in severe cases, damaged eyesight and fainting. Absorption of even small amounts over long periods can damage the eyesight.

TABLE 8-1
Continued

Product	*Properties*	*Uses*	*Risks*
METHYLENE CHLORIDE dichloromethane	Colorless liquid with a faint sweetish smell; highly volatile; miscible with water	Used as a solvent and degreasing agent	Inhalation or ingestion can induce dizziness, fatigue, headache, nausea and, in severe cases, can affect the heart and lead to fainting. When heated, toxic and corrosive gases are formed.
METHYL ETHYL KETONE 2-butanone MEK	Colorless liquid with acetone-like smell; volatile; miscible with water	Used as a solvent; used in dyes and paints, plastics, drugs, and cosmetics	Fumes irritate eyes and respiratory passages. Inhalation or ingestion can induce headache, dizziness, nausea and, in severe cases, fainting.
METHYL CHLOROFORM 1,1,1,-trichloroethane	Colorless liquid with a sweetish smell; highly volatile; sparingly soluble in water	Used as a solvent and degreasing agent; used in the drug and petroleum industries	Inhalation or ingestion of large amounts can induce headache, fatigue, dizziness and, in severe cases, can lead to fainting, affect the heart, and produce a drop in blood pressure. When heated, toxic and corrosive gases are formed.

Continued

Table 8-1
Continued

Product	*Properties*	*Uses*	*Risks*
STYRENE phenylethylene vinyl benzene	Colorless to yellowish liquid with sweetish or sharp smell; gives off vapor, especially when heated; partly soluble in water	Used in the manufacture of plastics and as a solvent; used in the reinforced plastics industry	Fumes irritate eyes and respiratory passages. Inhalation or ingestion can induce headache, fatigue, nausea, dizziness, coughing fits, difficulty breathing and, in severe cases, fainting. Absorption over long periods can impair reactions and memory.
TOLUENE methylbenzene toluol	Colorless liquid with sweet or sharp smell; volatile; sparingly soluble in water	Used as a solvent and in petrol and other petroleum distillates; used in the manufacture of drugs, explosives, TDJ, and other chemicals	Fumes irritate eyes and respiratory passages. Inhalation or ingestion can induce headache, fatigue, dizziness, nausea, loss of appetite, muscular weakness and, in severe cases, numbness, tingling of the limbs, and fainting. If both inhaled and ingested, pneumonia can result. Absorption of large amounts over long periods is thought to affect the liver and kidneys.

TABLE 8-1
Continued

Product	***Properties***	***Uses***	***Risks***
TRICHLORO-ETHYLENE ethylene trichloride	Colorless liquid with sweetish smell; volatile; partly soluble in water	Used as a solvent, dissolvent, and degreasing agent; used in the manufacture of other chemicals	Inhalation or ingestion can induce headache, dizziness, nausea and, in severe cases, can lead to fainting, affect the heart, and damage liver and kidneys. Inhalation over a long period of time is thought to damage the nervous system. When heated, toxic and corrosive gases are formed. Can react violently with strong alkalis, leading to formation of toxic and flammable gases.
XYLENE dimethylbenzene xylol	Colorless liquid; gives off vapor, especially when heated; sparingly soluble in water	Used as a solvent and in the manufacture of drugs, dyestuffs, fibers, and other chemicals; used in many petroleum distillates, such as motor fuel	Fumes irritate eyes and respiratory passages. Inhalation or ingestion can induce headache, dizziness, nausea, loss of appetite and, in severe cases, fainting and difficulty in breathing. If inhaled as well as ingested, can cause pneumonia. Absorption over long periods is thought to affect liver and kidneys.

ALDEHYDES

This section gives two types of indoor air quality information: (1) It provides recent information about the potential carcinogenicity of acetaldehyde and malonaldehyde, and (2) it provides information about mutagenicity of nine related low molecular weight aldehydes (acrolein, butyraldehyde, crotonaldehyde, glutaraldehyde, glyoxal, paraformaldehyde, propiolaldehyde, propionaldehyde, and valeraldehyde).

Acetaldehyde and malonaldehyde both meet OSHA's criteria for classification as potential occupational carcinogens. The potential for these substances to produce cancer in humans has not been quantified, but NIOSH recommends that reducing exposures to the lowest feasible concentration should lower the risk.

The carcinogenic potential of the nine related low molecular weight aldehydes has not been adequately quantified by appropriate experimentation. However, studies indicate that their chemical reactivity and mutagenicity are similar to those of acetaldehyde and malonaldehyde. Therefore, NIOSH is also concerned about occupational exposure to these nine aldehydes because they (in addition to acetaldehyde and malonaldehyde) may be used as substitutes for formaldehyde, a regulated carcinogen. As with the first two aldehydes, reducing exposure to these nine related aldehydes to the lowest feasible concentration should lower the risk.

Physical and Chemical Properties

Acetaldehyde is a two-carbon compound with a carbonyl group (HC=O). Its molecular weight is 44.05. This clear liquid boils at 69.5°F (20.8°C). Its pleasant, fruity odor at dilute concentrations becomes pungent and suffocating at high concentrations. Acetaldehyde is readily oxidized within the body to acetate and acetic acid by liver mitochondrial acetaldehyde dehydrogenase.

The chemical and physical properties of malonaldehyde are similar to those of acetaldehyde and other low molecular weight monoaldehydes and dialdehydes. Pure malonaldehyde is unstable and is precipitated as a sodium salt immediately before its use in bioassays. Recent studies of purity analysis by ultraviolet spectroscopy established that malonaldehyde (propanedial, sodium salt) is stable with no notable degradation for two years when stored at –4°F (–20°C). It is metabolized in vivo and in vitro by oxidation to malonic semialdehyde and by decarboxylation to acetaldehyde.

Production, Use, and Potential for Occupational Exposure

In one recent year, 280,000 tons of acetaldehyde were produced in the United States, using mainly the Hoechst-Wacker two-stage process for preparing acetaldehyde from ethylene. Acetaldehyde is primarily used as a substrate for acetic acid manufacture, although this use is declining as more economical chemical intermediates are replacing acetaldehyde. This chemical is also used in the synthesis of pyridine and pyridine bases, peracetic acid, pentaerythritol, 1,3-butylene glycol, and chloral. In addition, acetaldehyde is used in the silvering of mirrors, in leather tanning, as a denaturant for alcohol, in fuel compositions, as a hardener for gelatin fibers, in blue and casein products, as a preservative for fish, in the paper industry, and in the manufacture of cosmetics, aniline dyes, plastics, and synthetic rubber.

From the wide variety of applications, it is easy to believe that NIOSH's estimate of approximately 14,000 workers exposed to acetaldehyde is possibly understated. This figure is based on field survey data from workers potentially exposed during handling of the chemical and may be low because it omits any estimate of potential exposure in workplaces where acetaldehyde is used in trade-named or proprietary products.

Purified malonaldehyde is subject to polymerization and degradation, so it is usually generated shortly before use. Malonaldehyde or its enolic sodium salt is primarily used in research laboratories where it may be generated by acid hydrolysis of 1,1,3,3-tetramethoxypropane or 1,1,3,3-tetramethoxypropane. The annual production of malonaldehyde varies because it is produced on an as-needed basis by several U.S. chemical manufacturers. No figures are available for the number of U.S. workers exposed.

Table 8-2 lists the production estimates and potential number of workers exposed to the aldehydes discussed in this section, along with exposure limits.

Exposure Limits and Standards

This section describes some of the background for the standards and recommendations shown in Table 8-2.

Acetaldehyde. In recent rulemaking on air contaminants (54 *Federal Register* 2445) OSHA established a permissible exposure limit of 100 ppm (180 mg/m^3) of acetaldehyde in air as an eight-hour time-weighted average, with a short-term exposure limit of 150 ppm (270 mg/m^3). The PEL is based on the results of an acute toxicity study and is intended to protect workers against the risk of conjunctivitis and sensory irritation.

TABLE 8-2

Production, Exposure Potential, and Standards for Aldehydes

Aldehyde	Estimated Amount Produced (tons)	Number of Workers Potentially Exposed	Standards and Recommendations		
			NIOSH Recommendation (ppm)	OSHA PEL (ppm)	MSHA PEL (ppm)
Acetaldehyde	281,000	14,054	Ca	100 (180 mg/m^3), TWA 150 (270 mg/m^3), STEL	100 (180 mg/m^3), TWA 150 (270 mg/m^3), STEL
Acrolein	25,000	7,300	0.1 (0.25 mg/m^3), TWA 0.3 (0.8 mg/m^3), STEL	0.1 (0.25 mg/m^3), TWA 0.3 (0.8 mg/m^3), STEL	0.1 (0.25 mg/m^3), TWA 0.3 (0.8 mg/m^3), STEL
Butyraldehyde	750,000	5,392	—	—	—
Crotonaldehyde	5,000	148	2 (6 mg/m^3), TWA	2 (6 mg/m^3), TWA	2 (6 mg/m^3), TWA
Glutaraldehyde	1.5	353,905	0.2 (0.8 mg/m^3), CL	0.2 (0.8 mg/m^3), CL	0.2 (0.8 mg/m^3), CL
Glyoxal	—	44,937	—	—	—
Malonaldehyde	—	—	Ca	—	—
Paraformaldehyde	—	125,645	—	—	—
Propiolaldehyde	—	—	—	—	—
Propionaldehyde	50,000	1,557	—	—	—
Valeraldehyde	—	—	50 (175 mg/m^3), TWA	50 (175 mg/m^3), TWA	50 (175 mg/m^3), TWA

Note: Ca = Potential occupational carcinogen; TWA = time-weighted average; STEL = short-term exposure limit; CL = ceiling limit.

NIOSH has not established a recommended exposure limit for acetaldehyde. In testimony at the OSHA hearings on this rulemaking, NIOSH did not concur with the proposed PEL, citing acetaldehyde in a group of 53 chemicals that should be designated as potential occupational carcinogens for which a substantial level of risk remains at the proposed PEL.

MSHA has proposed a PEL of 100 ppm (180 mg/m^3) as an eight-hour TWA for acetaldehyde with an STEL of 150 ppm (270 mg/m^3). The OSHA Hazard Communication Standard requires that chemical manufacturers, distributors, employers, and importers provide appropriate hazard warnings for carcinogens or potential carcinogens listed in the latest editions of monographs by the International Agency for Research on Cancer (IARC).

Malonaldehyde. NIOSH, OSHA, and MSHA have not established exposure limits for malonaldehyde. In 1985, IARC concluded that there was inadequate evidence to establish the carcinogenicity of malonaldehyde in experimental animals and that there were no epidemiological data to evaluate the carcinogenic risk of malonaldehyde to humans.

Human Health Effects

Acute exposure of humans to 50 ppm acetaldehyde vapor for 15 minutes produced mild eye irritation, 200 ppm for 15 minutes produced bloodshot eyes and reddened eyelids, and 135 ppm for 30 minutes produced mild irritation of the upper respiratory tract. Acetaldehyde vapors, mist, and fumes that reach the eyes can cause painful but superficial injury to the cornea. Systemic effects resulting from chronic acetaldehyde exposure in the workplace have not been reported, but prolonged exposure to acetaldehyde may produce drowsiness.

Recommendations

NIOSH considers the OSHA classification appropriate for identifying potential occupational carcinogens. Potential occupational carcinogen is defined here as:

> any substance, or combination or mixture of substances which causes an increased incidence of benign and/or malignant neoplasms, or a substantial decrease in the latency period between exposure and onset of neoplasms in humans or one or more experimental mammalian species as the result of any oral, respiratory or dermal exposure, or any other exposure which results in the induction of tumors at a site other than the site of administration. This

> definition also includes any substance which is metabolized into one or more potential occupational carcinogens by mammals.

Officially the system is called the *Identification, Classification, and Regulation of Potential Occupational Carcinogens*, and it is also known as the OSHA carcinogen policy. In conformance with this policy, NIOSH recommends that acetaldehyde and malonaldehyde be considered potential occupational carcinogens. Excess cancer risk for workers exposed to the aldehydes has not been quantified, but the probability of developing cancer should be decreased by minimizing exposure. As a matter of prudent public health policy, employers should take reasonable precautions (such as appropriate engineering and work practice controls) to reduce exposures to the lowest feasible concentration.

Testing has not been completed to determine the carcinogenicity of the nine related low molecular weight aldehydes. However, the limited studies that have been conducted indicate that the chemical reactivity and mutagenicity of these chemicals are similar to those of acetaldehyde, malonaldehyde, and formaldehyde. Therefore, careful consideration should be given to reducing exposures to these nine related aldehydes.

Guidelines for Minimizing Worker Exposure

Actual steps taken to minimize worker exposure depend on specific situations too varied to cover in a single chapter, or even a single book. Here are some general guidelines that you can adapt to the particular conditions in your plant.

Monitoring exposure is a major part of the program to minimize exposure, and NIOSH recommends that employers monitor every employee who manufactures, transports, packages, stores, or uses aldehydes in any capacity. It is important to determine whether a potential exists for any worker to be exposed to the chemical. In work areas where exposures may occur, an initial survey should be taken to determine the extent of actual and potential worker exposure.

Sampling should be used to (1) identify the sources of emissions so that effective engineering or work practice controls can be instituted, and (2) ensure that controls already in place are operational and effective. The following list offers suggestions for the procedure:

- Samples should be collected over a full shift to determine exposures.
- When the potential for exposure is periodic, short-term sampling may be needed to replace or supplement full-shift sampling.
- Personal sampling is preferred to area sampling.

- If personal sampling is not feasible, and if area sampling results can be used to approximate individual worker exposure, area sampling can be substituted.

An excellent source of information is the *NIOSH Occupational Exposure Sampling Strategy Manual*, which gives guidance for developing efficient strategies to monitor worker exposures to toxic chemicals. The manual contains information about determining the need for exposure monitoring, the number of samples to be collected, and the appropriate sampling times.

Controlling Worker Exposure

Equipment maintenance and worker education are vital aspects of a good exposure control program. Workers should be informed about any materials that may contain or be contaminated with aldehydes, the nature of the potential hazard, and methods for minimizing exposure. Every attempt should be made to minimize exposure to aldehydes by using the following work practices and controls:

- Emergency procedures
- Product substitution
- Closed systems and ventilation
- Worker isolation
- Personal protective equipment (such as chemical protective clothing [CPC] and equipment, and respiratory protection devices)
- Proper decontamination and waste disposal
- Medical screening under certain circumstances

Emergency procedures. In the event of an accidental exposure to an aldehyde, quickly remove the exposed person to prevent continued exposure. Seek medical assistance and perform appropriate emergency procedures from the following list:

Eye exposure: If an aldehyde or an airborne vapor, mist, or fume containing an aldehyde gets into the eyes, immediately flush them with large amounts of water for a minimum of 15 minutes, lifting the lower and upper lids occasionally. Get medical attention.

Skin exposure: If an aldehyde or an airborne mist, vapor, or fume containing an aldehyde contacts the skin, gently wash the contaminated skin with soap and water. Get medical attention if irritation persists.

Inhalation: If aldehyde mists, vapors, or fumes have been inhaled, move the victim at once to fresh air. Begin cardiopulmonary resuscitation if respiration has stopped. Administer oxygen if possible. Get medical attention.

Ingestion: Airborne mists can settle on food, drink, cigarettes, breathing apparatus mouthpieces, and other objects that workers put into their mouths. If a worker is reacting to ingestion of an aldehyde, give the victim several glasses of water to drink, then induce vomiting by having the victim touch the back of his or her throat with a finger or by giving syrup of ipecac as directed on the package. Do not force liquids into an unconscious or convulsing person; do not force an unconscious or convulsing person to vomit. Get medical attention. Keep the victim warm and quiet until medical help takes over.

Chronic exposure: Evaluate chronic exposure to aldehyde mists, vapors, and fumes that results in dermatitis and conjunctivitis or chronic intoxication (e.g., symptoms similar to those of chronic alcoholism such as weight loss, anemia, delirium, apparent loss of intelligence, and psychological disturbances).

Rescue: If aldehyde mists, vapors, or fumes are sufficiently dense to incapacitate a worker, remove the worker from further exposure and implement appropriate emergency procedures (e.g., those listed on the Material Safety Data Sheet). Make all workers familiar with emergency procedures and the location and proper use of emergency equipment.

Product substitution. When feasible, substitute a less hazardous material for aldehydes. However, check thoroughly into comparisons of effects among aldehydes and various possible substitutes. Substitutes may have more serious adverse effects or may be more dangerous when combined with aldehydes already in workers' bodies.

Closed systems and ventilation. Engineering controls should be the principal method for minimizing respiratory and skin exposure to aldehydes in the workplace. Achieving and maintaining reduced concentrations of airborne aldehydes depend on adequate engineering controls such as closed-system operations and ventilation systems that are properly constructed and maintained.

Closed-system operations provide the most effective means for minimizing worker exposures to aldehydes. These systems should be used for producing, storing, transferring, packaging, and processing aldehydes. In quality control laboratories where production samples are prepared for analyses, exhaust ventilation systems should be designed to capture and contain vapors. Proper ventilation in the presence of aldehydes is so important that it justifies the cost of consultants and other specialists.

Worker isolation. Areas in which aldehydes are produced or used should be restricted to workers essential to the process or operation. If feasible, extensive use of automated equipment operated from a closed control room or booth should keep workers isolated from direct contact with aldehydes. The control room should be maintained at higher air pressure than the area surrounding the process equipment so that it does not draw air from around the equipment. When workers must enter the general work area to perform process checks, adjustments, maintenance, assembly line tasks, and related operations, they should take special precautions such as using personal protective equipment.

Personal protective equipment. Reliance on personal protective equipment (CPC, respiratory protection devices, and other equipment) is the least desirable method of controlling worker exposure to aldehydes. Such equipment should not be the primary control method during routine operations.

NIOSH recognizes that respirators may be required for providing protection in certain situations such as implementing engineering controls, certain short-duration maintenance procedures, and emergencies. Only the most protective respirators should be used for situations involving carcinogens. These respirators include:

- Any self-contained breathing apparatus equipped with a full facepiece and operated in a pressure-demand or other positive-pressure mode, and
- Any supplied-air respirator equipped with a full facepiece and operated in a pressure-demand or other positive-pressure mode in combination with an auxiliary self-contained breathing apparatus operated in a pressure-demand or other positive-pressure mode.

Any respiratory protection program must, at a minimum, meet the requirements of 29 CFR 1910.134. Respirators should be approved by NIOSH and MSHA. A complete respiratory protection program should include regular training and medical evaluation of personnel, fit testing, periodic environmental monitoring, and maintenance, inspection, and cleaning of equipment. The program should be evaluated regularly.

GATHERING INFORMATION BEYOND THE ALDEHYDE FAMILY

This chapter has provided an example of the information that can help employers design a safe working environment. Obviously this book cannot present the same amount of information for each of the hundreds of hazardous chemicals on the NIOSH and OSHA lists; the information is available, but employers are responsible for obtaining it. An excellent source is a toll-free telephone call to NIOSH at (800)-35-NIOSH. Table 8-3 gives several Internet sites that can be sources of helpful information as well.

TABLE 8-3
Helpful Internet Sites

Organization	*Internet Address*
Agency for Toxic Substances and Disease Registry (ATSDR)	http://atsdr1.atsdr.cdc.gov:8080/atsdrhome.html
Agricultural Health and Safety Center, University of California at Davis	http://www-oem.ucdavis.edu
Canadian Center for Occupational Health and Safety	http://www.ccohs.ca
Center for Environmental Health and Safety, Southern Illinois University	http://www.siu.edu/departments/environ
Centers for Disease Control and Prevention (CDC)	http://www.cdc.gov
Department of Defense (DOD), Safety and Health	http://www.acq.osd.mil/ens/sh
Department of Energy (DOE)	http://www.doe.gov
Duke Occupational & Environmental Medicine	http://dmi-www.mc.duke.edu/cfm/occ&env/index.html
Environmental Protection Agency (EPA)	http://www.epa.gov

TABLE 8-3
Continued

Organization	*Internet Address*
Florida Agricultural Informational Retrieval System (FAIRS)	http:/hammock.ifas.ufl.edu
Government Printing Office	http://www.acess.gpo.gov
National Institute for Occupational Safety and Health (NIOSH)	http://www.cdc.gov/niosh/homepage.html
National Institute of Environmental Health Sciences (NIEHS)	http://www.niehs.nih.gov
National Institutes of Health (NIH)	http://www.nih.gov
National Library of Medicine (NLM)	http://www.nlm.nih.gov
National Technical Information Service (NTIS)	http://www.fedworld.gov
Occupational Safety and Health Association (OSHA)	http://www.osha.gov
Swedish National Institute of Occupational Health (NIOH)	http://www.nioh.se/nioh.htm
Texas Institute of Occupational Safety and Health	http://pegasus.uthct.edu/TIOSH/TIOSH.html
U.S. Army, Logistics Management Institute, Army Industrial Hygiene	http://www.lmi.org/Armyih
University of Occupation and Environmental Health, Japan	http://www.nihs.go/jp
World Health Organization (WHO)	http://www.who.ch

CHAPTER 9

How Temperature and Humidity Affect Workers and Facilities

This chapter looks at air temperature in two ways: First it considers extremely hot working environments—the environments that can cause death or serious health problems. Later in the chapter there is a section that looks at temperatures whose extremes are limited to the discomfort ranges, and at the combined effects of temperature and humidity. Other conditions and health factors are discussed in this chapter. Temperature, moisture, mildew, and mold are often inseparable problems in plants of all sizes.

WHEN HOT WORKING ENVIRONMENTS ARE UNAVOIDABLE

Many manufacturing procedures require large amounts of heat, so it is not practicable to keep associated workers cool. Probably about half of the "hot environment" jobs keep workers outdoors and exposed to the sun. Basically those jobs are not covered in this book, except that rescue procedures are generally similar for indoor and outdoor hot environments. A few of the plants to which this chapter applies are those that include operations such as:

- Brick firing and ceramics
- Foundries
- Glass and glass products manufacture
- Rubber and rubber products manufacture
- Boiler rooms
- Bakeries and commercial kitchens
- Laundries
- Mines
- Smelters
- Steam tunnels

Discomfort and unpleasant conditions are not the major problem caused by the very high temperatures associated with these operations. The major problem is that workers suddenly exposed to working in a hot environment face additional, *and generally avoidable*, hazards to their safety and health—even to their lives. Let's look at the body's natural mechanisms for controlling its internal temperature.

How the Body Handles Heat

As warmblooded beings, humans maintain a fairly constant internal temperature, even when exposed to varying environmental temperatures. Control of body temperature is relatively easy when the temperature of the environment is slightly higher than the desired body temperature. The first step that the body takes to keep internal body temperatures within safe limits, and to get rid of excess heat, is to adjust the rate and amount of blood circulation through the skin. Blood temperature deviations from 98.6°F (37°C) trigger these changes in response to commands from the brain. When the brain senses rising blood temperature, it causes the heart to pump more blood, and the blood vessels expand to accommodate the increased flow. Microscopic blood vessels (capillaries) that thread through the upper layers of the skin begin to fill with blood. This sequence causes blood to circulate closer to the surface of the skin, where some of its heat is transferred to the cooler environment.

If increased blood circulation through the skin does not provide sufficient cooling, the brain continues to sense overheating and signals glands in the skin to shed sweat onto the skin surface. When the sweat evaporates, it cools the skin, helping blood circulating close to the surface to transfer more heat from the body.

The higher the temperature of the environment, the less efficient this cooling process becomes. When the air temperature is higher than the skin temperature, blood brought to the body surface cannot give up its heat. Under these conditions the heart continues to pump blood to the body surface, sweat glands exude liquids containing electrolytes onto the skin's surface, and evaporation of the sweat becomes the principal means of maintaining a constant body temperature.

High relative humidity works against the body's natural cooling system. Sweat cools the skin surface only to the extent that it evaporates, but the higher the humidity, the less the sweat evaporates. Therefore, the expression "It's not the heat, it's the humidity" is partially true. The problem *is* the heat, but without the high humidity the body would usually be able to maintain its internal temperature within a safe range.

These self-adjusting body functions, necessary for health and life, also have adverse effects on an individual's ability to work in a hot environment. The following are the consequences of having extra blood diverted to external body surfaces:

- Less blood goes to active muscles, the brain, and other internal organs.
- Strength declines.
- Fatigue occurs sooner than it would in lower temperatures.
- Alertness and mental capacity are diminished.
- Delicate or detailed work suffers from a loss of accuracy.
- Comprehension and retention of information are lowered.

Serious problems begin with the sequences just described. The following section examines safety and health problems that result from working in hot environments. Descriptions of the health problems include treatment for mitigating their effects.

Safety Problems

Certain safety problems are common to hot working environments. Heat tends to promote accidents; sweaty palms are slippery; less blood to the brain causes dizziness, loss of balance, and slow decisions; and safety glasses fog up. The hot environment is often due to molten metal, hot surfaces, steam, and other products, increasing the possibility of burns from accidental contact.

Aside from these obvious dangers, the general frequency of accidents appears to be higher in hot environments. One reason is that working in a hot environment lowers the mental alertness and physical performance of workers. Increased body temperature and physical discomfort promote irritability, anger, and other emotional states that sometimes cause workers to overlook safety procedures or to divert attention from hazardous tasks.

Health Problems

As already discussed, when the temperature is high enough, or the temperature-humidity combination is intolerable, the body's natural cooling mechanism is unable to keep the body's internal temperature from rising. Following are the most common heat-induced disorders, which should all include this caution: Persons who have heart problems, are on a low sodium diet, or have other medical problems should consult a physician about possible complications from working in hot environments.

Heat stroke. Heat stroke is the most serious health problem associated with working in hot environments. It occurs when the body's temperature regulatory system fails, and it partially or completely gives up shedding sweat onto the skin surface. One reason this condition is so serious is that it compromises the body's only effective means of removing excess heat—and there is little warning to the victim that he or she has passed to a crisis stage.

A heat stroke victim's skin is hot, usually dry, and red or spotted. Body temperature is usually 105°F (40.5°C) or higher, and he or she is mentally confused, delirious, perhaps in convulsions, or unconscious. Without quick and appropriate treatment, the person can die.

Any person with signs or symptoms of heat stroke requires immediate hospitalization. If a delay is necessary, as when waiting for transportation, first aid measures are important. First, move the victim away from the heat, soak his or her clothing thoroughly, and fan the person vigorously to increase cooling. Early recognition and treatment of heat stroke are the only means of preventing permanent brain damage or death.

Heat exhaustion. Heat exhaustion is caused by the loss of large amounts of fluid by sweating, sometimes with excessive loss of salt. It includes several clinical disorders whose symptoms are similar to the early symptoms of heat stroke. Unlike a person suffering from heat stroke, a worker with heat exhaustion still sweats but experiences extreme weakness or fatigue, giddiness, nausea, or headache. In the more serious cases, the victim may vomit or lose consciousness. The skin is clammy and moist, the complexion is pale or flushed, and the body temperature is normal or elevated only slightly.

Treatment starts with having the victim rest in a cool place and drink liquids. Victims of mild heat exhaustion usually recover with this treatment. Those with severe heat exhaustion may require extended care for several days. There are no known permanent effects.

Heat cramps. Heat cramps are painful spasms of the muscles that occur among those who sweat profusely in heat and drink large quantities of water but do not adequately replace the body's salt loss. Drinking large quantities of water tends to dilute the body's fluids while the body continues to lose salt. The low salt level soon causes painful cramps. Fatigued muscles, generally the ones that have been working the hardest, are usually the ones most susceptible to cramps, but other muscles can be affected. Cramps may occur during work or after work. Corrective action is to drink salted liquids.

Fainting. When a worker who is not accustomed to hot environments stands erect and immobile in the heat, the heart sometimes pumps an insufficient amount of blood to the brain. This happens because enlarged blood vessels in the skin and in the lower part of the body (a normal part of the body's attempts to control internal temperature) may cause the blood to pool in lower regions rather than return to the heart to be pumped to the brain.

Moving around stops the tendency to faint by preventing blood from pooling. If weakness continues, and the worker lies down away from the heat, he or she should recover in a short time.

Heat rash. Prickly heat, or heat rash, is likely to occur when heat is accompanied by high humidity. The person's skin remains wet because the humidity allows little or no evaporation of the sweat, and sweat ducts become plugged, causing the skin rash. When the rash is extensive or when it is complicated by infection, prickly heat can be very uncomfortable and can interfere with a worker's performance and efficiency. Corrective actions include resting in a cool place part of each day, drying the skin, and bathing regularly.

Transient heat fatigue. Transient heat fatigue refers to the temporary state of discomfort and mental or psychological strain arising from prolonged heat exposure. Workers unaccustomed to the heat are particularly susceptible and can suffer, to varying degrees, a decline in task performance, coordination, alertness, and vigilance. The severity of transient heat fatigue is lessened by a period of gradual adjustment to the hot environment (heat acclimatization).

Preparing Workers for a Hot Environment

Given that the product and/or process make it impossible to reduce the workplace heat, as when furnaces or sources of steam are in the work area, the next best way to reduce heat stress on workers is to allow an adjustment period. Humans are, to a large extent, capable of adjusting to heat. It typically takes five to seven days for the body to undergo a series of changes that will make continued exposure to heat more endurable.

On the first day of work in a hot environment, body temperature, pulse rate, and general discomfort will be high. With each succeeding daily exposure, all these responses will gradually decrease while the sweat rate will increase. When the body becomes acclimatized to the heat, the worker will be able to work with less strain and distress.

The worker adjustment program can consist of getting used to high temperatures gradually and/or getting used to spending more time in the hot environment. If a plant has several hot environment areas at different temperatures workers to be acclimatized should work first for a few days in the lowest of the hot temperatures and gradually move to the hottest areas. Workers should spend less time in hot environments until their bodies become acclimatized.

In general, heat disorders are more likely to occur among workers who have not been given time to adjust to working in the heat or among workers who have been away from hot environments and have become accustomed to lower temperatures. Workers' bodies can lose acclimatization even during a vacation or extended illness and should be given a readjustment period.

Permanently Lessening Stressful Conditions

Many industries have succeeded in reducing the hazards of heat stress by introducing engineering controls, training workers to recognize and prevent heat stress, and implementing work-rest cycles. One of the easiest and least costly ways to reduce heat exposure is to provide copious ventilation and introduce fresh air.

Heat stress depends partly on the amount of heat the worker's body produces while performing a job; the first step in lessening stress is to recognize that the body produces much more heat during hard, steady work than it does during intermittent or light work. Therefore, one way to reduce the potential for heat stress problems is to make the job easier and/or decrease its duration by providing adequate rest time. Mechanical aids can provide the muscle for lifting and other tasks, thereby reducing the number of calories burned (burning calories is how the body produces its heat). Another benefit of using mechanical aids is that remote control can be added to isolate workers from the heat by placing them behind shields or, best of all, in air-conditioned booths.

Rather than exposing workers to heat for extended periods of time, it is better to distribute their time in the hot environment into shorter, even, periods. Work-rest cycles are best, but if they are not practical, then management should arrange work-rotation cycles. Whether the workers rotating out of the hot environment are given a rest period or are given a time in a cooler environment, their bodies will have an opportunity to get rid of excess heat, slow down the production of internal heat, and provide greater blood flow to the skin.

Engineering controls. There are many ways that engineering controls can minimize exposure to or results of hot environments. For example:

- Improving the insulation on a furnace wall can reduce its surface temperature and the temperature of the area around it (and make the furnace more efficient by keeping its heat working instead of radiating).
- Exhaust fans located to remove moisture at its sources will lower the humidity of the entire area, thereby allowing workers' natural cooling methods, such as sweat, to work as intended.
- Open windows can allow excess heat to leave the work area. Best results can be obtained when open windows are high on one outside wall and low on an opposite outside wall.
- Exhaust fans, whether in windows or fixed in walls, remove more hot air than mere convection will.
- Fans, even if they do nothing more than circulate the air within an area, provide some relief.

Drinking water. In the course of a day's work in the heat, a worker may produce as much as 2 to 3 gallons of sweat. Because so many heat disorders involve excessive dehydration of the body, it is essential that water intake during the workday be about equal to the amount of sweat produced. Most workers exposed to hot conditions drink less fluid than needed because the thirst drive does not accurately reflect the body's needs. Therefore, workers should be encouraged, through training and reminders, not to depend on thirst to signal when and how much to drink. Instead, they should drink 5 to 7 ounces of fluids every 15 to 20 minutes. There is no optimum temperature of drinking water, but it must be palatable and convenient.

One of the adjustments the body makes when it becomes acclimatized to heat is that it loses much less salt in the sweat than does the body of a person who is not adjusted. The average American diet contains sufficient salt for acclimatized workers, even when sweat production is high. If, for some reason, salt replacement is required, the best way is to add a little extra salt to the food. Salt tablets are no longer recommended.

Protective clothing. Certain work clothing is sometimes a necessary part of the job, even if it worsens the heat problem. For example, tasks such as handling hot ingots in a foundry or hot trays in a bakery require insulated gloves. Some jobs require reflective clothing, insulated suits, or infrared-reflecting face shields. Aside from such requirements, the question of whether clothing helps or hurts depends on other factors and conditions.

If the air is hot but lower than skin temperature, clothing hinders the body's ability to transfer its excess heat to the air, especially when the air is moving and the clothing is solid (not open weave).

If the air temperature is higher than skin temperature, the benefits of clothing are even more dependent on other conditions because there are two conflicting processes to consider. First, we want to prevent heat in the air from further heating the body. Clothing can help here by being a partial shield. Second, the skin is cooled by evaporation of sweat. Clothing can interfere with this desirable process. Therefore, the choice when air temperature is higher than skin temperature depends on such factors as:

- How much higher the air temperature is
- How much the air is moving
- Whether the air is changing via ventilation, and the temperature of the incoming air
- The type of clothing—weight, fabric, weave, wicking tendency
- Individual preference

For extremely hot conditions thermally conditioned clothing is available. One such garment carries a self-contained air conditioner in a backpack, and another is connected to a source that pumps cool air into the suit. A different type of garment is a plastic jacket with pockets that workers can fill with dry ice or containers of regular ice.

Special Considerations

Sometimes plant construction is such that hot environmental conditions are exacerbated by a spell of hot outside weather. Workers acclimatized to their regular hot environment generally handle the extra heat for up to two days, but then the number of heat illnesses usually increases. Some reasons are progressive body fluid deficit, loss of appetite, and possible salt deficit if salted food is the main source of sodium. Even if the working environment is insulated from outside weather, a heat wave can trigger illnesses at work because workers do not get relief or proper sleep at home. Air-conditioning systems are sometimes inoperative due to overload of the electric company's sources.

During extended hot spells it is advisable to make a special effort to adhere rigorously to the preceding preventive measures and to have workers avoid any unnecessary or unusually stressful activity. Sufficient sleep and good nutrition are important for maintaining a high level of heat tolerance. Workers who may be at a greater risk of heat illnesses are the obese, the chronically ill, and older individuals.

When feasible, arrange to have the most stressful tasks performed early in the morning or at night. Avoid double shifts and overtime whenever possible, and extend rest periods to alleviate the increase in body heat load.

Consuming alcoholic beverages during prolonged periods of heat can cause additional dehydration. Persons taking certain medications, such as pills for blood pressure control, diuretics, or water pills, should consult their physicians to determine if any complications could occur during excessive heat exposure. Workers should increase their daily fluid intake during hot spells.

CONTROLLING TEMPERATURE AND HUMIDITY COMFORT ZONES

One of the reasons this book includes temperature in discussing indoor air quality is that workers include temperature in their judgment of air quality. Recent research suggests that indoor air quality is judged to be worse as temperatures rise above 76°F (24.4°C), *regardless of the actual air quality*. This is also a reminder to those responsible for indoor air quality—keeping the temperature within comfort zones can forestall air quality complaints.

Those responsible for controlling workplace temperature have to think of "temperature range" because there is no ideal temperature that is best for everyone or preferred by everyone. Table 9-1 lists temperature ranges that define the comfort zone for most individuals in light, mostly sedentary activity. It is assumed that workers are wearing "typical" clothing and therefore does not apply to situations such as a service operation that requires workers to go outside frequently in the winter. Rather than put on a coat each time, those workers will probably wear indoor clothing that is heavier than worn by those who remain indoors.

TABLE 9-1
The Dependence of Acceptable Temperature Ranges on the Relative Humidity

Relative Humidity	*Winter Temperature*	*Summer Temperature*
30%	68.5°F – 76.0°F (20.3°C – 24.4°C)	74.0°F – 80.0°F (23.3°C – 26.7°C)
40%	68.5°F – 75.5°F (20.3°C – 24.2°C)	73.5°F – 79.5°F (23.1°C – 26.4°C)
50%	68.5°F – 74.5°F (20.3°C – 23.6°C)	73.0°F – 79.0°F (22.8°C – 26.1°C)
60%	68.0°F – 74.0°F (20.0°C – 23.3°C)	72.5°F – 78.0°F (22.5°C – 25.6°C)

Relative Humidity and Dew Point

Because technical professionals frequently misunderstand the terms *relative humidity* and *dew point*, we briefly discuss them here. Air holds moisture, and the warmer the air is, the more moisture it can hold. *Absolute humidity* is the actual amount of moisture in the air. *Relative humidity*, usually implied by the word "humidity," is the amount of moisture the air is holding, expressed as a percentage of the maximum it can hold at that temperature. If you take a sample of air at a certain relative humidity and heat it, the relative humidity of that sample will decrease. Figure 9-1 is a graph that relates amount of moisture to air temperature and relative humidity.

Humidity ratio, the ordinate in the figure, is defined as pounds of moisture per pound of dry air. For example, if the temperature of a work area is 80°F and there is nearly 0.018 pound of moisture for every pound of dry air, the relative humidity in that area is an uncomfortable 80 percent. For comparison, look at some of the more acceptable humidity-temperature combinations in Table 9-1, such as 75°F and 40 percent relative humidity. Figure 9-1 shows that at these values a pound of dry air contains a little less than 0.008 pound of moisture. There is 2¼ times as much moisture in the 80°F air with 80 percent relative humidity. With that much moisture in the air, it is easy to see why sweat does not evaporate as easily and why workers in that room would be uncomfortable.

Dew point is the temperature below which moisture condenses out of the air. Although not scientifically rigorous, it is acceptable to say that dew point and 100 percent relative humidity are the same point. Therefore, if the 80°F air with 80 percent relative humidity cools to about 73°F, cold water pipes in the area will "sweat," and walls and other surfaces will probably become wet. Little or no sweat evaporates off workers' skin when conditions are at the dew point.

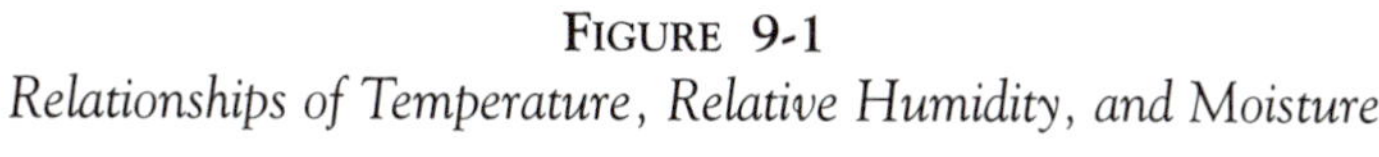

FIGURE 9-1
Relationships of Temperature, Relative Humidity, and Moisture

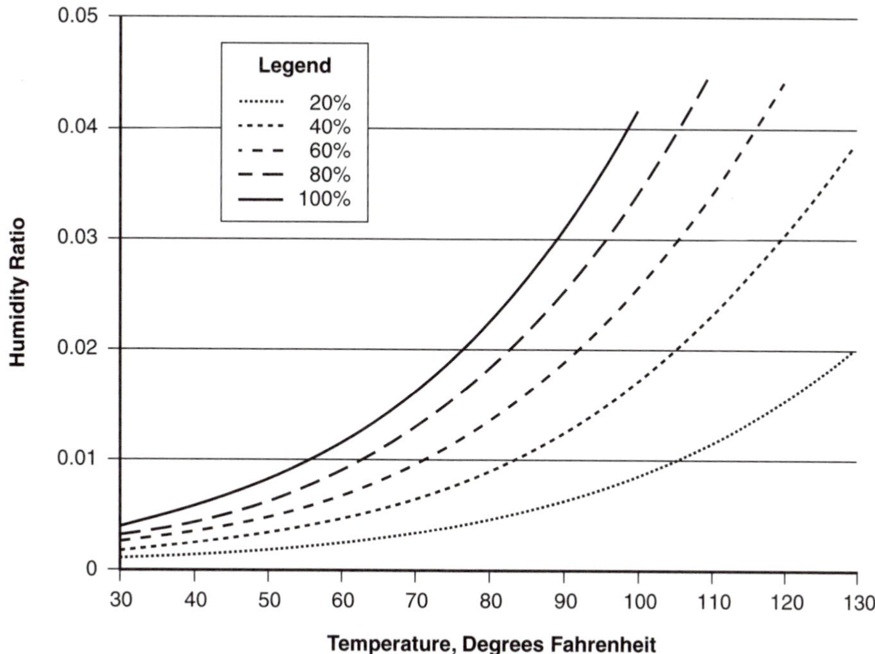

Recommended Level of Relative Humidity

There is considerable debate among researchers, IAQ professionals, and health professionals concerning recommended levels of relative humidity. In general, the range of humidity levels recommended by different organizations is 30 to 60 percent. As may be expected, there are tradeoffs. Some object to relative humidities below 30 percent because they feel discomfort from dryness. However, the lower the relative humidity, the lower the chances of growing mold and mildew. Concerns associated with dry air (mostly in terms of worker comfort) must be balanced against the risks associated with higher humidity (enhanced microbial growth). The fact is that if temperatures are maintained at the lower end of the comfort range (around 68° to 70°F) during heating seasons, relative humidity in most climates will not fall much below 30 percent (also within the comfort range) in occupied buildings. The problem is generally keeping relative humidity from exceeding its upper limit.

Moisture, Mold, and Mildew

Molds and mildew are fungi that grow on the surfaces of objects, within pores, and in deteriorated materials. They can cause discoloration and odor problems, deteriorate building materials, and lead to allergic reactions as well as more serious health problems in susceptible individuals. The following conditions are necessary for mold growth to occur on surfaces:

- Temperature range between 40°F (4°C) and 100°F (38°C)
- Mold spores (already present in most places)
- Nutrient base (most surfaces contain nutrients)
- Moisture

Unfortunately, the temperature range for most workplaces includes the range for mold and mildew growth. Spores are almost always present in outdoor and indoor air, and almost all commonly used construction materials and furnishings can provide nutrients to support mold growth. Dirt on surfaces provides additional nutrients; cleaning and disinfecting with nonpolluting cleaners and antimicrobial agents provides some protection against mold growth. However, it is virtually impossible to eliminate all nutrients. Moisture control is therefore an important strategy for reducing mold growth.

Mold growth does not require the presence of standing water; it can occur when high relative humidity or the hygroscopic properties (the tendency of materials to absorb and retain moisture) of building surfaces allow sufficient moisture to accumulate.

Temperature and relative humidity often vary within a room, whereas absolute humidity usually remains uniform. Therefore, if one side of a room is cooler than the other side, the cooler side will have a higher relative humidity. In other words, the highest relative humidity in a room is next to the coldest surface. It is called the "first condensing surface" because it will be where condensation first occurs if the temperature reaches the dew point. This explains why mold often grows on one patch of wall or only along the wall-ceiling joint. It is likely that the surface of the wall is cooler than the room air because there is a void in the insulation or because wind blows through cracks in the exterior of the building.

Steps to reduce moisture. Mold and mildew growth can be reduced if relative humidities near surfaces can be kept below the dew point. The three ways to control relative humidity below the dew point are (1) by reducing the moisture content of the air, (2) increasing air movement at the surface, and (3) increasing the air temperature (either the general space temperature or the temperature at critical building surfaces).

Either surface temperature or moisture content can be the dominant factor in causing a mold problem. A surface temperature–related mold problem may not respond well to increasing ventilation, whereas a moisture content–related mold problem may not respond well to increasing temperatures. Understanding which factor dominates will help in selecting an effective control strategy.

Consider evidence of mold and mildew in an old, leaky, poorly insulated building in a heating climate. Because the building is leaky, its high natural air exchange rate dilutes interior airborne moisture levels, maintaining a low absolute humidity during the heating season. Providing mechanical ventilation in this building in an attempt to control interior mold and mildew probably will not be effective. Increasing surface temperatures by insulating the exterior walls, and thereby reducing relative humidities next to the wall surfaces, would be a better strategy for controlling mold and mildew.

To reduce surface temperature–dominated mold and mildew increase the surface temperature through either or both of the following methods:

- Increase the temperature of the air near room surfaces either by raising the thermostat setting or by improving air circulation so that supply air is more effective at heating the room surface.
- Decrease the heat loss from room surfaces either by adding insulation or by closing cracks in the exterior wall to prevent wind-washing (air that enters a wall at one exterior location and exits at another exterior location without penetrating into the building).

If the problem is moisture content–dominated mold and mildew, you can reduce it by one or more of the following strategies:

- Source control (e.g., direct venting of moisture-generating activities such as showers) to the exterior
- Dilution of moisture-laden indoor air with outdoor air that is at a lower absolute humidity
- Dehumidification

Dilution is useful as a control strategy only during heating periods, when cold outdoor air tends to contain less moisture. Automobiles that automatically turn on the air conditioner when the driver turns on the defogger use a similar process to send dry air to defog the windows—they cool the air to lower its absolute humidity, and then heat it to lower the relative humidity. During cooling periods, dilution can fail because outdoor air often contains at least as much moisture as indoor air.

Identifying and correcting mold and mildew problems. Exterior corners are common locations for mold and mildew growth in heating climates, and in poorly insulated buildings in cooling climates. They tend to be closer to the outdoor temperature than other parts of the building surface for one or more of the following reasons:

- Poor air circulation (interior)
- Wind-washing (exterior)
- Low insulation levels
- Larger surface exposed to outside for heat loss

Sometimes mold and mildew growth can be reduced by removing obstructions to interior airflow, such as by rearranging furniture. Buildings with forced-air heating systems and/or room ceiling fans tend to have fewer mold and mildew problems than buildings with less air movement, other factors being equal.

Setback thermostats are commonly used to reduce energy consumption during the heating season, but the lower temperatures during unoccupied periods increase the probability of mold and mildew growth. Maintaining a room constantly at too low a temperature can have the same effect as a setback thermostat. Mold and mildew can often be controlled in heating climate locations by increasing interior temperatures during heating periods. Again there are tradeoffs to consider: Higher temperatures increase energy consumption and reduce relative humidity in the breathing zone, which can cause worker discomfort.

The problems of mold and mildew can be as extensive in cooling climates as in heating climates. The same principles apply: Either surfaces are too cold, moisture levels are too high, or both. A common example of mold growth in cooling climates can be found in rooms where conditioned "cold" air blows against the interior surface of an exterior wall. This condition, which may be due to poor duct design, diffuser location, or diffuser performance, creates a cold spot at the interior finish surfaces. A mold problem can occur within the wall cavity as outdoor air comes in contact with the cavity side of the cooled interior surface. It is a particular problem in rooms decorated with low-maintenance interior finishes (e.g., impermeable wall coverings such as vinyl wallpaper), which can trap moisture between the interior finish and the gypsum board. Mold growth can be rampant when these interior finishes are coupled with cold spots and exterior moisture. Possible solutions for this problem include:

- Preventing the hot, humid exterior air from contacting the cold interior finish (i.e., controlling the moisture content at the surface)
- Eliminating the cold spots (i.e., elevating the temperature of the surface) by relocating ducts and diffusers
- Ensuring that vapor barriers, facing sealings, and insulation are properly specified, installed, and maintained
- Increasing the room temperature to avoid overcooling

An obvious consideration that should not be neglected when looking at cooling climates is that increasing the room temperature decreases energy consumption. However, the tradeoff is that it can also cause worker comfort problems.

Localized cooling of surfaces commonly occurs as a result of "thermal bridges," elements of the building structure that have high heat conductivity (e.g., steel studs in exterior frame walls, uninsulated window lintels, and the edges of concrete floor slabs). Dust particles sometimes mark the locations of thermal bridges, because dust tends to adhere to cold spots. Using insulating sheathings significantly reduces the effect of thermal bridges in building envelopes.

In winter, windows are typically the coldest of the natural surfaces in a room. The interior surface of a window is often the first condensing surface.

Condensation on window surfaces has historically been controlled by using storm windows or "insulated glass" (e.g., double-glazed windows or selective surface gas-filled windows) to raise interior surface temperatures. The advent of higher performance glazing systems has led to a higher incidence of moisture problems in heating climate building enclosures, because buildings can now be operated at higher interior moisture levels without visible surface condensation on windows. In older buildings with less advanced glazing systems, visible condensation on the windows often alerts occupants to the need for ventilation to flush out interior moisture.

Another undesirable condition on one front, due to gains on another front, is concealed condensation. Thermal insulation in wall cavities increases interior surface temperatures in heating climates, reducing the likelihood of interior surface mold, mildew, and condensation. However, thermal insulation also reduces heat loss from the conditioned space into the wall cavities, decreasing the temperature in the cavities, and increasing the likelihood of concealed condensation. The first condensing surface in a wall cavity in a heating climate is typically the inner surface of the exterior sheathing, the "back side" of plywood or fiberboard. A general rule is: The higher the insulation value in wall cavities, the higher the potential for hidden condensation. Hidden condensation can be controlled by either or both of the following strategies:

- Reducing the entry of moisture into the wall cavities—controlling infiltration and/or exfiltration of moisture-laden air, or
- Elevating the temperature of the first condensing surface. In heating climate locations, this change can be made by installing exterior insulation while taking care to avoid significant wind-washing. In cooling climate locations, this change can be made by installing insulating sheathing to the interior of the wall framing and between the wall framing and the interior gypsum board.

PART III

MONITORING INDOOR AIR QUALITY

CHAPTER 10

Measuring and Monitoring Considerations

It is important to understand that measuring air quality is a complex operation, and it helps first to identify the purpose of measuring, and the category of instrument needed.[1] The first distinction to make among instrument choices is between *analytical devices* and *protection devices*, also called *protective monitors*. Analytical devices are usually used in laboratory settings or perhaps field laboratories. Instruments in this category are used by scientific and trained technical personnel. Whether the measurement is of automobile exhaust emissions, factory flues, dissolved gases in water, or even medical analysis of absorbed gas in the human body, highly skilled personnel and specialized procedures and equipment are required.

In contrast, protective devices are instruments that are designed to be used in a broad range of applications by ordinary workers with minimal specialized training. Most of the material in this chapter relates to protective devices.

1. *Training Program: Gas Detectors for Protection*, courtesy of Dynamation, Incorporated, Ann Arbor, Michigan.

CATEGORIES OF PROTECTIVE DEVICES

Protective devices can be categorized according to their applications, as follows:

Monitors designed to detect hazardous gas or vapors in an airstream intended for breathing. Workers receive their breathing air via pipes or hoses that end in some type of respirator or hood. The source of the air is almost always a compressor that feeds the hose either directly or by way of a tank. The breathing air is regulated by OSHA for industrial applications and National Fire Protection Association code for medical applications.

Typical hazards that require special attention include carbon monoxide, hydrocarbon vapors, lack of oxygen (usually caused by accidental introduction of inert gases such as nitrogen or argon), as well as excess humidity, rated by dewpoint.

A second major category of protective monitors can be thought of as "fixed" or permanently installed monitors, intended to protect workers in a specific location. They may also operate auxiliary equipment such as fans or mechanical vents. Many industrial processes produce hazardous gases as byproducts, such as chlorine from pulp mills, carbon monoxide from parking garages, and large quantities of carbon dioxide from breweries and mushroom growers. Chemical plants and petroleum refineries produce a variety of hazardous gas byproducts that can be a threat to their workers.

The reason permanently installed monitoring systems are often used here is that the effects are usually predictable. Smaller systems may incorporate a single sensor for one gas, whereas a large system in a chemical plant may incorporate hundreds of sensors for many different types of gases.

A third major category of gas monitors is the single-gas portable monitor, most often used for personal protection. In their most basic form, single-gas monitoring devices include passive badges or ambient absorption tubes that gradually change color on exposure to a specific gas or class of gases. These badges or gas tags are often worn by workers in manufacturing environments.

Single-gas electronic monitors designed for worker protection are usually very small and light, and may or may not include a digital display. Often these detectors are alarm-only devices carried in a shirt pocket or on the belt. In very noisy work areas a headphone or earpiece may be required to alert the user of an alarm.

Slightly larger devices with digital displays are often used to check for a specific gas. Firefighters respond regularly to home carbon monoxide detector alarms and must rely on a quality digital CO detector to evaluate the risk and locate the source. Models are available with a wide variety of sensors, and a larger, easy to read display is usually preferred.

> *The final major category of instruments for worker protection includes multigas monitors designed for use in confined spaces where multiple risks can occur.* One way to meet legislative requirements is with multiple instruments, each capable of detecting a specific gas. However, it is usually more convenient and economical to use a multigas monitor suitable for at least three gases. Most multigas monitors are equipped to detect oxygen, combustibles, and either carbon monoxide or hydrogen sulfide or both.

CATEGORIES OF GASES

It is helpful also to think of gases in categories. Any discussion of hazardous gases or the instruments designed to detect them usually involves three main categories:

- Oxygen
- Combustible gases
- Toxic gases

Oxygen

Oxygen is in a class of its own because it becomes hazardous only when the percentage of oxygen in the air falls outside a normal range centered at 20.9 percent. A low concentration can be caused by various oxygen-consuming functions such as bacterial action or even rusting iron products. They can lower the oxygen concentration by removing oxygen or by replacing it with another gas such as nitrogen or methane. At the other end of the concentration range, a major hazard from excessive oxygen is the danger of explosion. Although the oxygen itself does not burn, it provides the basis for most fires and explosions. High concentrations are usually caused by leaks from oxygen tanks or by misadjustment of their controls.

Combustible Gases

Combustible gases include every gas capable of being ignited. The combustible gas most often encountered is methane, also known as natural gas or city gas. Methane is a natural byproduct of decaying matter and is the gas most often used for calibrating gas detectors. Other gases used for calibrating include hexane and pentane. All combustible gases are characterized by upper and lower limits at which they will explode, called UEL and LEL, respectively. For example, methyl acetate can explode when at concentrations between 16 percent (UEL) and 3.1 percent (LEL). These limits depend somewhat on temperature, oxygen concentration, and other factors.

Toxic Gases

Toxic gases are those that produce physiological harm to humans. Because testing suspected toxic gases on humans is unacceptable, some gases are included on the list if animal studies demonstrate their toxicity, and the affected systems of the animals involved are similar to those of humans. Many combustible gases fall also into the category of toxic gases, but the level at which they become a toxic risk is much smaller than the level at which they are combustible. Toxic levels are usually measured in parts per million.

DISPERSION OF GASES

The molecular weight of a gas is important, mainly for determining where to find certain gases and how they will disperse. There are three conditions to consider:

- Gases whose molecular weight is lighter than that of air will rise.
- Gases whose molecular weight is heavier than that of air will settle.
- Gases whose molecular weight is about the same as that of air will disperse and mix with the air.

There is no sharp value dividing these categories, so you will find gases all along the continuum from lightest to heaviest. You will find some that, because they are only slightly lighter than air, will both rise slowly and disperse. (Information given in Appendix B includes relative weight.)

Another consideration is the degree of containment. For indoor situations, we consider that the air and its contaminants are held in the building and can rearrange themselves but are not free to blow away. But because no industrial facility is hermetically sealed, some exchanges with the outside can be expected. It helps to look at the extreme of a completely sealed building and then back off from there to estimate actual conditions. In a sealed room, with no ventilation or air exchange with other rooms or the outside, you can reasonably expect that, given sufficient time to stabilize, the atmosphere will stratify. Gases with the lowest molecular weight will be found at the top, and gases with the highest molecular weight will settle to the floor. It is very important to note that since this is a closed room, the heavy gas that sinks won't share the low levels with normal air—it will displace normal air, including valuable oxygen. Normal air will move up, to a level determined by the amount of the heavy gas present. In the same way, lighter gases will decrease the amount of normal air and oxygen at the highest levels. Unless there is a large difference in molecular weights, the displacement will not be complete. Rather, there will be a smooth change in concentrations.

In practice, rooms are not completely sealed, and it is likely that leaks (deliberate or accidental) will not be evenly distributed. If the upper-level leaks, for example at the top of a stairwell, are significantly larger than leaks at lower levels, the building will form a container that keeps heavy pollutants but allows lighter ones to escape, like helium balloons flying away. Buildings that are well sealed at upper levels but have loading docks with large open doors will trap lighter gases near the top floors but will allow heavy gases to spill out, especially if the lower floors are open to vertical airways such as stairwells.

For an example of why it is important to understand these actions of different gases, consider a building that is internally leaking small amounts of methane. The building is sealed quite well at upper levels. Methane is lighter than air, so it will rise to the upper levels, where it will displace oxygen. This dangerous methane-rich atmosphere cannot be detected by combustible sensors, because they require at least 17 percent oxygen to function! An investigator who measures methane at lower levels (where the concentration is lower) but none at upper levels (where the concentration is higher) might conclude that the problem is completely defined by the smaller methane concentration at lower levels if he or she doesn't understand these critical details.

GAS MONITORS, DATA RECORDERS, AND LIABILITY[2]

We now look at a few of the legal unknowns regarding liability concerns that affect decisions in all aspects of American life. Entire industries have developed around

2. The following thoughts and caveats have been provided courtesy of Dynamation, Incorporated, Ann Arbor, Michigan.

the concept of risk reduction and liability protection. New workplace legislation designed to protect workers has the side effect of increased opportunities for legal confrontation. Employers and property owners are now legally obligated to a host of new and specific responsibilities. Also of concern are the nonspecific, interpreted responsibilities implied by the heading "due diligence."

The narrow field of hazardous gas monitoring offers several examples of implied responsibility. The Confined Space Standard of 1993 requires monitoring for oxygen, combustible, and "potentially" toxic gases. Monitoring for oxygen and combustible gas is usually straightforward and easily accomplished with conventional technology. However, toxic monitoring is more involved because specific sensors for only a few of the most commonly encountered gases are available. The vast majority of monitors with toxic sensors are equipped to monitor carbon monoxide and/or hydrogen sulfide, the most commonly encountered gases. That leaves hundreds of additional toxic substances to be monitored, most of which form gases or give off vapors and have recommended exposure limits of less than 10 ppm.

What is the employer's or property owner's obligation to monitor for these toxic gases? It is small comfort to know that most of the hundreds of toxins are seldom encountered in the general workplace. Actually, a significant number, for example benzene and trichloroethylene, *are* commonly encountered.

Two of the main sources exerting pressure on the employer and the property owner are regulatory agencies and the ever-ready-to-sue public. What will it cost, in dollars and public relations, if a sewer worker is overcome by toxic vapors from something other than hydrogen sulfide? Maybe a good attorney can convince the judge that monitoring for commonly encountered gases constituted a sufficient degree of due diligence. But the adversarial attorney may point to Appendix E of the Confined Space Standard (recently revised to reinstate a reference to broad-range sensors as being best suited for sewers) and ask why that type of monitor wasn't used.

Perhaps, instead of being in a sewer, the worker is in a manufacturing facility and is overcome by a toxic such as trichloroethylene instead of the carbon monoxide that his monitor could have detected. No standard specifically requires monitoring for every toxic, but a real possibility exists that management may be held liable if every reasonable effort was not made to anticipate potential hazards and provide reasonable protection.

Some situations may require no more than a three- or four-gas monitor. Others may require a monitor that senses a broad range of toxics as well as carbon monoxide, hydrogen sulfide, oxygen, and combustibles. Still others may require testing with detector tubes or even a gas chromatograph. Supplied air, in tanks or by air hose, from an approved (Grade D) source may be the only solution in some situations. The most important point is that employers and property owners must be able to demonstrate they made every effort to comply with applicable laws, to evaluate and anticipate the potential risks, and to provide appropriate protection.

Another area of concern is the documentation of efforts to ensure appropriate protection. Prior to the final adoption of the Confined Space Standard, there was much expectation that instruments capable of recording exposure data might be required, but the published standard omitted any reference to such instruments.

One possible reason for this omission is that the permit required for allowing workers entry into a confined space records most of the pertinent data in an easily accessed form without requiring that complex computer files or printout records be cross-referenced to the entry. The standard mandates that this information be kept and reviewed within 12 months or annually.

GAS DETECTION GUIDELINES[3]

Table 10-1 gives information that applies specifically to confined spaces, but since confined spaces are another form of indoor environment, it contains information of general use for indoor air quality situations. All confined spaces are initially classified as permit-only spaces; they may be reclassified if appropriate criteria are met during evaluation testing. Tests of the atmosphere should be analyzed using equipment of sufficient sensitivity and specificity to identify and evaluate any hazardous atmospheres that may exist or arise so that appropriate permit entry procedures can be developed by a qualified professional.

Table 10-1 gives evaluation results and required verification testing as it pertains to gas detection, as indicated in the Federal Register 1910.146 and its appendices. Regarding the column headed *After Entry Gas Detecting*, use the following guidelines for toxic gases:

> For less than 15 minutes in confined space, use peak PEL value or if longer than fifteen minutes, use both peak and TWA, PEL values as described in subpart G, *Occupational Health and Environment Control*, or in subpart Z, *Toxic and Hazardous Substance;* i.e., for hydrogen sulfide: 15 PPM is peak and 10 PPM is TWA.

3. Provided by Dynamation, Incorporated, Ann Arbor, Michigan, with the following disclaimer, "Dynamation has studied part 1910.146 extensively and has compiled the table utilizing years of experience in confined space gas detection. However, any variation from official interpretation is without intent. It was developed as a reference guide to assist employers to understand gas detection requirements only. Please consult the Federal Register of January 14, 1993 for the complete confined space printed rules and regulations."

TABLE 10-1

Gas Detection Guidelines

Evaluation Results	*Permit Type*	*Pre-Entry Gas Detecting*	*After Entry Gas Detecting*	*Alarm Condition*
No actual or potentially hazardous conditions in confined space	Non-Permit	Alarm only acceptable for O_2, LFL, toxic, and vapors by lead person	Continuous monitoring with alarm only or direct reading detector	Evacuate and test with direct reading by supervisor
Only hazard is actual or potentially hazardous atmosphere where continuous forced ventilation alone can permit safe entry	Permit—No attendant and appropriate safety equipment required	Direct reading for O_2, LFL, toxic gases, and vapors	Periodic testing with pre-entry type detector	Evacuate and implement procedures to protect from hazard. Repeat pre-entry test
Actual or potential atmospheric and other hazards. Continuous ventilation (if used) will not permit safe entry	Permit—Attendant and appropriate safety equipment required	Direct reading for O_2, LFL, toxic gases, and vapors by supervisor	Continuous monitoring with direct reading pre-entry type detector	Follow safety permit instructions
Sewers—Actual or potentially hazardous atmospheres where forced ventilation if used will not ensure safety or it can be reasonably expected that conditions will change	Permit—Attendant and appropriate safety equipment required	Direct reading for O_2, LFL, H_2S. A broad range sensor recommended to detect toxic vapors by supervisor	Continuous monitoring for all gases with direct reading pre-entry type detector	Evacuate and test with specific gas device to identify hazardous substance

HOW TO CONDUCT AN ORGANIZED INVESTIGATION

As a first step in response to a complaint, do not set up a grand program to measure and monitor the indoor environment. Begin by looking at the following steps, selecting those that apply to your plant and the complaint in question, and rearranging them in an appropriate order.

- Examine the complaint file to see if any other complaints are similar or have some overlap.
- Determine the scope of the preliminary walk-through.
- Decide whether immediate action, such as evacuating the building or shutting down certain processes, should be initiated.
- Decide whether *any* action should be initiated right away.
- Decide whether it is time to bring in an outside consultant. You might repeat this decision several times as you gather data and learn facts about the problem.
- Identify what further information you will need before you can continue.

You might define the preliminary walk-through as part of these preinvestigation steps or as a separate step that automatically follows any air-quality complaint. It is only after you have considered the issues listed above and have completed the walk-through that you have a direction in which to move.

LAYING THE FOUNDATIONS

The following arrangements should always be at the ready, long before the first hint of an indoor air quality problem. Their very presence might be the reason certain problems never occur.

Communication

Ever since your first class in Management 101 you probably expected to hear that communication is a major factor in maintaining indoor air quality. You're right. *Effective communication systems help facility managers, staff, contractors, occupants, inspectors, regulators, and others to clarify their responsibilities and to cooperate in identifying potential IAQ problems.* That sentence is in italic because there is no end to what you can read into it. The range of people involved—from technically and actively involved to those who just want a clean atmosphere in which to work—shows that minimal efforts such as a memo on the bulletin board or an item in the company newsletter do not qualify as communication.

Of the many requirements that a good communication system must meet, there are two that especially apply here. First, the channel to and from (notice the "to and from") management should be tailored for the group(s) involved; a "Dear Fellow Employee" memo to everyone cannot convey information that actually applies to everyone. Second, any action required by the situation should be clearly explained, including specific facts about who, what, when, where, and how.

Some question why building occupants should be included as necessary links in the communication chain—they use the indoor air but do not control it. The answer is that, if the building occupants really do not have a part in controlling the indoor air, then the communication system is incomplete. Occupants can have two major roles in air quality management:

> *They can affect the air quality.* Occupants can smoke, bring in allergens, misuse equipment, block air flow, tamper with controls, open windows, and in other ways prevent the system from working as designed. They do these things mostly because of poor communication.
>
> *They are usually the first to know of a degradation in air quality.* Where indoor air quality is not monitored constantly, a problem might exist for a long time if occupants didn't report it. Appendix F presents forms that occupants can use to communicate important facts. Some of the forms can trigger an indoor air quality investigation; other forms provide valuable information to investigators.

On the surface, good communication offers the benefit of facilitating IAQ control. Beneath the surface, one of the benefits is the improved morale that always results from including employees in the loop, keeping them informed, and letting them know that management cares.

One tried and proven way to promote good communications is through a group that represents all of the interested parties in the building. Sometimes it is an independent and permanent group. In some organizations, it is a subcommittee that reports to the health and safety committee. When forming and empowering the group, consideration should be given to:

- What are its responsibilities?
- What is its authority?
- To what level of management should it report?
- What provisions should there be for reviewing and modifying its charter as the firm accumulates experience with it and as conditions change?
- What should the management structure be within the group?
- How is it to be funded?
- Where will its costs appear in the accounts?

An Indoor Air Quality Profile

An IAQ profile is a recording of building conditions from the perspective of indoor air quality. A review of construction and operating records, combined with an inspection of building conditions, helps to reveal potential indoor air problems and identify building areas that require special attention to prevent problems in the future. The profile becomes baseline data to facilitate later investigations.

The Underlying Rule

Indoor air quality problems you might encounter range from extremely dangerous situations, to simple annoying odors, to indoor temperatures that are less than ideal. The fixes vary from an adjustment by a facility engineer to a large-scale contractor involvement. Regardless of where on this continuum the solution lies, *the goal is always to resolve the complaint without causing other problems*. Corrective action that will cure or diminish a problem must take into consideration local laws, building codes, and environmentally safe practice. Changes to the overall design and operation of the building may necessitate the involvement of a registered professional engineer or other registered or certified professionals.

PREPARING FOR AN AIR SAMPLING PROGRAM

If it has been decided that air sampling should be used, the first step should be to collect information using the investigation and preparation activites just described. Investigators should develop the Indoor Air Quality Profile, along with a sampling strategy based on a comprehensive understanding of how the building operates and the nature of the complaints. A plan for interpreting the results will also be needed.

It would be desirable to have the investigation include certain routine air quality measurements, to obtain a "snapshot" of current conditions. These measurements should be limited to those that are indicative of common indoor air quality concerns, such as temperature, relative humidity, air movement, and possibly carbon dioxide. Although unusual readings could indicate a problem, they should always be interpreted in perspective, based on site-specific conditions.

Measurement of specific chemical or biological contaminants can be expensive. Before spending time and money to obtain measurements of indoor air pollutants, investigators should pause to examine the following:

- How will the results be used (that is, will results be compared to standards or guidelines, to levels in complaint-free areas, or to previous readings)?
- What substances, known to be present, will be measured?
- What unknown substances will possibly be detected?
- Where will samples be taken?
- When will samples be taken? If there is a pattern, the time of highest concentration should certainly be one of the test times.
- What sampling and analysis method will provide the most useful information?

How Will the Results Be Used?

Without a doubt, air sampling will generate numbers. To avoid being swamped with data, investigators should first develop a plan that includes what numbers to expect, what numbers might appear unexpectedly, and what each possible set of numbers will mean. One important part of the plan should be a list of actions that will be taken in response to certain numbers. Be prepared for the possibility that the numbers may not help solve the problem. Another possibility, which often happens, is that the complaints will be resolved, or will cease, without sampling or with inconclusive sampling results.

The reason for first establishing how the results will be used is that the design of an air sampling strategy should fit the intended use of the measurements. Some examples of potential uses of measurements include:

Comparing different locations. You might obtain useful information by comparing the air in two or more different areas of the building or by comparing indoor to outdoor conditions. The objective may be to:

1. Determine if a control approach has the desired effect of reducing pollutant concentrations.

2. Determine if changes made have been effective in improving ventilation.
3. Establish baseline conditions for comparisons in other areas or at other times. For example:
   - Record concentrations in outside air.
   - Record concentrations in areas where no symptoms are reported.
   - Record expected background range for typical buildings without perceived problems.
4. Test a hypothesis about the source of a problem. For example, record the emissions from a certain piece of equipment at various times and under various conditions.

Testing for "indicator" compounds. Indicators can be associated with particular types of building conditions such as:

1. Peak carbon dioxide concentrations of over 1000 parts per million are an indicator of insufficient ventilation.
2. Carbon monoxide over several parts per million indicates unacceptable presence of combustion byproducts. If carbon dioxide concentrations are high at the same time, it is likely that the same machinery is causing both problems.

Comparing results to standards. A frequent use of measurement results is to compare the measured concentrations to guidelines, standards, or regulations. Sources of these standards include:

- OSHA PELs (Occupational Safety and Health Administration Permissible Exposure Limits)
- NIOSH RELs (National Institute for Occupational Safety and Health Recommended Exposure Limits)
- ACGIH TLVs (American Conference of Governmental Industrial Hygienists Threshold Limit Values)
- EPA National Ambient Air Quality Standards
- World Health Organization Air Quality Guidelines
- Canadian Exposure Guidelines for Residential Air Quality

There are no widely approved procedures to define whether indoor air quality test results are acceptable. However, a generally accepted rule is that test results which are within standards or guidelines should be the basis for declaring a building safe. In the other direction, it is usually correct to conclude that contaminant concentrations above the standards or guidelines are a clear indication of a problem. Investigations have shown that, even when concentrations are well within guidelines, occupants may report health and comfort problems. In fact, it is extremely rare for occupational standards to be exceeded, or even approached, in public and commercial buildings, including those experiencing indoor air quality problems.

Where specific exposure problems are suspected, more detailed diagnostic testing may be needed to locate or understand major sources, confirm the exposure, and to develop appropriate remedial actions. For example, controlling microbial or pesticide contamination may involve surface or bulk sampling. Surface sampling involves wiping a measured surface area and analyzing the swab to see what organisms are present; bulk sampling involves analyzing a sample of suspect material. Specialized skills, experience, consultants, and equipment may be needed to obtain, analyze, and interpret such measurements.

What Substances Should Be Measured?

Experienced investigators strongly recommend *not* launching an all-out program to measure "everything." Measuring "indicator" compounds such as carbon dioxide or carbon monoxide can be a cost-effective strategy that yields the most information for the least expense. An inappropriate concentration of these and other indicator compounds is frequently present when there is an air quality problem. Such measurements can help the investigator understand the nature of the problem and to circumscribe the problem and its physical area. Repeating these measurements at different times and under different conditions can provide a considerable amount of information before the full investigation begins.

Air sampling for specific pollutants works best as an investigative tool when it is combined with other types of information-gathering. When both of the following conditions are present, it is probably prudent to begin a program of chemical sampling:

1. If symptoms or observations strongly suggest that a specific pollutant or a specific source is the cause of the complaint; and
2. If sampling results are important in determining an appropriate corrective action.

Where Should Air Samples Be Taken?

Of course, the room or area where complaints arise should be one location for air sampling. Interpretation of the information gathered there becomes greatly facilitated when combined with information gathered from other locations such as complaint-free areas in the building, and outdoors. If there is a time pattern to the complaints, then measurements in the complaint area should be taken when it is expected that pollution will be at its highest, lowest, and at other times.

The conditions experienced by building occupants are best simulated by sampling the air in an area away from the influence of individuals. However, it is also informative to examine the air surrounding a particular person, such as one who remains at a desk for most of the workday. Samplers placed on the desk when the individual is elsewhere can provide a good indication of the quality of the air to which that person is exposed.

There are several strategies available for locating air sampling sites. One approach that usually meets with success is to first divide the building into homogeneous areas based on key factors identified in the building inspection and interviews. Following are some suggestions for dividing a building:

- *Control zones*. Depending on the building's HVAC design, the smallest control zone could be a part of a room, a room, a floor, or even an entire building.
- *Types of HVAC zones*. Some buildings' HVAC systems might be designed so that all interior areas are controlled as a zone, and all perimeter areas (with outside walls) as another zone.
- *Complaint and noncomplaint areas*. Most investigators will automatically divide the building into these areas for some measurements.
- *Relationship to major sources of pollution*. For example, when any of the chemicals from tobacco smoke are suspected, the investigators can divide the building into smoking areas, areas that share a hallway, areas that share a ventilation branch, areas with no apparent connection, areas with direct outside connection, and others.
- *Complaint types*. Separate measurement programs might be established for areas in which workers report different types of reactions to pollutants.

After dividing the building for the investigation, it is easier to plan test sites and other parts of the testing procedure. One mistake to avoid is to treat the building divisions as unchangeable. Sometimes early tests reveal surprises that indicate a different way of dividing the building would yield more information.

When Should Air Samples Be Taken?

Almost all investigations attempt to obtain worst-case conditions as one set of measurements. When the time pattern is unknown, it might be necessary to set up recorders or data loggers whose results can be examined. The time pattern often reveals clues that lead to the source of the problem. Following are some of the possibilities:

- *Pollutant concentrations are at their lowest in the morning, and become progressively worse during the workday*. This is probably the most common situation, and suggests that the plant's equipment is emitting pollutants as it operates. Of course, there are other possible explanations, such as allergens being brought in by workers.
- *Pollutant concentrations are highest in the morning, and lessen during the workday*. If windows and doors are closed after the shift, and opened during the day, this pattern suggests that some tank or piping inside the plant is leaking.
- *Investigation might show that concentration patterns change during the workday and then return to a certain arrangement after the shift*. The first question to ask is whether the HVAC system is turned off or set to a different cycle after the shift. It is also possible that worker activity or opening of doors (especially by vertical shafts such as stairwells and elevators) causes the air to circulate one way during a shift and another way after the shift.

In addition to worst-case sampling, it can be helpful to obtain samples during average or typical conditions as a basis of comparison. Any identifiable condition, such as worst case, best case, quiescent times, peak activity, break times, and others can yield valuable information.

Just as a tooth stops hurting when you arrive at the dentist's office, symptoms or odors that occur intermittently will seldom be present during an indoor air quality investigation. Generally it is not cost-effective to take air samples when air quality seems satisfactory, unless the purpose is to establish a baseline for future comparisons.

Intermittent and unpredictable air quality problems are difficult to solve. Sometimes investigators have to ask appropriate building staff or other occupants to document changes over time. If outside investigators are used, and are not on-site when an odor episode occurs, the building engineer should inspect the air handler and intake area while other staff members document the status of several potential sources.

Another technique is to manipulate building conditions to *create* worst-case conditions. For example, arrange for the trash truck to idle at the loading dock, or close outdoor air dampers to minimum settings. Chemical smoke and tracer gases can be used to assess where emissions may travel in response to various building conditions. Of course, all strategies that raise pollutant levels should be used in such a way or at such a time that occupants will not be placed in danger.

CHAPTER 11

Principles of Air Sampling and Analysis

This chapter provides a general guide to making measurements that are often needed in the course of developing an indoor air quality profile or investigating an IAQ complaint. We highlight the more practical methods and note some inappropriate tests to avoid. Most of the instruments discussed here are relatively inexpensive and readily available.

In addition to the instruments described in this chapter, there are hundreds of others for special purposes, higher accuracies, and less common applications. (Some of them come with an instruction manual the size of a small telephone book.) Keeping abreast of the variety of new and existing instruments can involve attending trade shows, studying magazine advertisements, and talking to vendors.

HOW TO SELECT INSTRUMENTS

The growing interest in indoor air quality is stimulating the development of instruments for IAQ research as well as for building investigations. Most instruments that are optimum for one purpose are entirely inappropriate for another purpose, and therefore it is necessary to keep the end use in mind when selecting instruments. Following is a list of general characteristics that will help in selecting the right instrument, avoiding those that are either inadequate or have too many features (and therefore are too expensive for the intended application).

Criteria Related to Ease of Use

- *Portability*. Weight, selection of power sources, ease of recharging batteries, operating time on a charge, auxiliary equipment required, battery saver features
- *Direct-reading vs. analysis required*. If analysis required, is it self-contained?
- *Ruggedness*. Ability to withstand physical abuse, operating temperature range, standby temperature range, sealed from environment
- *Time required for each measurement*. Shortcuts for repetitive measurements, remote control
- *Training required*. Self-explanatory layout, conversions, pocket for manual

Criteria Related to Quality Assurance

- Availability of service and customer support
- Maintenance and calibration requirements
- Mean time before failure
- Self-check features

Output

- Time-averaged vs. instantaneous readings
- Sensitivity, selectivity
- Compatibility with computer or data logging accessories

Cost

- Single use vs. reusable
- Training requirements
- Update arrangements
- Purchase vs. rent or lease

USING SAMPLING DEVICES

The next few paragraphs take a wide-angle look at the major categories of sampling devices that are used for evaluating indoor air quality. Most monitoring programs use one or more of these instruments.

Vacuum Pump

With this instrument, a vacuum pump with a known airflow rate draws air through collection devices, such as a filter (catches airborne particles), a sorbent tube (attracts certain chemical vapors to a powder such as carbon), or an impinger (bubbles the contaminants through a solution in a test tube).

When adapting an industrial hygiene sorbent tube sampling method to indoor air quality, investigators should consider two important questions:

- Are the emissions to be measured from a product's end use the same as those of concern during manufacture?
- Is it necessary to increase the air volume sampled? Such an increase may be needed to detect the presence of contaminants at the low concentrations usually found in nonindustrial settings. For example, an investigator might have to increase sampling time from 30 minutes to 5 hours in order to detect a substance at the low concentrations found during IAQ investigations.

In situations where standard sampling methods are changed, qualified industrial hygienists and chemists should be consulted to ensure accuracy and precision.

Direct-Reading Meter

Direct-reading meters estimate air concentrations through one of several detection principles. These instruments may report specific chemicals (such as carbon dioxide by infrared light), chemical groups (such as certain volatile organics by photo ionization potential), or broad pollutant categories (such as all respirable particles by scattered light). Detection limits and averaging time developed for industrial use may or may not be appropriate for indoor air quality applications.

Detector Tube Kit

Detector tube kits generally include a hand pump that draws a known volume of air through a chemically treated tube that reacts with certain contaminants. The length of color stain resulting in the tube correlates to chemical concentration.

Personal Monitoring Devices

Also known as dosimeters, personal monitoring devices are badges and other sensors worn by individuals to measure their exposure to one or more chemicals. "Active" personal monitors are those that include a pump; "passive" monitors do not include a pump.

BASIC INDICATORS OF VENTILATION AND COMFORT

As mentioned in Chapter 10, a few of the indicator measurements go a long way toward indicating air quality in most situations. This section looks at some simple measurements that, for many investigations, provide all the information necessary.

Temperature and Relative Humidity

Air quality complaints are sometimes vaporous, such as "the air is stuffy" or "it saps my energy," and the resultant investigation might find that temperature and/or relative humidity are the only numbers that are outside their normal ranges. In fact, investigators who do not consider temperature and humidity to be indoor air quality problems may find nothing to explain the complaints.

Actually, temperature and relative humidity measurements are excellent indicators of thermal comfort. Workers' sense of thermal comfort (or discomfort) results from an interaction of temperature, relative humidity, air movement, clothing, activity level, and individual physiology.

An inexpensive way to take measurements is with a standard thermometer (accuracy of ±1° Fahrenheit is recommended) and a sling psychrometer. The latter uses two thermometers—one with the bulb kept dry and the other wet—which the investigator swings through the air. Moving rapidly through the air has no effect on the dry bulb, but it cools the wet bulb by evaporating the water around it. The lower the relative humidity, the more the wet bulb thermometer will cool;

the investigator then consults a table to convert the difference in temperature readings to relative humidity.

There are also electronic instruments available for measuring temperature and relative humidity. (Sling psychrometers and other instruments are described in Moffat, D.W., *Plant Monitoring and Inspecting Handbook,* Prentice Hall, Englewood Cliffs, New Jersey.) A thermo-hygrometer measures both quantities. Regardless of the method used, it is important to follow the manufacturer's instructions to maintain accuracy. For example, electronic humidity meters usually require frequent calibration. Allow time for the readings to stabilize; some take several minutes.

Indoor relative humidity can vary throughout the day as outdoor conditions change. Therefore, a single reading might not reveal much about conditions that may have triggered worker complaints. Frequent readings throughout the day or, still better, a recording sensor that takes continuous readings, will give an investigator much more useful data for analyzing complaints.

Temperature and humidity measurements, in addition to explaining workers' thermal comfort reports, provide indirect indications of HVAC functioning and the potential for airborne contamination from biological or organic compounds. Although there is disagreement as to ideal temperature and humidity conditions, Chapter 9 gave the ranges that most groups prefer. Comparison of indoor and outdoor temperature and humidity readings taken during complaint periods can indicate whether thermal discomfort might be due to extreme conditions beyond the design capacity of HVAC equipment or the building envelope.

Conditions can change considerably throughout a room, so it is important that one set of measurements be taken at the location of complaints. Measure, also, next to thermostats to confirm calibration. One of the clues to watch for is unreasonably large variations in the readings throughout a room—frequently indicating a room air distribution or mixing problem. If readings are highly variable over time, there may be HVAC control or balance problems.

Tracking Air Movement

Chemical smoke can be helpful for evaluating HVAC systems, tracking potential contaminant movement, and identifying pressure differentials. Air, and therefore smoke, moves from areas of higher pressure to areas of lower pressure as long as there is an airway between them. The airway can be large, such as a door, or small, such as a crack in the wall or space around pipes that run between the areas.

If the smoke is used at the same temperature as the air, there will be no convection effect, and all movement observed will be the result of air currents.

Investigators can learn about airflow patterns by observing the direction and speed of smoke movement. Puffs of smoke released at the shell of the building (by doors, windows, or gaps) will indicate whether the HVAC system is maintaining interior spaces under positive pressure relative to the outdoors.

Chemical smoke is available in various types of dispensers such as bottles, guns, pencils, and tubes. It is recommended that the dispenser allow users to release controlled quantities, directed at specific locations. Releasing a number of small puffs is often more informative than releasing a large amount in a single puff.

Check the manufacturer's warning. Most tell you to avoid direct inhalation of chemical smoke because it can be irritating; investigators should protect themselves, and use the smoke when others are not present. Do not release chemical smoke directly on smoke detectors.

The technique of tracking with smoke consists of releasing puffs in different places. For example, observation of a few puffs released in the middle of an area can help investigators visualize air circulation within the room or cubicle. Dispersal of the smoke in several seconds suggests good air circulation; smoke should not stay essentially still for several seconds. Poor air circulation may contribute to sick building syndrome complaints or may contribute to comfort complaints even if there is sufficient overall air exchange.

Puffs of smoke released near diffusers, grilles, and other HVAC vents give a general idea of airflow. Being able to describe the airflow (in or out, vigorous, sluggish, or no flow) is helpful for evaluating the supply and return system and for determining whether ventilation air actually reaches the breathing zone.

Tracking the smoke can reveal some surprising information about the ventilation system. Investigators responding to complaints at one company found that, indeed, air in the breathing zone did not contain the desired levels of oxygen and carbon dioxide. Since air in the ducts met their standards, they measured the flow into that office area (see the next section). Although their measurements showed a sufficient supply of new air, they stepped up the flow. However, the complaints continued. Investigators resorted to smoke tracking, and learned that fresh air came from the ducts and went directly into the return inlets, without mixing with the room air. Workers were receiving no benefit from the fresh air being brought into the area. "Short-circuiting" occurs when air moves directly from supply diffusers to return grilles, instead of mixing with room air in the breathing zone. When a substantial amount of supply air short-circuits, occupants may not receive adequate supplies of outdoor air and source emissions may not be diluted sufficiently.

For a variable air volume system, be sure to take into account how the system is designed to modulate. It could be on during the test, but off for much of the day.

Measuring Airflow

Following smoke movement gives a good indication of where and how the air is flowing, but it does not show *how much* air is flowing. Measurements of airflow allow investigators to determine if the HVAC system is operating according to design, and to identify potential problem locations. Building investigations often include measurements of outdoor air quantities, exhaust air quantities, and airflows at supply diffusers and return grilles.

Airflow quantities can be calculated by the formula:

Airflow quantity = (Airflow velocity)(Cross-sectional area of airstream)

For example, if air is moving at 100 feet per minute in a 24 × 12 inch duct, the airflow quantity is:

Airflow quantity = (100 feet per minute)(2 square feet duct area)
= 200 cubic feet per minute

Cross-sectional area is easy to measure in a straight run of rectangular ductwork, but can be complicated at other locations such as mixing boxes or diffusers. The most practical method in those locations is to base an estimate on experience.

Air velocity measurements are typically made with a pitot tube or an anemometer, but you may find that the measurements vary considerably with probe location. For example, turbulence around mixing vanes makes it difficult to take measurements at supply diffusers. An acceptable method is to average the results of several measurements, assuming that, in a closed system, readings above average will match with readings below average. ASTM Standard Practice D 3154 provides guidance on making such measurements.

To measure total airflow at locations such as grilles, diffusers, and exhaust outlets, you can obtain a hood that gathers all the airflow and funnels it to the reading device. Hoods are available in various sizes. Some of them can operate in any position, so they can check exhaust outlets and other locations regardless of whether they are on a wall, floor, ceiling, or in any other plane.

Estimating Outdoor Air Quantities

Outdoor air quantities can be evaluated by measuring airflow directly. Investigators often estimate the proportion of outdoor air using techniques such as thermal mass balance (temperature) or carbon dioxide measurements. Estimating outdoor air quantity using temperature measurements is referred to as "thermal balance" or sometimes "thermal mass balance."

Thermal balance. Using this test requires the following conditions:

1. Airstreams of return air, outdoor air, and mixed air (supply air before it has been heated or cooled) must be accessible for separate measurement. Some panels read out supply, return, outdoor, and/or mixed air temperature. Some systems are already equipped with an averaging thermometer that is strung diagonally across the mixed air chamber; the temperature is read out continuously on an instrument panel.
2. There must be at least a several degree temperature difference between the building interior and the outdoor air.
3. Total airflow in the air handling system can be estimated either by using recent balancing reports or by pitot tube measurements in ductwork. As an alternative, the supply air at each diffuser can be estimated (such as by using a flow measuring hood) and the results can be added together to calculate total system airflow.

Temperature measurements can be made with a simple thermometer or an electronic sensor. Several measurements should be taken across each airstream and averaged. It is generally easy to obtain a good temperature reading in the outdoor air and return airstreams. The best average temperature reading of the mixed airstream requires a large number of measurements taken upstream of the point at which the stream is heated or cooled, although this might be difficult or impossible in some installations. The following formulas give quantity of outdoor air, where temperatures are in Fahrenheit and volume is in cubic feet per minute.

$$\text{Outdoor air (percent)} = 100\left[\frac{T_{\text{return air}} - T_{\text{mixed air}}}{T_{\text{return air}} - T_{\text{outdoor air}}}\right]$$

$$\text{Outdoor air (cfm)} = \left[\frac{\text{Outdoor air (percent)}}{100}\right]\text{Total airflow (cfm)}$$

Carbon dioxide measurements. CO_2 readings can be taken at supply outlets or air handlers to estimate the percentage of outdoor air in the supply airstream. Percentage of outdoor air is calculated from the following formula, and then used in the preceding formula to calculate quantity of outdoor air. The numbers are critical, so concentrations should be measured, not assumed.

$$\text{Outdoor air (percent)} = 100\left[\frac{C_S - C_R}{C_O - C_R}\right]$$

Where: C_S = ppm CO_2 in supply air (if measured in a room), or
= ppm CO_2 in the mixed air (if measured at an air handler)
C_R = ppm CO_2 in the return air
C_O = ppm CO_2 in the outdoor air

Carbon Dioxide as a Ventilation Indicator

Carbon dioxide is a normal, though small, component of the atmosphere, and exhaled breath from building occupants is an important source of indoor carbon dioxide. Therefore, CO_2 concentrations can, under some test conditions, provide a good indication of the adequacy of ventilation. Comparison of peak CO_2 readings between rooms, between air handler zones, and at various heights above the floor, will help to identify and diagnose several building ventilation deficiencies.

CO_2 can be measured with either a direct-reading meter or a detector tube kit. All applicable data, such as occupancy, air damper settings, weather, and other conditions, should be noted for each period of measurements. Carbon dioxide measurements for ventilation should be collected away from anything that could affect the reading (for example, the sampling device should be held away from exhaled breath). Individual measurements should be short-term. As with many other measurements of indoor air conditions, it is advisable to take one or more readings in "control" locations to serve as baselines for comparison. Readings from outdoors and from areas where there apparently are no indoor air quality problems are frequently used as controls. Outdoor air samples should be taken near the outdoor air intake, because that will measure the air that is brought into the building.

To provide maximum help to investigators, measurements to evaluate the adequacy of ventilation should be taken at several times during the day. Two measurement times that should always be included are the time when concentrations are expected to be at their highest and the time when they should be at their lowest. If few occupants enter or leave the area during the day, and if there is no rapid exchange of air, carbon dioxide levels will typically rise during the morning, fall during the lunch period, then rise again and reach a peak in mid-afternoon.

Peak carbon dioxide concentrations above 1000 ppm in the breathing zone indicate ventilation problems. Peak concentrations below 1000 ppm *generally* indicate that ventilation is adequate for human occupancy. However, there are several reasons for not jumping to the conclusion that, because of low CO_2 readings, there is no air quality problem. Buildings with CO_2 concentrations well below 1000 ppm have been found to have serious problems. For example, if strong contaminant sources are present, good ventilation alone may not be sufficient; source control might be needed to supplement the ventilation. Other reasons to continue checking include the fact that humans (even with computers) make errors, and varying CO_2 concentrations over time can cause low readings that may be misleading.

Carbon dioxide readings higher than about 1000 ppm suggest another set of possible conditions that may alone, or in combination with other conditions, cause the air quality complaints. Some possibilities are: increased occupant population, air exchange rates below ASHRAE guidelines, poor air distribution, and poor air mixing. A higher average CO_2 concentration in the general breathing zone (at least two feet from exhaled breath) than in the air entering return grilles is an indication of poor air mixing. Smoke tubes and temperature profiles will help identify air circulation patterns.

If carbon dioxide measurements taken before the occupied period begins are higher than outdoor readings taken at the same time, there may be an operating problem with the HVAC system. Potential problems include:

- Ventilation is turned off too early the evening before.
- Combustion byproducts from a nearby roadway or parking garage are drawn into the building.
- Combustion byproducts from an internal source continue to add CO_2 to the building's atmosphere.
- A gas-fired heating appliance in the building has a cracked heat exchanger.

Outdoor CO_2 concentrations above 400 ppm may indicate an outdoor contamination problem from heavy and/or stationary traffic or other combustion sources. Note, however, that detector tubes cannot provide accurate measurements of CO_2 in hot or cold weather.

VOLATILE ORGANIC COMPOUNDS

Defined as carbon-containing compounds, there are hundreds of VOCs in many indoor air atmospheres, at concentrations ranging from the trace level to well over recommended levels. Outgassing from paint, caulk, and adhesives are just a few of the sources. Formaldehyde is a VOC that is mentioned frequently. Volatile organic compounds may present an indoor air quality problem when individual organics or mixtures exceed normal background concentrations.

Total Volatile Organic Compounds (TVOCs)

Several direct-reading instruments are available that provide a low sensitivity "total" reading for different types of organics. Such estimates are usually presented in parts per million and are calculated with the assumption that all chemicals detected are the same as the one used to calibrate the instrument. A photoionization detector (discussed later in this chapter) is an example of a direct-reading instrument used as a screening tool for measuring TVOCs.

A laboratory analysis of a sorbent tube can provide an estimate of total solvents in the air. Although methods examined here report "total volatile organic compounds" (TVOCs) or "total hydrocarbons" (THC), analytical techniques differ in their sensitivity to the different types of organics.

Each method of measurement is useful for certain purposes, but generally should not be compared to measurements from other methods. Direct-reading instruments do not provide sufficient sensitivity to differentiate normal from problematic mixtures of organics. However, instantaneous readouts may help to identify "hot spots," sources, and pathways. TVOCs or THC determined from sorbent tubes provide more accurate readings, but this method cannot distinguish peak exposures. A direct-reading instrument can only identify peak exposures that happen to occur during the measurement period.

High concentrations of individual volatile organic compounds may also cause IAQ problems. Individual VOCs can be measured in indoor air with a moderate degree of sensitivity (that is, measurement in parts per million) through adaptations of existing industrial air monitoring technology. Examples of "medium sensitivity" testing devices include XAD-4 sorbent tubes (for nicotine), charcoal tubes (for solvents), and chromosorb tubes (for pesticides). After a sufficient volume of air is pumped through these tubes, the investigator sends them to a laboratory for extraction and analysis by gas chromatography. Variations use a passive dosimeter (charcoal badge) to collect the sample, or a portable gas chromatograph onsite for direct injection of building air. These methods may not be sensitive enough to detect many trace level organics present in building air.

"High sensitivity" techniques have recently become available to measure trace organics—VOCs in the air in concentrations of parts per billion. Sampling may involve Tenax and multiple sorbent tubes, charcoal tubes, evacuated canisters, and other technology. Analysis involves gas chromatography followed by mass spectroscopy.

With measurement results in hand, investigators face the difficult task of identifying contaminated areas of the plant. However, there are no real regulations that state explicitly the division between safe and unsafe concentrations of various VOCs. Guidelines for public health exposure (as opposed to occupational exposure) for a few VOCs are available in the World Health Organization (WHO) Air Quality Guidelines for Europe. These guidelines reflect carcinogenic as well as noncarconogenic effects. Occupational exposure standards exist for many other VOCs. However, no general-application safety factor for applying these occupational limits to general IAQ is currently endorsed by EPA and NIOSH.

Measurement of trace organics may indicate the presence of dozens to hundreds of trace VOCs, although their significance is difficult to determine. It may be helpful to compare levels in complaint areas to levels in outdoor air or in noncomplaint areas.

Formaldehyde

There is considerable information available on this aldehyde because it is one VOC that has been studied extensively. Reasons for the high level of interest include its toxicity and the fact that it is used in many manufacturing processes. Small amounts of formaldehyde are present in many indoor environments.

Symptoms that may signal an elevated formaldehyde concentration include itching of the eyes, nose, or throat. When new materials are brought into the plant, or other materials are suspect, it is often helpful perform a sampling test. Several measurement methods are available; two methods that are generally acceptable and are commonly used for IAQ screening involve impingers and sorbent tubes. Sensitivity and sampling time are important considerations when selecting a method. Many methods allow detection of concentrations well below 0.1 parts per million. Measuring short-term peaks (around a two-hour sample time) is ideal for evaluating acute irritation. Dosimeters may accurately record long-term exposure, but will miss these peaks.

Various guidelines and standards for formaldehyde exposure have been prepared. Several organizations have adopted 0.1 ppm as a guideline that provides reasonable protection against irritational effects in most people. Hypersensitivity reactions may occur at lower levels of exposure. Worst-case conditions are created by minimum ventilation, maximum temperatures, and high source loadings.

BIOLOGICAL CONTAMINANTS

Human health, and life itself, can be affected by exposure to both living and nonliving biological contaminants. The term *bioaerosols* describes airborne material that is or was living, such as mold, bacteria, parts of living organisms (such as insect body parts), and animal feces.

Tests for bioaerosols cannot generally be justified as routine parts of an air investigation program. Such tests should usually be limited to the following situations:

- Situations in which a walk-through investigation or human profile study suggests microbiological involvement
- Situations in which no other pollutant or physical condition can account for the symptoms that are present

One reason for downplaying sampling or testing is that the results are too often misleading. A less expensive, more reliable, and preferred procedure is to inspect the building's sanitary conditions. When sampling is included in the investigation, it should be accompanied by observations of sanitary conditions and a determination as to whether any health problems may be caused by biological contamination.

The many biological contaminants found in indoor environments make it impossible for a single technique to be effective for sampling. A variety of specific approaches can be used to retrieve, enumerate, and identify each kind of microorganism from water, surfaces, and air. Other specific methods are used for materials such as feces or insect parts. The utility of these techniques depends on their use by professionals who have a thorough understanding of the sample site and the target organism.

When air sampling is to be conducted, there are several possible approaches. The most common type of air sampler uses a pump to pull air across a gelatinous, nutrient agar, which is then incubated. Any bacterial or fungal colonies that subsequently grow can be counted and identified by a qualified microbiologist. The microbiologist uses different types of agar and incubation temperatures to culture different types of organisms. Only living organisms or spores in the area are counted by this method.

Settling plates, which are simply collection plates opened to room air and then incubated, are sometimes used to identify which bioaerosols are present in different areas under investigation. The drawbacks to this technique are that it does not indicate the quantity of bioaerosols present, and that only the bioaerosols that are heavy enough to fall out onto the agar will be recorded.

Quantities and types of bioaerosols can vary greatly over time in any given building, making sampling results difficult to interpret. Comparison of relative numbers and types of microorganisms between indoors and outdoors or between complaint areas and background sites can help to establish trends. However, no tolerance levels or absolute guidelines have been established. Low bioaerosol results by themselves are not considered proof that a problem does not exist, for a variety of reasons:

- The sampling and identification techniques used may not be suited to the type(s) of bioaerosols that are present.
- Biological growth may have been inactive during the sampling period.
- The analysis technique used may not reveal nonliving bioaerosols (such as animal parts or feces) than can cause health reactions.

AIRBORNE DUST

There are a number of mechanisms by which particles and fibers get into the air. They might be suspended, stirred up by air movement, or stirred up by convection. Their effects on workers range from harmless, through nuisance, to serious health problems.

Many collection and analytical techniques are available. Dust can be collected by using a pump to draw air through a filter. The filter can then be weighed (gravimetric analysis) or examined under a microscope. Direct readouts of airborne dust are also available (such as using a meter equipped with a "scattered light" detector).

With few exceptions, IAQ measurements for airborne dust will be well below occupational and ambient air guidelines. Results showing unusual types or elevated amounts of particles or fibers can reveal potential exposure problems.

COMBUSTION PRODUCTS

Combustion products are complex because they are usually combinations of particles, gases, and sometimes heat. They are released by internal combustion engines, furnaces, tobacco smoke, and other sources. Components include airborne and potentially harmful gases such as carbon monoxide and nitrogen oxides. Depending on the plant layout, and provisions for providing the equipment with air, the source may deplete the oxygen and simultaneously replace it with a harmful gas.

Direct-reading meters, detector tubes, and passive dosimeters are among the techniques most commonly used to measure carbon monoxide and nitrogen oxides. Comparing the results with occupational standards will reveal whether an imminent danger exists. Any readings that are above outdoor concentrations or background building levels may indicate a mixture of potentially irritating combustion products, especially if susceptible individuals are exposed.

PHOTO IONIZATION DETECTORS[1]

This section uses two specific instruments as examples to show some features and operating characteristics of PIDs. There are other brands that operate on the same basic principles. Their purpose is to measure total gases and vapors in the air.

The instruments operate by drawing gases and vapors past an ultraviolet lamp, which produces a positive charge in the gases and vapors. This charge creates an electric current which varies with the amount of gas present. A PID changes the electric current into a reading in parts per million on the instrument readout.

Calibration

To maintain reading accuracy, PIDs must be calibrated with a gas or vapor at a known concentration. Benzene and isobutylene are most often used as calibration vapors because their vapors ionize efficiently in ultraviolet (UV) light. Toluene is sometimes substituted for benzene, because it is much less toxic. PIDs are calibrated by the factory using one or more calibration gases at a range of concentrations. Gas concentrations and actual meter readout are plotted on graphs to produce a calibration curve for each chemical.

Users perform a two-step field calibration process to verify that the instrument is working properly. First, the zero setting is checked by attaching either a

1. From the paper *Photo Ionization Detectors (PIDS)*, courtesy of RAE Systems, Sunnyvale, California.

zero gas filter or a small cylinder of clean breathing air. Zero gas filters remove dust, gases, and vapors that may be present in the surrounding air. They are often less expensive and easier to transport than cylinders of breathing air.

The second field calibration step compares the actual PID meter readout to a known concentration of calibration gas, usually 100 ppm isobutylene. After the concentration of gas is entered in the PID memory, a small cylinder of it is attached to the instrument to take an actual reading for comparison.

Field Measurements

Correct interpretation of PID readings requires that average background readings be recorded for comparison with work area readings. Therefore, field measurements should begin by recording background readings in an area outside, but similar to, the work area to be monitored. For example, the parking lot might be a good background location for monitoring a hazardous waste site, and an office area might be a good background location for monitoring a chemical plant. The key to selecting a background location is to remember that its air should be similar to the air of the work area to be monitored, minus any contaminants that might be in the work area. This means that the background location should be free of vehicle and heavy equipment exhaust, bug sprays, and vapors from natural vegetation. Such gases, vapors, and chemicals cause elevated readings in the background location. The background reading is important because monitoring results are reported in terms of "ppm (or other units) above background."

PIDs have no ability to differentiate among the air contaminants that cause a reading. They cannot differentiate between mixtures and pure compounds. They simply record the amount of current produced by gases and vapors in the ionization chamber. After calibration, a PID will record total gas and vapor concentrations as compared with the response to the calibration gas.

Ionization Potentials

Not all gases and vapors can be electrically charged (ionized) by ultraviolet light. It depends on the *ionization potential* of the gas or vapor compared to the electron volt (eV) sensitivity of the PID lamp. Typical sensitivities are 9.5 eV, 10.0 eV, 10.2 eV, 10.6 eV, 11.7 eV, and 11.8 eV. If the ionization potential of a gas or vapor is less than the lamp voltage, the PID should be able to detect that compound. For example, benzene has an ionization potential of 9.24 eV, which is lower than the voltage rating of all UV lamps, so benzene can be detected by a PID. However, methane cannot be measured accurately because its ionization potential of 12.98 eV is higher than any of the UV lamp ratings. Table 11-1 gives ionization potentials for several compounds that might be found in industrial indoor air.

TABLE 11-1
Ionization Potentials of Selected Compounds

Acetaldehyde	10.21	1-Bromobutanone	9.54
Acetic Acid	10.37	1-Bromo-2-chloroethane	10.63
Acetone	11.41	Bromochloromethane	10.77
Acetylene dichloride	9.80	Bromoethane	10.28
Acrolein	10.10	Bromoethene	9.80
Acrylonitrile	10.91	Bromoform	10.48
Allene	9.83	1-Bromo-3-hexanone	9.26
Allyl alcohol	9.67	Bromoethyl ethyl ether	10.08
Allyl chloride	10.20	Bromomethane	10.53
Aminoethanol	9.87	1-Bromo-2-methylpropane	10.09
Ammonia	10.15	2-Bromo-2-methylpropane	9.89
Aniline	7.70	1-Bromopentane	10.10
Arsine	9.89	1-Bromopropane	10.18
Benzaldehyde	9.53	2-Bromopropane	10.08
Benzene	9.25	1-Bromopropene	9.30
Benzenethiol	8.33	2-Bromopropene	10.06
Benzyl chloride	10.16	3-Bromopropene	9.70
Benzonitrile	9.71	2-Bromothiophene	8.63
Benzotrifluoride	9.68	o-Bromotoluene	8.79
Bromobenzene	8.98	m-Bromotoluene	8.81
1-Bromobutane	10.13	p-Bromotoluene	8.67
2-Bromobutane	9.98	1,3-Butadiene	9.07

Continued

TABLE 11-1
Continued

2,3-Butadione	9.23	iso-Butyl ethanoate	9.95
n-Butanal	9.83	iso-Butyl mercaptan	9.12
2-Butanal	9.73	iso-Butyl methanoate	10.46
n-Butane	10.63	1-Butyne	10.18
2-Butanone	9.53	2-Butyne	9.85
iso-Butanol	10.47	n-Butyraldehyde	9.86
s-Butanol	10.23	Carbon disulfide	10.13
t-Butanol	10.25	Carbon tetrachloride	11.28
2-Butanol	10.10	Chlorobenzene	9.07
1-Butene	9.58	Chlorobromomethane	10.77
cis-2-Butene	9.13	1-Chloro-2-bromoethane	10.63
trans-2-Butene	9.13	1-Chlorobutane	10.67
3-Butene nitrile	10.39	2-Chlorobutane	10.65
n-Butyl acetate	10.01	1-Chlorobutanone	9.54
s-Butyl acetate	9.91	1-Chloro-2,3-epoxy propane	10.60
n-Butyl alcohol	10.04	Chloroethane (ethyl chloride)	10.97
n-Butyl amine	8.71	Chlorethene	10.00
iso-Butyl amine	8.70	2-Chloroethoxyethene	10.61
s-Butyl amine	8.70	1-Chloro-2-fluorobenzene	9.16
t-Butyl amine	8.64	1-Chloro-3-fluorobenzene	9.21
n-Butyl benene	8.69	1-Chloro-2-fluoroethene (cis)	9.87
iso-Butyl benzene	8.68	1-Chloro-2-fluoroethene (trans)	9.87
t-Butyl benzene	8.68	Chloroform	11.37
Butyl Cellosolve	8.68	o-Chloroiodobenzene	8.35
n-Butyl mercaptan	9.15	1-Chloro-2-methylbenzene	8.72

Table 11-1
Continued

1-Chloro-3-methylbenzene	8.61	Cyclopentane	10.52
1-Chloro-4-methylbenzene	8.78	Cyclopentanone	9.26
Chlorobutadiene	8.79	Cyclopentene	9.01
Chloromethyl ethyl ether	10.08	Cyclopropane	10.06
Chloromethyl methyl ether	10.25	2-Decanone	9.40
1-Chloro-2-methylpropane	10.66	Diborane	11.00
Chloroprene	8.80	Dibromochloromethane	10.59
1-Chloropropane	10.82	1,1-Dibromoethane	10.19
2-Chloropropane	10.78	Dibromomethane	10.49
3-Chloropropene	10.04	1,2-Dibromopropane	10.26
2-Chlorothiophene	8.68	1,2-Dichlorobenzene	9.07
o-Chlorotoluene	8.83	1,3-Dichlorobenzene	9.12
m-Chlorotoluene	8.83	1,4-Dichlorobenzene	8.94
p-Chlorotoluene	8.70	Dichlorodifluoromethane	11.75
Cumene (isopropyl benzene)	8.75	1,1-Dichloroethane	11.06
Crotonaldehyde	9.73	1,2-Dichloroethane	11.04
Cyanoethene	10.91	cis-Dichloroethene	9.65
Cyanogen bromide	10.91	trans-Dichloroethene	9.66
3-Cyanopropene	10.39	Dichloromethane	11.35
Cyclobutane	10.50	1,2-Dichloropropane	10.87
Cyclohexane	9.98	1,3-Dichloropropane	10.85
Cyclohexanone	9.14	1,1-Dichloropropanone	9.71
Cyclohexene	8.95	2,3-Dichloropropene	9.82
Cyclo-octatetraene	7.99	Dicyclopentadiene	7.74
Cyclopentadiene	8.55	Dibutyl amine	7.69

Continued

TABLE 11-1
Continued

Diethoxymethane	9.70	3/5-Dimethyl-4-heptanone	9.04
Diethyl amine	8.01	2,2-Dimethyl-3-pentanone	8.98
Diethyl ether	9.53	2,2-Dimethyl-propane	10.35
n-Diethyl formamide	8.89	Dimethyl sulfide	8.69
Diethyl ketone	9.32	Di-n-propyl disulfide	8.27
Diethyl sulfide	8.43	Di-n-propyl ether	9.27
1,2-Difluorobenzene	9.31	Di-isopropyl ether	9.20
1,4-Difluorobenzene	9.15	Di-n-propyl amine	7.84
Difluorodibromomethane	11.18	Di-n-propyl sulfide	8.30
Difluoromethylbenzene	9.45	p-Dioxane	9.13
Di-iodomethane	9.34	Epichlorohydrin	10.60
Di-iosobutyl ketone	9.04	Ethane	11.65
Di-isopropylamine	7.73	Ethanal	10.21
1,1-Dimethoxyethane	9.65	Ethanol	10.62
Dimethoxymethane	10.00	Ethene (ethylene)	10.52
Dimethyl amine	8.24	Ethyl acetate	10.11
2,3-Dimethylbutadiene	8.72	Ethyl amine	8.86
2,2-Dimethyl butane	10.06	Ethyl amyl ketone	9.10
2,2-Dimethyl butane-3-one	9.18	Ethyl benzene	8.76
2,3-Dimethyl butane	10.02	Ethyl bromide	10.29
2,3-Dimethyl-2-butene	8.30	Ethyl butyl ketone	9.02
3,3-Dimethyl butanone	9.17	Ethyl chloride (chloroethane)	10.98
Dimethyl disulfide	8.46	Ethyl chloroacetate	10.20
Dimethyl ether	10.00	Ethyl ethanoate	10.10
Dimethylformamide	9.45	Ethyl disulfide	8.27

TABLE 11-1
Continued

Ethylene chlorohydrin	10.90	2-Furaldehyde	9.21
Ethylene dibromide (EDB)	10.37	Furan	8.89
Ethyleneimine	9.94	Furfural	9.21
Ethylene oxide	10.56	Hexachloroethane	11.22
Ethyl formate	10.61	n-Hexane	10.18
Ethyl iodide	9.33	n-Heptane	10.07
Ethyl mercaptan	9.29	2-Heptanone	9.33
Ethyl methanoate	10.61	4-Heptanone	9.12
Ethyl isothiocyanate	9.14	1-Hexene	9.46
Ethyl methyl sulfide	8.55	Hexanone	9.34
Ethyl propanoate	10.00	Hexamethylbenzene	7.85
Ethyl trichloroacetate	10.44	Hydrazine	8.93
Mono-fluorobenzene	9.20	Hydrogen cyanide	13.91
Mono-fluoroethene	10.37	Hydrogen selenide	9.88
Mono-fluoromethanal	11.40	Hydrogen sulfide	10.46
Fluorotribromomethane	10.67	Hydrogen telluride	9.14
o-Fluorotoluene	8.92	Iodine	9.28
m-Fluorotoluene	8.92	Iodobenzene	8.73
p-Fluorotoluene	8.79	1-Iodobutane	9.21
Freon 11	11.77	2-Iodobutane	9.09
Freon 12	12.31	Iodoethane (ethyl iodide)	9.33
Freon 13	12.91	Iodomethane (methyl iodide)	9.54
Freon 14	16.25	1-Iodo-2-methylpropane	9.18
Freon 22	12.45	1-Iodopentane	9.19
Freon 113	11.78	1-Iodopropane	9.26

Continued

Table 11-1
Continued

2-Iodopropane	9.17	Methyl amine	8.97
o-Iodotoluene	8.62	Methyl bromide	10.53
m-Iodotoluene	8.61	2-Methyl-1,3-butadiene	8.85
p-Iodotoluene	8.50	2-Methylbutanal	9.71
Isoamyl acetate	9.90	2-Methylbutane	10.31
Isoamyl alcohol	10.16	2-Methyl-1-butene	9.12
Isobutane	10.57	3-Methyl-1-butene	9.51
Isobutyl amine	8.70	3-Methyl-2-butene	8.67
Isobutyl acetate	9.97	Methyl-n-butyl ketone	9.34
Isobutyl alcohol	10.47	Methyl butyrate	10.07
Isobutyl formate	10.46	Methyl Cellosolve	9.60
Isobutyraldehyde	9.74	Methyl chloroacetate	10.35
Isopentane	10.32	Methyl chloride	11.28
Isoprene	8.85	Methyl chloroform	11.25
Isopropyl acetate	9.99	Methylcyclohexane	9.85
Isopropyl alcohol	10.16	4-Methylcyclohexene	8.91
Isopropyl amine	8.72	Methylcyclopropane	9.52
Isopropyl benzene	8.75	Methyl dichloroacetate	10.44
Isopropyl ether	9.20	Methyl ethanoate	10.27
Isovaleraldehyde	9.71	Methyl ethyl ketone	9.53
Mesitylene	8.40	Methyl ethyl solfide	8.55
Mesityl oxide	9.08	2-Methyl furan	8.39
Methanol	10.85	Methyl iodine	9.54
Methyl acetate	10.27	Methyl isobutyl ketone	9.30
Methyl acrylate	10.72	Methyl isobutyrate	9.98

TABLE 11-1
Continued

Methyl isopropyl ketone	9.32	Pentachloroethane	11.28
Methyl isothiocyanate	9.25	1,3-Pentadiene (cis)	8.59
Methyl methacrylate	9.74	1,3-Pentadiene (trans)	8.56
Methyl methanoate	10.82	Pentafluorobenzene	9.84
Methyl mercaptan	9.44	Pentamethylbenzene	7.92
2-Methylpentane	10.12	n-Pentanal	9.82
3-Methylpentane	10.08	2,4-Pentanedione	8.87
2-Methylpropane	10.56	2-Pentanone	9.39
2-Methylpropanal	9.74	3-Pentanone	9.32
2-Methyl-2-propanol	9.70	1-Pentene	9.50
2-Methylpropene	9.23	Perchloroethylene	9.32
Methyl-n-propyl ketone	9.39	Perfluoro-2-butene	11.25
Methyl styrene	8.35	Perfluoro-1-heptene	10.48
Morpholine	8.88	n-Perfluoropropyl iodide	10.36
Naphthalene	8.10	n-Perfluoropropyl iodomethane	9.96
Nitric oxide	9.25	Phenol	8.69
Nitrobenzene	9.92	Phenyl ether	8.09
Nitrotoluene	9.43	Phenyl isocyanate	8.77
n-Nonane	10.21	Phosphine	9.96
5-Nonanone	9.10	Pinene	8.07
n-Octane	10.24	Propadiene	10.19
3-Octanone	9.19	n-Propanal	9.95
4-Octanone	9.10	Propane	11.07
1-Octene	9.52	1-Propanethiol	9.20
n-Pentane	10.35	n-Propanol	10.51

Continued

TABLE 11-1
Continued

Propanone	9.69	Tetrahydropyran	9.26
Propenal (acrolein)	10.10	1,2,4,5-Tetramethylbenzene	8.03
Propene	9.73	2,2,4,4-Tetramethyl-3-pentanone	8.65
Prop-1-ene-2-ol	8.20	Thioethanol	9.29
Prop-2-ene-1-ol	9.67	Thiomethanol	9.44
Propionaldehyde	9.98	Thiophene	8.86
n-Propyl acetate	10.04	1-Thiopropanol	9.20
n-Propyl alcohol	10.20	Toluene	8.82
n-Propyl amine	8.78	Tribromoethene	9.27
n-Propyl benzene	8.72	1,1,1-Trichlorobutanone	9.54
Propylene	9.73	1,1,1-Trichloroethane	11.25
Propylene dichloride	10.87	1,1,2-Trichloroethane	11.00
Propylene oxide	10.22	Trichloroethene	9.45
n-Propyl ether	9.27	Trichloromethyl ethyl ether	10.08
n-Propyl formate	10.54	Triethylamine	7.50
Propyne	10.36	1,2,4-Trifluorobenzene	9.37
Pyridine	9.32	1,3,5-Trifluorobenzene	9.32
Styrene	8.47	Trifluoroethene	10.14
Tetrachloroethene	9.32	Trifluoroiodomethane	10.40
1,1,2,2-Tetrachloroethane	11.10	Trifluoromethylbenzene	9.68
1,2,3,4-Tetrafluorobenzene	9.61	Trifluoromethylcyclohexane	10.46
1,2,3,5-Tetrafluorobenzene	9.55	1,1,1-Trifluoropropene	10.90
1,2,4,5-Tetrafluorobenzene	9.39	1,2,3-Trimethylbenzene	8.48
Tetrafluoroethene	10.12	1,2,4-Trimethylbenzene	8.27
Tetrahydrofuran	9.54	1,3,5-Trimethylbenzene	8.39

TABLE 11-1
Continued

2,2,4-Trimethyl pentane	9.86	Vinyl ethanoate	9.19
2,2,4-Trimethyl-3-pentanone	8.82	Vinyl fluoride	10.37
n-Valeraldehyde	9.82	Vinyl methyl ether	8.93
Valeric acid	10.12	Water	12.58
Vinyl acetate	9.19	Xenon difluoride	11.50
Vinyl benzene (styrene)	8.47	Xenon tetrafluoride	12.90
Vinyl bromide	9.80	o-Xylene	8.56
Vinyl chloride	10.00	m-Xylene	8.56
4-Vinylcyclohexene	8.93	p-Xylene	8.45

Relative Response

Because some gases are ionized more efficiently than others, due to their chemical structure, PIDs are more sensitive to some gases and vapors than others. In general, aromatic hydrocarbons are ionized more efficiently than aliphatics. Ketones are ionized more efficiently than aldehydes, and aldehydes more efficiently than alcohols. It is critical to know the relative response of the instrument to the compounds being measured in order to interpret the instrument readout properly. Relative responses for pure compounds are explained differently from responses for compound mixtures.

Measuring pure compounds. The instrument response to each chemical depends on the type of UV lamp and the calibration gas used. Instrument manufacturers should provide information—typically a table showing the relative response of a PID to a standard concentration, such as 100 ppm of various gases and vapors. Users of the instrument can note variations between different UV lamps and between calibrations gases with the same lamp.

The relative response to a chemical is expressed as a calibration factor. When the calibration factor is 1.0 for a particular chemical, the PID will show exactly the same reading as the actual airborne concentration. Calibration factors below 1.0 mean that the readings will be higher than the actual concentrations, and factors above 1.0 mean the readings will be lower than actual concentrations. Users can determine actual concentrations by multiplying the instrument reading by the calibration factor for the compound being measured.

Measuring mixtures. PIDs have no ability to separate gases. They cannot identify individual chemicals within a mixture, so the relative responses to mixtures is different from responses to pure compounds. The only way to determine the relative response of a PID to a particular gas or vapor mixture would be to expose the instrument to known concentrations of that mixture and record the response. A relative response thus recorded could be used only when the air being monitored contained exactly the same chemicals, in exactly the same ratio. Therefore, it is not feasible to calibrate a PID for a variety of mixtures.

However, PIDs can be used to make qualitative estimates of total selected airborne gases and vapors. If the chemicals present are known, two or more UV lamps can be used to make qualitative estimates of the percent composition of different gases and vapors. For example, consider air that contains the following gases:

	Ionization Potential
Acetone	9.69 eV
Methyl ethyl ketone	9.53 eV
Chloroform	11.37 eV

With a 10.6 eV lamp, the PID can detect the acetone and the methyl ethyl ketone, but not the chloroform, because chloroform has an ionization potential higher than the 10.6 eV lamp. An 11.7 eV lamp can detect all three vapors, so the difference between the responses with the two lamps will represent the approximate concentration of chloroform vapor.

Interferences

A number of factors may cause interferences that affect the accuracy of PID readings. Identifying potential interferences in each monitoring location is a prerequisite to properly interpreting the results of PID monitoring.

Linearity of responses. PIDs are designed to detect reasonably low concentrations of gas and vapor. When calibrating a typical PID, the instrument response to the calibration gas will be linear up to approximately 400–500 ppm. Higher concentrations of the calibration gas can overload the ionization chamber and give reduced responses. When high concentrations enter the ionization chamber, the UV light does not reach all of the molecules at once. The ionization chamber is small and only a limited number of molecules can fit in the light path at one time. Therefore the instrument response to very high concentrations may be high at first and then fall off gradually until the PID stabilizes at a lower reading.

Dust and dirt. Dust and dirt in the sampled air can obscure the UV lamp window and cause inaccurate readings. A dust filter helps prevent this problem. Periodic cleaning of the lamp window may be necessary when using a PID in dusty environments.

Humidity. PIDs are sensitive to humidity. High humidity can cause water vapor to condense on the lamp window, producing incorrectly low readings. The water vapor deflects and scatters the light. Water vapor filters can help prevent this problem to some extent.

Hot atmospheres. Heated gases and vapors can cause inaccurate readings by condensing on the lamp window in a PID. If readings show sudden changes when monitoring warm air, condensation should be suspected. The lamp window may need to be cleaned after exposure to warm air to remove residues left by condensation.

Non-ionizable compounds. Gases and vapors with ionization potentials higher than the lamp, although not detected, can sometimes interfere with PID readings. These compounds can condense on the lamp window, deflecting, scattering, or blocking the UV light. High concentrations (>5,000 ppm) of methane gas have been shown to produce this effect.

Corrosive gases and vapors. Corrosive atmospheres can cause permanent fogging or etching of the lamp window in PIDs. These atmospheres may also damage the electrodes within the ionization chamber. The only solution is to avoid using PIDs in corrosive conditions. If an instrument is exposed to corrosive gases or vapors, it should be thoroughly cleaned and recalibrated immediately after use.

CHAPTER 12

Conducting an Investigation to Resolve Complaints

It is too easy for an indoor air quality investigation to get out of hand. Books, this one included, describe monitoring programs, measuring instruments, and many other parts of an investigation—but with the intention that investigators will choose procedures that are appropriate for the air problem at hand. Reasons for the investigation range from a complaint of an identifiable odor in one area to building-wide complaints about a problem from an unknown source. Setting up a program more sophisticated than necessary to solve a given problem not only violates fiscal sense, but can also confuse the issue—making the original problem harder to solve.

If a model shop is located next to an office area, and people at their desks complain of a smell of overheated oil, an investigator's first step should be to look into what the model shop workers are doing—not to launch an extensive air measurement program. Should a milling machine operator in the model shop be cooling the cut with oil, chances are the oil is getting hot enough to give off the odor smelled in the office. Checking relative air pressures in the two areas would show whether model shop air is being drawn into the office area, and would cost less than monitoring and analyzing the office air.

An important issue to settle early in the investigation is whether to proceed with in-house personnel, turn the job over to a consultant, or work with a combination of the two. Putting cost considerations aside, it is desirable to have employees gain experience in the indoor air quality field, but not at the risk of doing an inadequate job when the skills and/or equipment are not available in-house. Typically the extent of the problem is not known at the time of the complaint, so present staff responds. Hopefully the early stages of the investigation will show whether outside help will be needed, but sometimes that decision will be reversed as the investigation progresses.

This chapter offers suggestions for resolving complaints. Most investigations begin with the general and theoretical, then shift toward focused and practical. As the next section shows, a workable method is to pattern the investigation after a management brainstorming session—start by accepting all possibilities without evaluating them, and follow that with more detailed examinations.

CONDUCTING AN INVESTIGATION

Indoor air quality investigations usually follow the standard form of decision pathways: information-gathering, hypothesis formation, hypothesis testing, corrective action. If the testing results do not support the hypothesis, then hypothesis formation and hypothesis testing form a loop that the investigation may traverse several times. The goal is to understand the problem and its cause well enough to be able to decide on a corrective action. Throughout the process, the investigator must be on the lookout for multiple causes that might require multiple corrective actions.

Preliminary Walk-Through

Nothing takes the place of personally walking the site of the complaint and collecting information. Note the following:

- *What and who occupies the area.* Note the workers, their clothing, their habits (smoking, if allowed, their hygiene, etc.). Note machinery and equipment. Note incoming and outgoing material. Note furnishings, draperies, and other items.
- *HVAC system.* Take layout drawings and records on the walk-through, and note if anything has been changed since pre-complaint days. Look at louvre

settings, control locations, control access, and other variables. Sometimes complaints are solved at this point; for example, an investigator might note that cartons have been stacked so as to affect ventilation outlets or returns.

- *Pollutant pathways*. Mentally review how pollutants arising in other parts of the plant could flow to the complaint area. Make preliminary plans as to what pressure checks and chemical smoke tests should be made.
- *Contaminant sources*. Observe what operations in the area could be the source of contaminants. Although the walk-through is intended as an early look at the complaint area, this is a good time to speculate on possible involvement of other areas, and maybe to extend the walk-through.
- *Indicators of reaction to discomfort*. Look for occupant activities that indicate workers sense inadequate ventilation or poorly controlled temperature. Such indicators include desktop fans, heaters, or humidifiers; supply diffusers blocked with tape or cardboard; popped-up ceiling tiles; and interference with thermostat settings.

The walk-through may result in solving the problem, providing enough information to give direction to the rest of the investigation, or yielding no information. Usually it gives enough information to tell investigators what steps, such as monitoring air, checking pollutant pathways, or examining HVAC systems, will most likely solve the problem.

Hypotheses are potential explanations of the IAQ complaint. One view is that you start with the premise that "everything is a possibility," and the early stages of an investigation are a series of eliminations that quickly narrow the number of possibilities to a small percentage of the original list. The question, "Is the hypothesis consistent with the facts collected so far?" should be asked repeatedly.

Two other questions should circulate through your mind as you walk the area looking for information to start the investigation:

1. If there are obvious pollutant sources, do they appear to be adequately controlled?

- Are pollutant indicators present, such as odors, excessive dust, or staining?
- Are there sanitation problems such as debris near outdoor air intake, visible mold growth, or major water damage, that could be introducing air contaminants?
- Are there any conditions or activities occurring in or near the building that could be related to the complaints in timing, location, or health effects?

2. Are there deficiencies in the HVAC system that serves the complaint area?

- Does equipment that serves the area (such as thermostats, diffusers, fans, dampers, and filters) appear to be operating, clean, and in good condition?
- Do operating procedures exist, does the staff follow them, and are there signed check-off lists?
- Do records show that the system was commissioned (set, tested, and balanced) after construction?
- Do records show that system components are regularly inspected, calibrated, and adjusted?

Two often-overlooked sources of clues are workers and records of work in the area. Do not give workers the opportunity to say, "Why didn't they ask? We all know that the headaches started the day they moved the partitions." Information on conditions before and after the complaints can be the most important information to collect. The cut-off date is not the day the first complaint was registered, but the day workers first noticed the effect. That may be a hard date to pin down, because workers probably weren't thinking of poor air quality when they first felt the symptoms. After the symptoms have continued, it is hard to remember when they first started. In addition, it sometimes takes a while for the contaminant to build up in the area and then in workers to a level that would cause symptoms.

After you form a hypothesis, it is often more efficient to experiment with the results of a temporary physical change than to launch into a measurement program. Have a suspected operation stopped for a few days or, if possible, move it to another location. See if using fans to supplement the HVAC system mitigates the problem.

The hypothesis, plus common sense, often tell you exactly what to do next. If the complaint is limited to a single room, and the walk-through shows that there are no possible sources in the room, a reasonable hypothesis is that pollution from another source is getting into the room. The hypothesis tells you to first check pollutant pathways into the room. On the other hand, if the complaint involves a recognizable odor (lacquer thinner, for instance), the hypothesis "lacquer thinner is being used someplace" tells you that the easiest next step is to find out where it is used, and *then* track down pathways the fumes could follow.

Some investigators enforce a hard rule against taking any measurements during the walk-through, so nothing detracts them from concentrating on the overall picture. Others find that using heatless chemical smoke devices, digital thermometers, hygrometers, and direct reading meters (typically for measuring carbon dioxide) or detector tubes with a hand pump, to take occasional measurements helps them develop a feel for the building. Any instrument used should be calibrated and in working order.

Collecting Additional Information

As preliminary stages of the investigation gradually blend into later stages, there may be more visits—maybe many more—to the complaint area. A lot depends on who is investigating, how accurate the plant records are, and other factors. If the investigator is a veteran plant engineer, and plant records of modifications are up to date, he or she can gather sufficient information in fewer visits to the area.

If information from the walk-through does not lead to a reasonable hypothesis, try collecting more information about the occupants, HVAC system, pollutant pathways, and/or contaminant sources. Detailed or sophisticated measurements of pollutant concentrations or ventilation quantities may be required at this time. Outside experts might be needed. The following activities can be confined to a localized complaint area, be extended to include adjacent areas, or even encompass the entire building. Sometimes it can initially look as if the activities should focus on a certain area, but then early results make it clear that different boundaries should be drawn.

Collect information about the building. Look especially for information about changes that have been made, even changes to other parts of the building. A change in a distant part of the ducts might affect back pressure, which might cause a change in ventilation in the complaint area. It helps to also know how reliable the information has been in the past as far as being up to date.

Collect information about history related to the complaint. Has the same complaint been filed before? In the same area? In different areas? Is there a pattern, such as always after plant shut-down for vacation? Were they always filed by the same person? What was (were) the outcome(s) of the investigations?

Identify known HVAC zones and complaint areas. Begin to identify potential sources and pollutants (that is, special use areas near the complaint location). Having a copy of mechanical and floor plans can be helpful at this stage, especially if they are known to be up to date.

Notify building occupants of the upcoming investigation. Communicate the Ws: why, where, when, what.

Identify key individuals needed for access and information. A person familiar with the HVAC system is needed in most investigations. Individuals who filed or joined in the complaint should be available to help clarify exactly what the complaint is. Those in charge of potential sources can help answer questions and should be consulted if modifying or suspending an operation is a possibility. It may be necessary to prepare a list of telephone numbers and other ways to reach key individuals.

Closing In on the Problem

This section presents a systematic procedure for moving from clues to the source of the problem. Some investigations will use all the steps here, and others can jump directly to an applicable part of this procedure.

Define the Complaint Area. Use the spatial pattern of complaints to define the complaint area. Building locations where symptoms or discomfort occur are usually good rooms or zones to give particular attention during initial investigations. However, pollutant dispersion may make it necessary to revise the complaint area, even to locations far removed from the complaint area, as the investigation progresses. Back-tracking along pollutant pathways may be a technique that is a better substitute for concentrating on the complaint area.

Table 12-1 gives suggestions that can speed your move from clues to sources. The table applies to industrial plants in general; changes and additions made by investigators in each plant will tailor the table and add to its utility.

Looking for Timing Patterns

The timing of symptoms and complaints can indicate potential causes and provide directions for further investigation. Review the data for cyclic patterns of symptoms that may be related to HVAC or other activities in and around the building. Table 12-2 offers suggestions based on timing patterns.

Looking for Symptom Patterns

Investigations often fail to prove that any particular pollutant or group of pollutants could be the cause of the problem. Such causal relationships are difficult to establish, and there is little information available about the health effects of many chemicals. Typical indoor levels are much lower than the levels at which toxicology has found specific effects. Therefore, it may be more useful to look for patterns of symptoms than for specific pollutant and health effect relationships.

TABLE 12-1
Defining the Complaint Area

Spatial Pattern	***Suggestions***
Individual(s)	Check for drafts, radiant heat (gain or loss), and other localized temperature control or ventilation problems near the affected individual(s).
	Review the local pollutant source(s) near the affected individual(s).
	Consider that common background sources may affect only susceptible individuals.
	Consider the possibility that individual complaints may have different causes that are not necessarily related to the building (particularly if symptoms differ among the individuals).
Localized area (affecting individual rooms, zones, or air handling systems)	Check ventilation and temperature control within the complaint area.
	Review pollutant sources common to all portions of the localized area.
	Check HVAC system for zone, branches or components that might handle the localized area separately.
Widespread, no apparent spatial pattern	Check ventilation and temperature control for entire building.
	Check outdoor air quality, especially near intakes.
	Review sources that are spread throughout the building, such as cleaning materials.
	Consider explanations other than air contaminants.

Investigators who are not medically trained cannot make a diagnosis and should not attempt to interpret medical records. Also, confidentiality of medical information is protected by law in most jurisdictions and is a prudent practice everywhere.

TABLE 12-2
Search Directions Based on Timing Patterns

Timing Pattern	*Suggestions*
Symptoms begin and/ or are worse at the start of the occupied period	Review HVAC operating cycles. Emissions from building materials, or from the HVAC system itself, may build up during unoccupied periods.
Symptoms worsen over course of occupied period	Consider that ventilation may not be adequate to handle routine activities or equipment operation within the building.
Intermittent symptoms	Look for daily, weekly, or seasonal cycles or weather-related patterns.
	Check linkage to other events in and around the building.
Single event—may no longer be present	Consider spills, accidents, and other unrepeated events as sources.
Recent onset of symptoms	Ask staff and occupants to describe recent changes or events (such as remodeling, renovation, redecorating, HVAC system adjustments, new procedures, leaks, or spills.
Symptoms relieved on leaving the building, either immediately, overnight, or (in some cases) after extended periods away from the building	Consider that the problem is likely to be building-related, though not necessarily due to air quality. Other stressors, such as lighting or noise, may be involved.
Symptoms never relieved, even after extended absence from building	Consider that the problem may not be building-related.

Table 12-3 lists some common symptom groups that can be related to indoor air quality, along with possible sources or causes of these symptoms. Use this table with the understanding that it is far from exhaustive; possible patterns are limitless.

TABLE 12-3
Suggested Directions to Search, Based on Symptom Patterns

Symptom Pattern	*Suggestions*
Thermal discomfort	Check HVAC condition and operation.
	Measure indoor and outdoor temperature and humidity; see if extreme conditions exceed design capacity of HVAC equipment.
	Check for drafts and stagnant areas.
	Check for excessive radiant heat gain or loss.
Common Symptom Groups: Headache, lethargy, nausea, drowsiness, dizziness	If onset was acute, sudden, and/or severe, arrange for medical evaluation; problem may be carbon monoxide poisoning.
	Check combustion sources for uncontrolled emissions or spillage. Check outdoor air intakes for nearby sources of combustion fumes.
	Consider evacuation and/or medical evaluation if problem isn't corrected quickly.
	Consider other pollutant sources.
	Check overall ventilation; see if areas of poor ventilation coincide with complaints.
Congestion; swelling, itching, or irritation of eyes, nose, or throat; dry throat; may be accompanied by non-specific symptoms (e.g., headache, fatigue, nausea)	May be allergic, if only a small number affected; more likely to be irritational response if large number are affected.
	Urge medical attention for allergies.
	Check for dust or gross microbial contamination due to sanitation problems, water damage, or contaminated ventilation system.

TABLE 12-3
Continued

Symptom Pattern	*Suggestions*
Congestion *(Cont.)*	Check outdoor allergen levels, pollen count.
	Check closely for sources of irritating chemicals such as formaldehyde or those found in some solvents.
Cough; shortness of breath; fever, chills and/or fatigue after return to the building	May be hypersensitivity pneumonitis or humidifier fever. A medical evaluation can help identify possible causes.
	Check for gross microbial contamination due to sanitation problems, water damage, or contaminated HVAC system.
Diagnosed infection	May be Legionnaire's disease or histoplasmosis, related to bacteria or fungi found in the environment.
	Contact local or state health department for guidance.
Suspected cluster of rare or serious health problems such as cancer, miscarriages	Contact local or state health department for guidance.
Other Stressors: Discomfort and/or health complaints that cannot be readily ascribed to air contaminants or thermal conditions	Check for problems with environment, ergonomic, and not-related psychosocial stressors.

Collecting Information About the HVAC System

When the HVAC system is the cause of IAQ complaints, it is usually for one of two reasons. First, the quantity or distribution of outdoor air might be inadequate to serve the ventilation needs of building occupants. Second, the distribution system might be introducing outdoor contaminants or transporting pollutants within the building.

The investigation should begin with components of the HVAC system that serve the complaint area and surrounding rooms, but may have to expand if connections to other areas are discovered. Your goal is to understand the design and operation of the system well enough to answer the following questions:

- Are the components that serve the immediate complaint area functioning properly?
- Is the HVAC system adequate for the building's occupancy and usage?
- Are the ventilation or thermal comfort functions working properly?
- Should the definition of the complaint area be redefined on the basis of the HVAC layout and operating characteristics?

A basic evaluation of the system may include limited measurements of temperature, humidity, airflow, and carbon dioxide, along with smoke tube observations. When more complex measurements are needed, they may include extensive or sophisticated measurements of the same variables such as repeated CO_2 measurements taken at the same location under different operating conditions, and continuous temperature and relative humidity measurements recorded with a data logger. A detailed engineering study may be needed if the investigation turns up problems such as:

- Low airflows
- HVAC controls not working
- HVAC controls working according to inappropriate strategies
- Building operators do not understand, or are unfamiliar with, the HVAC system.

Conversations with the facility staff can provide important current information about equipment operation, maintenance schedules, breakdowns, and other incidents. They may provide inspection reports or other written records. Staff members who are familiar with building systems in general and with the specific features of the building under investigation can be very helpful in identifying conditions that may explain the indoor air quality complaints. Some facilities have extensive preventive maintenance programs. On the other hand, discussion could reveal that facility staff are not operating the building according to its design because:

- They do not understand the HVAC system design logic.
- They have been directed to run the system at the lowest possible cost.
- They do not have staff to operate the building properly.

- The system has not been modified to accommodate changes in space, occupancy, or other factors.

Table 12-4 summarizes the preceding steps, and adds detail to some of them. The table is not all-inclusive; it will become valuable to each plant as investigators add items of special application to their plants.

SAMPLING INSTRUMENTATION, METHODS, AND LEVELS

This section continues the recommendation that a major air measurement program be delayed until preliminary screening indicates that it is needed. It gives screening methods for specific contaminants.

Low Contaminant Levels

Indoor air quality investigations in industrial plants usually look for low concentrations of contaminant levels, often in parts per million. Such small concentrations require instruments with sufficient sensitivity. Personnel must be trained in selecting, setting up, and using the instruments, and in recording the data. Investigators must choose sampling procedures that can detect and measure concentrations of toxic materials which are much lower than are normally found in industrial investigations. Few procedures have been validated for these lower level contaminants.

Present OSHA sampling and analytical procedures were developed to meet precision and accuracy requirements for airborne contaminants in the range of OSHA Permissible Exposure Limits (PELs) and American Conference of Governmental Industrial Hygienists (ACGIH) Threshold Limit Values (TLVs). These procedures are used for sampling eight-hour Time-Weighted Averages (TWAs) and Short Term Exposure Limits (STELs) of 15/30 minutes.

Extensive air monitoring may not be warranted in many investigations, because the main problem is often that inadequate fresh air was brought in and/or distributed.

General and Specific Screening

To further reduce expenditures on a major monitoring program, use screening techniques to determine the potential sources that may require more sensitive and

TABLE 12-4
HVAC Information to Collect

Strategies	***Tools***
Review existing documentation on HVAC design, installation, and operation	Collect: Design documents, testing and balancing reports
	Operating instructions, control manufacturer's installation data
Talk to facilities staff	Ask facilities staff to record their observations of equipment cycles, weather conditions, and other events using Log of Activities and System Operations
Inspect system layout, condition, and operation	Use: Zone and room records
	HVAC Checklist, short or long form
	Thermometer and hygrometer to measure temperature and humidity
	Manometer or equivalent to measure pressure differentials 0-2″ and 0-10″ water gauge (w.g.) to measure at fans and intakes 0-0.25″ w.g. with pitot tube to check airflow in ducts Chemical smoke to observe airflow patterns A device to assess airflow from diffusers Anemometer or velometer for rough quantitative Flow hood for accurate quantitative Carbon dioxide measurement devices Detector tubes with a hand pump Direct reading meter

Continued

TABLE 12-4
Continued

Strategies	*Tools*
Use additional instruments as appropriate	Instruments often used by professional IAQ consultants include:
	A hygrothermograph to log temperature and humidity
	Tracer gas and measurement equipment
	A device to measure airborne particulates
	Measurement devices for carbon monoxide and other contaminants of interest

accurate evaluation. Collect screening samples using detector tubes or direct reading instruments. Increase instrument sensitivity by using higher flow rates or longer sampling times. Low-range detector tubes are available from manufacturers. Table 12-5 presents contaminant types, screening methods, concentration range, and validated testing methods.

Much of the specialized equipment is available through the OSHA Health Response Team (HRT), OSHA Calibration Laboratory (OCL), or regional offices.

In addition to general screening, there are many chemicals for which optional screening is based upon professional judgment. The next few paragraphs mention the most common ones.

Screening results may indicate a need for validated sampling procedures to further quantify employee exposures. Much of the information on validated sampling and analytical methods is in the OSHA *Chemical Information Manual* or in the OSHA *Analytical Methods Manual*.

Acetic acid. Low levels of acetic acid have been found from off-gassing of silicone caulking compounds and in hospitals where x-ray developing equipment is improperly ventilated. Use detector tubes (0–10 ppm) to evaluate compla ints of eye, nose, and throat irritation.

Asbestos. Screening is not a routine procedure. Any requested screening should be done in accordance with the proper OSHA standards.

TABLE 12-5
Sampling and Analytical Methods

Contaminant	*Concentration Range*	*Screening Method*	*Validated Method**
Bioaerosols	0–1,000 cfu/m^3	Viable Biological Sampler	
Carbon Dioxide	0–2,000 ppm	Detector Tube, Ion Chromatograph	Sampling bag, Gas Chromatograph/ Thermal Conductivity Detector. OSHA ID172
Carbon Monoxide	2–50 ppm	Detector Tube, Meter	Sampling bag, meter
Formaldehyde	0.04–1 ppm	Detector Tube	Coated XAD-2, Gas Chromatograph/Nitrogen Phosphorus Detector. OSHA-52
Nitric Oxide	0–25 ppm	Detector Tube	Triethanolamine Tube with Oxidizer, Differential Pulse Polarographic. OSHA ID190
Nitrogen Dioxide	0–5 ppm	Detector Tube	Triethanolamine-Molecular Sieve Tube, Ion Chromatograph. OSHA ID182
Particulates	0–40,000 particles/cc	Light Scattering Meter	
Ozone	0–0.1 ppm	Detector Tube, Chemiluminescent Meter	

Continued

TABLE 12-5
Continued

Contaminant	*Concentration Range*	*Screening Method*	*Validated Method**
Radon	4–200 pci/l	Radon Cartridge, Electrect	

* Referenced OSHA procedures can be found in the OSHA *Analytical Methods Manual* or the OSHA *Chemical Information Manual*. See individual manufacturer's literature for information on interferences to the screening or sampling method.

Carbon dioxide. Carbon dioxide measurement is a useful screening technique that is often helpful in determining whether adequate quantities of outside fresh air have been introduced and distributed into the building. Use low-level detector tubes (0–2000 ppm) or portable infrared spectrometers to screen for indoor carbon dioxide levels. Table 12-6 gives NIOSH recommendations.

These levels are guidelines only. A carbon dioxide level that exceeds 1000 ppm does not necessarily mean that the building is hazardous and should be evacuated. Rather, 1000 ppm is a level that should be used as a guideline that helps maximize comfort for all occupants.

Carbon monoxide. This gas is a serious hazard that should be detected with low-level detector tubes (2–200 ppm). It also can be detected with direct-reading CO monitors.

Formaldehyde. This gas frequently results from off-gassing of insulation, building materials, carpets, drapes, glues, and adhesives. Use low-level (0.04–1 ppm) detector tubes to evaluate complaints of eye, nose, and throat irritation.

Ozone and nitrogen oxides. Generally evaluated with detector tubes. It is best to also collect outdoor samples since ambient levels of ozone may reach levels of one to three times the Permissible Exposure Level of 0.1 ppm during air temperature inversions.

If a more accurate or continuous ozone evaluation is required, a chemiluminescent monitor that is specific for ozone and can measure in the range of 0.01 to 10 ppm is available from OSHA's Health Response Team. Their mission is to provide

TABLE 12-6
NIOSH Recommendations for Carbon Dioxide

CO_2 *Concentration*	*Indication*
250–350 ppm	Normal outdoor ambient conditions
600 ppm	Minimal air quality complaints
600–1,000 ppm	Less clearly interpreted
1,000 ppm	Inadequate ventilation; complaints such as headache, fatigue, and eye and throat irritation will be more widespread; 1,000 ppm should be used as an upper limit for indoor levels.

technical assistance in the fields of industrial hygiene and specialized engineering to OSHA's National, Regional, and Area Offices in support of Agency objectives.

Radon. A rapid, easy-to-use screening method for measuring radon gas concentrations is available from the Salt Lake City Analytical Laboratory. Use it for deciding if additional measurements are required or if remedial actions should be undertaken.

If required, additional longer term quantitative procedures are available from the Health Response Team. The HRT or regional offices may be contacted if sampling is to be initiated, and for interpretation of the results. If the results of screening samples are less than four picocuries per liter of air (pCi/l), EPA recommends that follow-up measurements probably are not required. Readings larger than four pCi/l indicate that follow-up measurements should be performed.

Airborne particles. Use a particle counting instrument capable of measuring concentrations as low as 2000 particles per cubic centimeter of air for comparing particulate concentrations in different areas. These measurements should help investigators determine where additional ventilation or air filtration is necessary for eliminating or minimizing employee complaints.

Airborne microorganisms. The American Conference of Governmental Industrial Hygienists recommends a preassessment of the extent of microbial contamination prior to initiation of air sampling. Before biological sampling, several precautions must be taken, including making arrangements for preparing culture media for sampling, specialized shipping procedures, and making arrangements for analysis by a laboratory familiar with the handling and processing of biological samples. Contact the Directorate of Technical Support for information about laboratories experienced in analysis of microbial samples and with knowledge of the health effects.

If sampling for airborne microbials is deemed necessary, equipment is available from OSHA's Health Response Team.

Legionella pneumophila. It is often present in hot water tanks, washing systems, and pools of stagnant water. However, health effects are not observed until the contaminants become aerosolized within the building confinements. Identification of predominant taxa, or at least fungi, is recommended in addition to determining the number of colony-forming units per cubic meter (cfu/m^3) of air. During growing seasons, outdoor fungus spore levels can range from 1000 to 100,000 cfu/m^3 of air.

Contamination indicators include:

- 1,000 viable colony-forming units in a cubic meter of air
- 1,000,000 fungi per gram of dust or material
- 10,000 bacteria or fungi per milliliter of stagnant water or slime

Levels in excess of these do not necessarily imply conditions that are unsafe or hazardous. The type and concentrations of the airborne microorganisms, as evaluated by experts, will determine the hazard to occupants.

Miscellaneous airborne contaminants. Use a portable infrared spectrometer to evaluate a wide variety of potential air contaminants including acetic acid, ammonia, carbon dioxide, carbon monoxide, nitric oxide, nitrogen dioxide, sulfur dioxide, and a number of volatile organic compounds. The spectrometer can be connected to a strip-chart recorder to obtain a hard copy showing variations of concentration during the day. Take care in interpreting the results, because the instrument is not always specific for one compound.

Equipment not generally available in the field, such as particulate analyzer, infrared spectrometer, and airborne biological sampler are available through the Health Response Team. These instruments are provided with written descriptions, operating manuals, and methods of analysis.

PERSONAL SAMPLING FOR AIR CONTAMINANTS

Indoor air quality specialists recommend not starting an investigation with general air monitoring or sampling. A more efficient procedure is to screen the area in question and then evaluate the potential for employee exposure. This evaluation should use observation and screening samples, and should not overexpose employees in an attempt to simulate worst-case conditions. On the basis of information so gathered, a full-shift air sampling program might be initiated.

Screening with portable monitors, gravimetric sampling, or detector tubes can provide data for evaluating the following:

- Processes, such as electronic soldering
- Exposures to substances with exceptionally high PELs in relatively dust-free atmospheres such as ferric oxide and aluminum oxide
- Intermittent processes with substances without STELs
- Engineering controls, work practices, or isolation of process
- The need for CSHO protection

Take a sufficient number of samples to obtain a representative estimate of exposure. Contaminant concentrations vary seasonally, with weather, with production levels, and in a single location or job class. The number of samples taken depends on the error of measurement and differences in results. Consult the *NIOSH Occupational Exposure Sampling Strategy Manual* for specific instructions.

If air sampling and monitoring have previously been conducted in the plant, review the records.

Bulk samples are often required to assist the Salt Lake City Laboratory in properly analyzing field samples. Some test results that fall into these categories include those of silica, portland cement, asbestos, mineral oil and oil mist, chlorodiphenyl, hydrogenated terphenyls, chlorinated camphene, fugitive grain dust, and explosibility testing.

General Sampling Procedures

Some of the following steps apply to all situations; other steps should be modified to fit conditions in your particular plant. You should follow most of the steps in the order given, but a few of them can be completed in a different order without affecting the results.

- Determine the appropriate sampling technique, prepare and calibrate the equipment, and prepare the filter media. Screen the sampling area.
- Select the employee to be sampled and discuss the purpose and procedure. Major points to bring out are when and where the sampling equipment will be removed, and the importance of not removing or tampering with the equipment.
- Instruct the employee to notify the supervisor or the CSHO if the sampler requires temporary removal. Turn off or remove sampling pumps before the employee leaves a potentially contaminated area (as when he or she goes to lunch or on a break).
- Place the sampling equipment on the employee so that it does not interfere with work performance. Attach the collection device (filter cassette, charcoal tube, etc.) to the shirt collar or as closely as practical to the nose and mouth (in a hemisphere forward of the shoulders with a radius of approximately six to nine inches. The inlet should always be in a downward vertical position to avoid gross contamination. Position the excess tubing so as to not interfere with the employee's work.
- Turn on the pump and record the starting time. Observe the pump operation for a short time after starting to make sure it is operating correctly.
- Check pump operation and condition at least every two hours. More frequent checks may be necessary if the environment is such that the filter could clog. Ensure that the sampler is still assembled properly and that the hose has not become pinched or detached from the cassette or the pump. For filters, observe for symmetrical deposition, finger prints, large particles, etc. Record the flow and other pertinent data.
- Take photographs as appropriate, and detailed notes concerning visible airborne contaminants, work practices, potential interferences to movements, and other information that will assist in determining appropriate engineering controls.
- Prepare blanks during the sample period for each type of sample collected. For any given analysis, one blank will suffice for up to 20 samples collected, except for asbestos which requires a minimum of two field blanks. These blanks may include opened but unused charcoal tubes.
- Before removing the pump at the end of the sampling period, check the flow rate to ensure that the rotameter ball is still at the calibrated mark (if there is a pump rotameter). If the ball is no longer at the mark, record the pump rotameter reading.
- Turn off the pump and record the ending time.
- Remove the collection device from the pump and seal it with an OSHA-21

form as soon as possible. Do not place the seal around the cassette's perimeter; place it lengthwise so it covers the inlet and the outlet ports.
- Prepare the samples for mailing to the Salt Lake City Analytical Laboratory for analysis.
- Recalibrate pumps after each day of sampling (before charging).
- For unusual sampling conditions, such as wide temperature and pressure differences from the calibration conditions, call the regional technical support section.

CHAPTER 13

Implementing Regulatory Compliance

Several organizations provide requirements, standards, advisories, and recommendations for working conditions, including indoor air quality. Of all the federal, state, local, and industry association organizations that supply input, the Occupational Safety and Health Administration (OSHA) is the one that comes to mind first and has the largest effect. It operates through federal offices in about half the states, and through approved state OSHAs in the other half. Accordingly, this chapter begins with a look at some OSHA statistics and facts.

OSHA'S INFLUENCE

One effect of the regulations and enforcement can be seen in the occupational injury and illness incidence rates per 100 full-time workers. Table 13-1 summarizes these figures for selected years. Numbers for series of years are averages.

TABLE 13-1
Occupational Injury and Illness Incidence per 100 Full-Time Workers

Calendar Years	Total Cases
1973	11.0
1974	10.4
1973-1979	9.70
1990-1993	8.65

OSHA Enforcement

The act that created OSHA gave it responsibilities and powers to enforce its regulations. Those responsibilities and powers carry through to the 25 states that have approved state OSHAs as well. Statistics for various size businesses and for each standard industrial classification can fill a book this size; the following sums for businesses with over 250 employees can be considered typical.

From October 1994 through September 1995 federal OSHA issued 238 citations for violations of standards (state-plan OSHAs typically issue 3⅓ times as many citations as does federal OSHA). Penalties were as high as nearly 3½ million dollars. Of those citations, 42 were related to indoor air quality and their penalties totaled $5,139,251.50. The average penalty for these 42 citations was $122,363.13. When you include these numbers with the other incentives (such as wanting to be fair to employees, good public relations, etc.) for complying with clean air standards, there is sufficient motivation for management to take all necessary steps.

Employee Rights

The Occupational Safety and Health Act of 1970 encouraged employers and employees to work together to reduce workplace hazards and to implement safety and health programs. In the belief that subjecting oneself to hazardous conditions should not be the price of going to work, the act gave employees several rights and responsibilities. It allowed state plans to grant additional rights, but required that they at least preserve the federal OSHA rights. The following list, not in any special order, summarizes those rights that relate directly to indoor air quality, and those that relate indirectly because they are general in nature. Employees have the right to:

- Have access to relevant employee exposure and medical records.
- Request that the OSHA area director conduct an inspection if the employee believes hazardous conditions or violations of standards exist in the workplace.
- Have an authorized employee representative accompany the OSHA compliance officer during the inspection tour.
- Observe any monitoring or measuring of hazardous materials and see the resulting records.
- Object to the abatement period set by OSHA for correcting violations.
- Submit a written request to the National Institute for Occupational Safety and Health (NIOSH) for information on whether any substance in the workplace has potentially toxic effects in the concentration being used.
- Be notified by the employer if the employer applies for a variance from an OSHA standard, to testify at a variance hearing, and to appeal the final decision.
- Withhold employee names from employer if a written and signed complaint is filed.
- File a discrimination complaint if punished for exercising rights or for refusing to work when faced with an imminent danger of death or serious injury if there is insufficient time for OSHA to inspect.

Reporting hazards. Employers cannot simply ignore hazards because, if a hazard is not being corrected, an employee is expected to contact the OSHA office having jurisdiction. When the complaint is in writing, and OSHA determines that there are reasonable grounds for believing that a violation or danger exists, the office conducts an inspection. Workers are encouraged to point out hazards, describe accidents or illnesses that resulted, describe past complaints, and inform the inspector if working conditions were changed just for the inspection. Employees have a right to talk privately to the compliance officer on a confidential basis.

When shortages of personnel, materials, or equipment prevent an employer from complying fully with a new safety and health standard in the time provided, the employer may apply for a temporary variance from the standard. As part of the process, the employer must inform workers of the application and must post a summary of the application wherever notices are normally posted in the workplace. Employees must be informed that they have the right to request a hearing on the application. Notices of variance applications are published in the *Federal Register*, inviting all interested parties to comment. Employees, employers, and other interested groups are encouraged to participate in the variance process.

When facing imminent danger. Although there is nothing in the OSHA law that specifically gives an employee the right to refuse to perform an unsafe or unhealthful job assignment, OSHA's regulations, which have been upheld by the U.S. Supreme Court, provide that an employee may refuse to work when faced with an imminent danger of death or serious injury.

Empoyee Responsibilities

On the other side of the rights/responsibilities list, employees have the responsibility to assist in making the workplace safe and healthful, including the following:

- Read the OSHA poster at the jobsite.
- Comply with all applicable OSHA standards.
- Follow all lawful employer safety and health rules and regulations; wear or use prescribed protective equipment while working.
- Report hazardous conditions to the supervisor.
- Report any job-related injury or illness to the employer, and seek treatment promptly.
- Cooperate with the OSHA compliance officer conducting an inspection if he or she inquires about safety and health conditions in the workplace.
- Exercise rights under the OSH Act in a responsible manner.

Using the OSHA Consultation Service to Prevent Citations

Using a free consultation service, employers can find out about potential hazards at their worksites, improve their occupational safety and health management programs, and even qualify for a one-year exemption from routine OSHA inspections. This service is completely separate from the inspection effort; no citations are issued, no penalties will be proposed, and no possible violations will be reported to the inspection staff. The employer, firm name, any information provided and any unsafe or unhealthful working conditions also are confidential and not reported to the inspection staff.

As an employer requesting a consultation, you have just one obligation—a commitment you make prior to the visit. You commit to correcting serious job safety and health hazards in a timely manner.

The consultation is voluntary, so you must initiate it with a phone call or a letter. Although OSHA encourages a complete review of your firm's safety and health situation, you may choose to limit the visit to one or more specific problems.

Three steps in an OSHA consultation

1. *The Opening Conference*. The consultation is a "practice session" because it allows employers to learn how an inspection is conducted, but without the threat of citations and penalties. It begins with an opening conference at which the consultant explains his or her role, and the obligation you incur.

2. *The Walk-Through*. Together you and the consultant will examine conditions in your workplace. OSHA encourages maximum employee participation because better informed and more alert employees can easily work with you to identify and correct potential injury and illness hazards. Talking with employees during the walk-through helps the consultant identify and judge the nature and extent of specific hazards.

The consultant will study and discuss the applicable OSHA standards for the entire workplace or for the specific operations you designate. He or she will point out other unsafe conditions that might not be cited under OSHA standards, but nevertheless may pose safety or health risks. Consultants may suggest and even provide other measures, such as self-inspection and safety and health training you and your employees can use to prevent future hazardous situations.

A comprehensive consultation also includes (1) appraisal of all mechanical and environmental hazards and physical work practices, (2) appraisal of the present job safety and health program or establishment of one, (3) a conference with management on findings, (4) a written report of recommendations and agreements, and (5) training and assistance with implementing recommendations.

3. *The Closing Conference*. As with an actual inspection, a consultation visit ends with a closing conference at which the consultant reviews detailed findings. Recommendations will be discussed, along with descriptions of what you are doing right. You can discuss problems, possible solutions, and abatement periods for eliminating or controlling any serious hazards identified during the walk-through.

It rarely happens, but consultants have found "imminent danger" situations during walk-throughs. You must then take immediate action to protect all employees. In certain other situations which would be judged a "serious violation" under OSHA criteria, you and the consultant are required to develop and agree to a reasonable plan and schedule to eliminate or control that hazard. OSHA defines imminent danger as a condition where there is reasonable certainty a danger exists that can be expected to cause death or serious physical harm immediately, or before the danger can be eliminated through normal enforcement procedures. A serious violation is a violation where there is a substantial probability that death or serious physical harm could result.

After the closing conference, the consultant will send you a detailed written report explaining the findings and confirming any abatement periods agreed upon. Consultants may contact you from time to time to check your progress. Ultimately, OSHA requires hazard abatement so that each consultation visit achieves its objective of effective employee protection. Although consultants do not report possible violations to OSHA enforcement staff, there have been rare instances of employers failing to eliminate or control identified serious hazards or imminent danger situations according to the plan and within the agreed-upon limits or an agreed-upon extension. The situation then must be referred from consultation to an OSHA enforcement office for appropriate action.

Regular OSHA Inspections

Federal OSHA, as well as state-plan OSHAs, maintain staffs of inspectors, or compliance officers, who have the legal authority to visit and inspect any place of business with no prior notification. In fact, anyone who knows of a scheduled visit, and notifies the company, is subject to a fine of up to $1,000 and/or a six-month jail term.

Their schedule is basically random, although OSHA has established some priorities for arranging inspections.

- *Imminent danger situations*. Top priority is given to locations where an imminent danger situation is known or suspected to exist.
- *Catastrophes and fatal accidents*. Second priority is given to investigation of fatalities and accidents resulting in hospitalization of five or more employees.
- *Employee complaints*. Third priority is given to formal employee complaints of alleged violations of standards or of unsafe or unhealthful working conditions. Because such complaints are kept confidential, employers do not know when a complaint against them has been filed.
- *Programmed inspections*. Next in priority are programmed inspections aimed at specific high hazard industries, occupations, or health substances, or other industries identified in OSHA's current inspection procedures. Industries are selected on the basis of factors such as the injury incidence

rates, previous citation history, employee exposure to toxic substances, or random selection.

- *Follow-up inspections*. A lower priority is given to follow-up inspections, which are held to determine if previously cited violations have been corrected.

Variances

There are sometimes valid reasons for allowing an employer to continue operating even though not in full compliance with a certain standard. Variances can be either temporary or permanent. Valid reasons include:

- Shortages of material or equipment
- Lack of professional or technical personnel
- Present facilities or methods provide employee protection "at least as effective" as that required by OSHA.

Employers who operate plants in state-plan states and also in states under federal OSHA jurisdiction may apply directly to federal OSHA for a single variance applicable to all the plants in question. OSHA will work with the state-plan states involved to determine if a variance can be granted which will satisfy state as well as federal requirements.

OSHA *may* grant an employer additional time, by way of a temporary variance, to achieve compliance. However, such variance will not be granted to an employer who simply cannot afford to pay for the necessary alterations, equipment, or personnel. The variance allows the needed time, or one year, whichever is shorter, and it is for two six-month periods.

How to apply for a temporary variance

An application for a temporary variance must include the following information:

- Identification of standard or portion of standard that cannot be met
- Reasons employer cannot comply
- Documentation, with dates, of measures already taken to comply
- Certification that a) workers have been informed of the variance application, b) a copy has been given to the employees' authorized representative, c) a summary of the application has been posted wherever notices are normally posted, and d) employees have been informed that they have the right to request a hearing on the application.

How to apply for a permanent variance

A permanent variance, referred to as an alternative to a particular requirement or standard, is also possible. It may be granted to employers who prove their conditions, practices, means, methods, operations, or processes provide a safe and healthful workplace as effectively as would compliance with the standard. OSHA weighs the employer's evidence and arranges a variance inspection and hearing where appropriate.

Part of the application process includes informing employees of the permanent variance and of their right to request a hearing. Any time after six months from issuance of a permanent variance, the employer or employees may petition OSHA to modify or revoke it. OSHA may modify or revoke the permanent variance on its own accord.

One additional step in the temporary or permanent variance process is for the employer to apply for an interim order, which allows operations to continue under existing conditions until the variance decision is made. If the interim order is granted, the employer and other concerned parties are informed, and the terms of the order are published in the *Federal Register*.

Variances are not retroactive. An employer who has been cited for a standards violation may not avoid penalties by applying for a variance. However, employers may file a variance application while a citation is outstanding.

HOW TO KNOW YOUR OWN PLANT[1]

As suggested by the consultation just described, OSHA's objective is not to catch employers disregarding regulations, but to have employers create and maintain safe and healthful working conditions. However, employers cannot depend on consultation closing conferences for learning the volume of information with which they should be familiar. In addition, NIOSH and OSHA are constantly determining and evaluating hazards. They revise standards, sometimes making them stricter and sometimes loosening them. Such new determinations are especially frequent in the indoor air quality field.

Some companies keep a person or a staff whose responsibility is to keep abreast of regulations and to revise the plan's self-evaluation procedures. However, there are consulting firms that are experts in all areas of regulatory compliance. This chapter is based on the services of one such consultant.

1. Courtesy of J. J. Keller & Associates, Inc., Neenah, Wisconsin.

Compliance Audits

The full title of one manual is *Compliance Audits: Essential Checklists for OSHA, EPA, and Other Key Agencies*. It is a 200-page combination of descriptive information and extensive checklists, plus its purchase includes a year of updates.

Descriptive material includes details on establishing a safety committee, managing a worker safety and health program, and conducting your own audit and inspection. Specific procedures are given for conducting the audit of your preparedness, including forms to fill out. Recommendations include keeping a copy of the forms to provide the necessary documentation for a variety of purposes, including corporate reviews, regulatory inspections, post-accident investigation, and insurance company requests.

Typically there are separate checklists for different areas of the plant. Items that derive directly from certain regulatory citations usually include their numbers so you can go to the source. Following is a small sampling:

- Is local ventilation sufficient to disperse fumes? .178(g)(2)
- Are all lines to a confined space containing inert, toxic, flammable, or corrosive materials valved off and blanked or disconnected and separated before entry?
- Are all exposures from dust, fumes, etc. controlled? .94
- Do you have emergency eye wash and shower facilities within the immediate work area where employees are exposed to injurious corrosive materials? .151(c)
- Are workers who wear respirators trained and certified in their use?

Regulatory agencies rely heavily on your reports. A consultant's guidance materials should include clear explanations of critical record keeping requirements that demand your compliance. The guidance should remind you to document key activities such as training sessions. You will appreciate other reminders, such as to check notebook entries for general understandability (the writer may have left the company by the time the notes are needed), and to include emergency phone numbers and means of contacting key personnel.

Selecting a Consultant

In choosing a consultant, look for ability to write plain English explanations that will help you make sure your permit applications contain information necessary for complying with regulatory standards. There should be an emphasis on OSHA's heavily enforced topics such as the Confined Space Standard (1910.146) where toxic fume accumulations are a constant concern.

Check that a potential consultant's knowledge and experience include indoor air quality. Further, be sure that the knowledge and experience with IAQ extend to topics peculiar to you plant, such as respirators, hazard communications, tobacco smoke, or HVAC.

The consultant should give you a clear understanding of critical record-keeping requirements that demand your compliance. Equally important are the reports you should prepare to document your successful self-inspection (see Chapter 14).

Training Employees

Typical of OSHA's new and revised standards are the recent *Hazard Communications* (29 CFR 1910.1200; 1926.59) and *Personal Protective Equipment* (29 CFR 1910.132–.138) standards. Both rules have training components that are affected by the revisions. A good way to keep abreast of such changes is to subscribe to a consultant's safety training newsletter. Through such a periodical you should receive a summary of new or revised regulations and contacts for more information. Among the several references in one newsletter is, "For more information: See the February 9, 1994 *Federal Register* or contact Akio Konoshima, OSHA, (202) 219-8151."

Other subjects in a typical training newsletter might include:

- Make your case study an effective learning tool
- Five arguments to convince top management to train
- How do you train a group you know disagrees with you?
- Focus on respirators (always a prominent topic in indoor air quality)
- Working with computer terminals

Industrial Safety Report

Another way to keep abreast of regulatory changes is to subscribe to a monthly report. The particular one reviewed here discusses topics such as proposed regulations, enforcement activities, state-level activity, updates on regulated chemicals, training programs, recent penalties, rule clarification, and safety issues.

EVALUATING WORKERS' HEALTH[2]

Meeting the standards for indoor air quality should certainly be the goal of every employer, but the final test is workers' health. By contracting with a testing organization, you can design tests and services to meet OSHA medical surveillance and other regulatory requirements, or as part of a preventive health program. Uniform, cost effective programs are available to specific groups of employees or to all employees.

OSHA Medical Compliance Programs

OSHA has established guidelines for medical surveillance when working with specific exposures. You can contract with a health evaluation organization that will deliver health testing programs to meet or exceed the regulation or criteria requirements. Some take fully equipped vans to your facilities for testing employees. Table 13-2 summarizes some of the requirements.

Table 13-3 shows recommended tests to comply with requirement programs. Those items with double check marks would be considered if test results are outside of normal limits. The health questionnaire obtains background information which is vital to interpretation of test results. It includes items such as social history, family medical history, personal medical history, and review of systems. The last item in the table—respirator certification—is critical, and should be the responsibility of an Occupational Physician.

Some tests shown, such as for hearing, may not apply directly to indoor air quality. However, having the results on record can often help in unexpected ways; a trained medical examiner might make a connection between hypersensitivity to an indoor air quality problem and an apparently unrelated health problem. A nonexistent air quality deficiency might be blamed for headaches which are actually due to a health problem.

2. Courtesy of Health Evaluation Programs, Inc., Park Ridge, Illinois.

TABLE 13-2
Health Compliance Tests

Objective	*Description*
Basic Health Program	One firm's basic health program makes an overall evaluation of body systems. The findings, along with a visit to workers' personal physicians, provide a sound basis for a preventative health program.
Respirator	One example of why OSHA does not have specific tests for determining whether one is capable of wearing a respirator is that standards must be different for part-time and for emergency wearers. An evaluation consultant can design a minimum test profile.
Asbestos	Tests for asbestos workers are the same as that for respirator wearers, but an occupational physician examination is recommended.
Benzene	Compliance for possible benzene exposure requires a review of the blood systems. These tests also can warn of early signs of other health problems.
Coke Oven	Workers who have coke oven exposure and are over 45 or have at least five years in the industry must be tested twice a year with sputum and urine cytology tests.

TABLE 13-3
Examples of Health Screening Programs

	Basic Health Program	**OSHA Compliance for:**			
		Respirator	**Asbestos**	**Benzene**	**Coke Oven**
Health Questionnaire	✔				
Chest X-ray	✔	✔	✔	✔	✔
EKG	✔				
Pulse & Blood Pressure	✔	✔	✔	✔	✔
Blood Test, Hematology	✔			✔	
Blood Test, Chemistries	✔			✔	✔
Urinalysis	✔				✔
Hearing	✔				
Vision	✔				
Ocular Tension	✔				
Lung Function	✔	✔	✔	✔	✔
Physical Measurements	✔	✔	✔	✔	✔
Industrial Health Questionnaire		✔	✔	✔	✔
Physical Examination	✔✔	✔✔	✔	✔	✔✔
Skin Examination					✔
Cytology, urine, sputum					✔
Respirator Certification		✔	✔	✔	✔

CHAPTER 14

Recording, Reporting, and Posting Requirements

The twin objectives to an indoor air quality program are: (1) to have clean air and (2) to document the fact that your plant has met clean air requirements. This chapter looks at written reports that document your clean air program and its results. Some of the reports are legal necessities, such as those required by OSHA; some are required by management, and some provide technical information helpful to facilities engineers.

OSHA REQUIREMENTS

In order to give OSHA backing for its enforcement functions, Congress put upon employers the responsibility for maintaining certain records to document their efforts at compliance. This section gives you first the official words of Congress and, second, the guidance provided by OSHA regarding indoor air quality.

Congressional Law

On December 29, 1970, Congress passed Public Law 91-596 and cited it as *Occupational Safety and Health Act of 1970*. The following is taken from Section 8. It is worth examining because it gives the legal wording which, while all-encompassing, never intended to give details. This same Act created OSHA and gave them authority and responsibility for creating definitive requirements that employers must follow. In business terms it is the equivalent of a policy directive by top management—to be acted on by middle management and recast as definitive directions to those reporting to the middle managers.

> (c) (1) Each employer shall make, keep and preserve, and make available to the Secretary or the Secretary of Health, Education, and Welfare, such records regarding his activities relating to this Act as the Secretary, in cooperation with the Secretary of Health, Education, and Welfare, may prescribe by regulation as necessary or appropriate for the enforcement of this Act or for developing information regarding the causes and prevention of occupational accidents and illnesses. In order to carry out the provisions of this paragraph such regulations may include provisions requiring employers to conduct periodic inspections. The Secretary shall also issue regulations requiring that employers, through posting of notices or other appropriate means, keep their employees informed of their protections and obligations under this Act, including the provisions of applicable standards.
>
> (2) The Secretary, in cooperation with the Secretary of Health, Education, and Welfare, shall prescribe regulations requiring employers to maintain accurate records of, and to make periodic reports on, work-related deaths, injuries and illnesses other than minor injuries requiring only first aid treatment and which do not involve medical treatment, loss of consciousness, restriction of work or motion, or transfer to another job.
>
> (3) The Secretary, in cooperation with the Secretary of Health, Education, and Welfare, shall issue regulations requiring employers to maintain accurate records of employee exposures to potentially toxic materials or harmful physical agents which are required to be monitored or measured . . . Such regulations shall provide employees or their representatives with an opportunity to observe such monitoring or measuring, and to have access to the records thereof. Such regulations shall also make appropriate provision for each employee or former employee to have access to such records as will indicate his own exposure to toxic materials or harmful physical agents. Each employer shall promptly notify any employee who has been or is being exposed to toxic materials or harmful physical agents in concentrations or at levels which exceed those prescribed by an applicable occupational safety and health standard . . . and shall inform any employee who is being thus exposed of the corrective action being taken.

Details from OSHA

The following is selected from *All About OSHA*, a booklet that OSHA published in "semi-non-legalese" to help business managers understand and comply with the requirements. Its language is formal enough to cover all situations and exceptions.

> Employers of 11 or more employees must maintain records of occupational injuries and illnesses as they occur. Employers with 10 or fewer employees are exempt from keeping such records unless they are selected by the Bureau of Labor Statistics (BLS) to participate in the Annual Survey of Occupational Injuries and Illnesses. The purposes of keeping records are to permit BLS survey material to be compiled, to help define high-hazard industries, and to inform employees of the status of their employer's record. Employers in state plan states are required to keep the same records as employers in other states.
>
> OSHA recordkeeping is not required for all employers in retail trade, finance, insurance, real estate, and service industries—Standard Industrial Classification (SIC) 52–89 (except building materials and garden supplies, SIC 52; general merchandise and food stores, SIC 53 and 54; hotels and other lodging places, SIC 70; repair services, SIC 75 and 76; amusement and recreation services, SIC 79; and health services, SIC 80). A few regularly exempt employers will have to maintain records if they are selected to participate in the Annual Survey of Occupational Injuries and Illnesses. They will be notified in advance and supplied with the necessary forms and instructions. Exempt employers, like nonexempt employers, must comply with OSHA standards, display the OSHA poster, and report to OSHA within 48 hours any accident that results in one or more fatalities or the hospitalization of five or more employees.
>
> Employers must keep injury and illness records for each establishment. An establishment is defined as a "single physical location where business is conducted or where services are performed." An employer whose employees work in dispersed locations must keep records at the place where employees report for work. In some situations, employees do not report to work at the same place each day. In that case, records must be kept at the place from which they are paid or at the base from which they operate.
>
> Recordkeeping forms are maintained on a calendar year basis. They are not sent to OSHA or any other agency. They must be maintained for five years at the establishment and must be available for inspection by representatives of OSHA, HHS, BLS, or the designated state agency. Only two forms are needed for recordkeeping.
>
> **OSHA No. 200, Log and Summary of Occupational Injuries and Illnesses.** Each recordable occupational injury and illness must be logged on this form within six working days from the time the employer learns of it. If the log is

prepared at a central location by automatic data processing equipment, a copy current to within 45 calendar days must be present at all times in the establishment. A substitute for the OSHA No. 200 is acceptable if it is as detailed, easily readable and understandable as the OSHA No. 200.

OSHA No. 101, Supplementary Record of Occupational Injuries and Illnesses. The form OSHA No. 101 contains much more detail about each injury or illness. It also must be completed within six working days from the time the employer learns of the work-related injury or illness. A substitute for the OSHA No. 101 (such as insurance or workers' compensation forms) may be used if it contains all required information.

Annual Survey. Employers selected to participate in the annual statistical survey receive in the mail, soon after the close of the year, form OSHA No. 200S for this purpose. Each employer selected must complete this report, using form No. 200 as the source of information, and return it to BLS. Small business employers (employers with 10 or fewer employees) selected for the survey are notified at the beginning of the year and are supplied with a form OSHA No. 200.

Many specific OSHA standards have additional recordkeeping and reporting requirements. [This book cannot include those specific requirements because of the variety of standards that might apply to any individual company.]

A copy of the totals and information following the fold line of the last page of OSHA No. 200 for the year must be posted at each establishment wherever notices to employees are customarily posted. This copy must be posted no later than February 1, and kept in place until March 1. Even though there were no injuries or illnesses during the year, zero must be entered on the totals line, and the form posted.

Employers wishing to set up a recordkeeping system different from the one required by OSHA regulations may apply for a recordkeeping variance. Petitions for recordkeeping variances must detail and justify the employer's intended procedures and must be submitted to the regional commissioner of BLS for the area in which the workplace is located. Similarly, in state plan states, only BLS can grant a variance from recordkeeping requirements.

As with applications for variances from standards, an employer filing for a recordkeeping variance must give a copy of the application to the employees' authorized representative. The employer also must post a summary of the application wherever notices are normally posted. Employees have 10 working days to submit to BLS their own written data, views, or arguments.

Keeping Employees Posted

Employers are responsible for keeping employees informed about OSHA and about the various safety and health matters that might affect them. Federal OSHA and states with their own occupational safety and health programs require that each employer post certain materials at a prominent location in the workplace. These include:

- Job Safety and Health Protection workplace poster (OSHA 2203 or state equivalent) informing employees of their rights and responsibilities under the act. Besides displaying the workplace poster, the employer must make available to employees, upon request, copies of the act and copies of relevant OSHA rules and regulations. Any official edition of the poster is acceptable.
- Summaries of petitions for variances from standards or recordkeeping procedures.
- Copies of all OSHA citations for violations of standards. These must remain posted at or near the location of alleged violations for three days, or until the violations are corrected, whichever is longer.
- Log and Summary of Occupational Injuries and Illnesses (OSHA No. 200). The summary page of the log must be posted no later than February 1, and must remain in place until March 1.

All employees have the right to examine any records kept by their employers regarding exposure to hazardous materials, or the results of medical surveillance.

Occasionally, OSHA standards or NIOSH research activities will require an employer to measure and record employee exposure to potentially harmful substances. Employees have the right (in person or through their authorized representative) to be present during the measuring as well as to examine records of the results. Under these substance-specific requirements, each employee or former employee has the right to see his or her examination records, and must be told by the employer if exposure has exceeded the levels set by standards. The employee must also be told what corrective measures are being taken.

In addition to having access to records, employees in manufacturing facilities must be provided information about all of the hazardous chemicals in their work areas. Employers are to provide this information by means of labels on containers, material safety data sheets, and training programs.

USING COMPUTERS TO PREPARE REPORTS

Many commercial sources provide computer software and other items that will simplify the less-than-pleasant task of preparing reports. Some equipment connects to measuring instruments and places their outputs directly into formatted reports. Computer software usually asks on-screen questions; you fill in values and other answers and the computer prints the report.

Air Quality Monitor and Summary Report Writer[1]

Several air quality monitors provide you with a range of sensitivities, accuracies, types of gases sensed, and other variables. This section describes one instrument that shows results on LCD readouts, sends results to a printer for a summary report, and stores measurements in its memory. This report is not intended to satisfy official requirements; rather it provides information useful for analysts, management, and company records.

This instrument monitors and records carbon dioxide, temperature, relative humidity, and toxic gas. Plug-in sensors are available for monitoring CO, H_2S, SO_2, Cl_2, NO_2, NO, HCN, NH_3, O_2, O_3, ETO, H_2, and HCl.

For unattended operation, you can set programmed sampling times. Each time it takes a sample, an internal positive displacement pump draws air through the temperature and relative humidity sensors. Air is then drawn through a nondispersive, infrared CO_2 sensor, and the values are displayed on the readout. These measurements are automatically stored in the datalogging memory, which can store 16,000 samples in the standard memory. Table 14-1 shows memory capacity in days as a function of sampling interval and memory size, assuming monitoring of four parameters.

TABLE 14-1
Memory Capacity in Days

Sampling Interval, Minutes	*16K Standard Memory*	*32K Optional Memory*	*64K Optional Memory*	*128K Optional Memory*
1	2.8	5.5	11.1	22.2
5	13.9	27.8	55.5	111.1
15	41.6	83.3	166.6	333.3
60	166.6	333.3	666.6	1333.3

1. Courtesy of Metrosonics Inc., General Products Division, Rochester, New York.

There are several reports that can be displayed, printed, or saved to disk. The Summary Report identifies, for each parameter, average value, minimum value, maximum value, and date and time. A Time History Report details actual values of each parameter for each averaging interval, including actual time, over the entire test. Data is automatically formatted for transfer to common database programs, such as Quattro Pro, for later analysis. Figure 14-1 shows a typical Summary Report.

The comment and numeric code fields allow an operator to specify various types of information such as size of area, number of occupants, other pollutants noted, probe height, and other items.

Automatic Indoor Air Quality Report[2]

With OSHA instituting and proposing new regulations, the need for documentation will constantly increase and become more complex. As expected, suppliers respond by releasing new versions of software that simplify the task considerably. This section summarizes the features of one such product; there are many others on the market, and new ones will be released.

- *An indoor air quality report can be generated automatically.* This feature utilizes any combination of CO, CO_2, %RH, airflow, and temperature to produce a comprehensive IAQ report. You will be asked to enter the name of the engineer or hygienist, name and address of location, dates, and other background information. The software will then use stored data to create reports that include graphs, tables, and text from the ASHRAE 62-1989 ventilation standard. You can edit any part of the report to tailor it to specific requirements.
- *Graphs can be exported to other databases and word processors.* You can export graphics in the PCX format into dBase, QuattroPro, WordPerfect, and other programs that will perform additional analysis, data processing, and documentation.
- *A printer driver allows numerous options for use with color laser printers, bubblejets, and other printers.*

As mentioned, there are several suppliers who offer software that will simplify the task of documenting test results and preparing reports. Following are suggested questions to ask before selecting the one that is best for your application.

---

2. Courtesy of Solomat, A Neotronics Company, Norwalk, Connecticut.

FIGURE 14-1
Typical Summary Report

```
REPORT PRINTED 06/20/97 AT 10:55:21
SAMPLE RATE: 15 minutes
ALARM LOGGING STATUS: Off

  TEST LOCATION:  Manufacturing Control
  OPERATOR NAME:  Smith, Edward
OPERATOR NUMBER:  012-34-5678
     DEPARTMENT:  Plant Engineering
COMMENT FIELD 1:  _______
COMMENT FIELD 2:  _______
  NUMERIC CODES:  _______  _______  _______  _______  _______

              OVERALL STATISTICS FOR TEST NUMBER 1

TEST STARTED 6/16/97 AT 17:06:16
TEST DURATION: 2 DAYS 15:44:40

CARBON DIOXIDE:
     ALARMS:     LOWER:        0 ppm        UPPER: 5000 ppm

    MINIMUM:    420 ppm      OCCURRED    6/18/97 AT 15:52:15
    AVERAGE:    490 ppm
    MAXIMUM:    696 ppm      OCCURRED    6/19/97 AT 08:07:27

TEMPERATURE:
     ALARMS:     LOWER:       32.0 F        UPPER: 140.0 F

    MINIMUM:     68.7 F      OCCURRED    6/18/97 AT 06:52:19
    AVERAGE:     71.8 F
    MAXIMUM:     76.7 F      OCCURRED    6/16/97 AT 17:07:47

RELATIVE HUMIDITY:
     ALARMS:     LOWER:        0.0%         UPPER: 100.0%

    MINIMUM:     45.7 %      OCCURRED    6/16/97 AT 20:22:14
               AVERAGE:       51.5 %
    MAXIMUM:     57.8 %      OCCURRED    6/19/97 AT 07:52:36
```

Is it compatible with the computer(s) on which you will use it? Prepare a list of basic facts about your computer(s), such as type (PC or Macintosh), DOS version, Windows version (if you will run under windows), amount of RAM (volatile memory), space available on hard drive, and size(s) of floppy disk drive(s).

Will installation on your computer be simple and straightforward? How long will it take, and are any special skills required?

Is it designed for user-friendly operation? Are on-screen questions self-explanatory? Does a single use proceed straight through or does it require zig-zagging back and forth? Does data have to be entered repeatedly in different parts of the program? Are menu selections clearly written?

Are complete manuals included? Have manuals, which you can have open on your desk, been replaced by on-screen help? On-screen help might be more convenient for short answers, but when help consists of details or a multiple-step procedure, you have to write it all down; having it already printed in a manual would be more convenient.

Does the software "know" OSHA limits and requirements? If not, does it allow you to enter RELs, PELs, and other limits?

Is it upgradable? You can expect the rules to change as regulators learn more about pollutants, so find out what the software supplier will do about keeping you up to date.

Can you annotate the report? The value of a set of data can be greatly enhanced by adding comments along with the regular notations. You might find software that allows you to choose whether the comments appear on the screen only, in the printed report only, or both.

How can errors be corrected? When entering the company name and other basic information, some software allows you to go back at any time and make changes such as correcting spelling errors. Some allow you to go back only a certain amount, such as one line, and some require you to abandon the inputs and start over.

What security is provided? If data security is a problem, does the software provide for passwords and other measures that keep unauthorized people from using it?

What choices do you have in formatting a report? Some possibilities include: rearranging the order of items; controlling the number of decimal places;

placing commas in large numbers; highlighting out-of-range figures; and rounding numbers. It can be convenient to be able to print more than one kind of report—for management, for engineering, for a customer, for an inspector, etc.—from the same set of data.

PART IV

MAINTAINING INDOOR AIR QUALITY

CHAPTER 15

Describing Your Overall Air Quality Position

You might be planning an air quality program because there have been worker complaints, because you hear OSHA inspectors' footsteps getting closer, or because it's right to provide clean air for your employees. Regardless of the reason, a good place to start is with an overall assessment of the building's air quality position. This chapter suggests ways to establish a building IAQ profile. Preparing to write the profile forces you to gather information that should always be at hand. From there, the sequence branches according to whether the information has uncovered an air quality problem. If it has not, you are ready to write the building IAQ profile; if it has, it is best to pause and clean up the problem before finalizing the profile. The suggestions include cycling through the plan, with each cycle going into a higher level of detail, as demonstrated by the layout of the following section.

STARTING A PLANT IAQ PROFILE

A good way to describe your building's air quality situation is to develop a building IAQ profile to document everything that could relate to or affect indoor air quality, including the structure, function, and occupancy of the building, as well as weather and other outdoor conditions. If your plant consists of more than one building, it would be appropriate to prepare a profile for each one. Material discussed in this chapter applies to any building, whether it is the only building or one of several that make up the plant.

A building IAQ profile will give you baseline information for future comparisons and alerts. It will help you identify potential problem areas and give you a way to assign priorities for budgeting maintenance and modification projects. It can be one of several profiles that management uses for assembling a valuable overall profile that includes lighting, security, and other systems. Kept up to date, this overall profile can be one of the most important and useful documents in the company's files.

Be sure to budget sufficient time and resources to make a complete IAQ building profile. If the budget forces you to abandon the project before completion, all work done is a waste; a partial profile is of negligible value. When the need for a building profile is not urgent, some building managers have assigned the tasks as fill-in for engineers between projects. This method takes a little longer, and total worker-hours are higher because there has to be some refresher time to get oriented before starting each segment. In addition, the same workers might not be able to follow through from beginning to final report.

Seed Questions

The following are seed questions because they will grow and bring forth other questions, which will lead to still another generation of questions, and so on. In fact, the list shows that questions are already growing out of each seed. One danger you must avoid is forcing all questions that your profile answers to be descendants of questions in this list; because each plant is different, it is important for you to be free to create new questions or families of questions.

- When the building was conceived and built, how was it intended to be used? Find out what was to go into the building (how many people, what kinds of materials, what equipment) and what was to come out (tangible and/or intangible products).

- Is the building presently being used as intended? If not, was it ever so used? Obtain a history of the building, and determine how its usage evolved.
- What structural and other major changes have been made? The HVAC system is always a major concern in IAQ questions; has it been modified, expanded, or changed in any way since it was originally installed? Whenever it was changed, was it reset and retested to reflect new demands? Was it ever (or always) completely satisfactory?
- Was the building commissioned? If it was, compare the original information to current conditions. Is there a commissioning file that was kept up to date?
- Assuming that the building currently has no air quality problem, what changes may be needed to prevent future problems? To answer this question when there are current problems, make a list of what conditions will be changed by solving each problem. What probable and what possible changes in occupancy and usage do you foresee, to prevent future IAQ problems?

Measurements

Although one purpose of creating the IAQ building profile is to determine what information you need and what measurements you should take, it is usually necessary to include some basic measurements in the profile report. A good set of basic measurements provides information that you cannot deduce from plans and specifications. The profile's value will be enhanced by inclusion of such measurements as:

- *Airflow.* Take enough measurements to determine overall airflow patterns. Areas with no air movement are especially important. Note both vertical and horizontal airflow patterns, and record outside weather conditions, especially wind, along with each measurement.
- *Temperature.* Try for enough data to reveal two types of variations: time and spatial. Record other variables, such as the fact that all room occupants were at a meeting while certain readings were being taken. Outside weather conditions, especially temperature, provide important data for interpreting the indoor measurements.
- *Relative humidity.* Time variations are most meaningful, especially if they correlate with other measurements such as temperature and outside weather conditions.
- *Carbon dioxide.* As mentioned in previous chapters, carbon dioxide measurements are a good indicator of other conditions and problems. When

taking these measurements, be sure to note oxygen-consumers such as number of workers, level of worker activity, and starting or stopping of equipment that uses oxygen.

- *Pressure differentials*. Concentrate on obtaining some measurements in each zone of the air handling system. One of the main purposes of recording pressure differentials is to help explain some of the airflow patterns you measured—air flows from areas of relatively high pressure to areas of relatively low pressure.

Few buildings have been adequately commissioned, so the system may never have delivered the airflows shown on the design drawings. Should future IAQ complaints develop into litigation, the value of a thorough IAQ profile enhanced by actual dated measurements will be priceless.

Personnel Necessary for IAQ Profile

With the exception of a few simple requirements of the IAQ profile, such as taking physical measurements of room and window areas, you will need personnel with a variety of technical skills. The following list suggests the skills needed for most buildings; you might add to the list according to the special needs of your particular building.

- Basic understanding of HVAC systems, and your system in particular.
- Ability to read architectural and mechanical plans and to understand manufacturer's catalog data and owner's manuals.
- Ability to identify, and preferably be familiar with, office equipment.
- Ability to identify, and preferably be familiar with, factory and other equipment in the building.
- Ability to work cooperatively with building occupants and gather information about space usage.
- Ability to collect information about HVAC and other air handling systems, evaluate equipment condition, and determine maintenance schedules.
- Ability to understand the practical meaning of information contained in the Material Safety Data Sheets (MSDSs).
- Ability to take the measurements listed in the preceding section.

An Overall Procedure

Following is a recommended sequence for developing an IAQ building profile. There are three main steps, then an evaluation of whether those steps revealed an

IAQ problem. If there is a problem, place profile development on hold while resolving the problem. The final step is to produce the IAQ management plan, and possibly an accompanying report. As with other recommendations, you should adjust these to optimize them for the special conditions of your plant.

1. Collect and review pertinent records. This step is more dependent on the special conditions in your plant than the other two steps are. In addition, this step will reveal problems, such as owner's manuals that cannot be located, which will cause extra work in digging out the needed information. Subparts of this step include:

- Review design, construction, and operating documents.
- Determine if the building was commissioned; obtain and study the documentation.
- Check HVAC maintenance and operation records against equipment lists.
- Review complaint record.

You will have to start files to organize the information that will go into the IAQ profile. Typical outputs from the first step include:

- Description of the HVAC system design and operation; set of operating instructions; manuals.
- Sets of maintenance and calibration records for all building systems that have an effect on air quality.
- History of major changes in occupancy, room division, or structure.
- Information about complaint locations and history.

2. Conduct a walk-through inspection. While not a white-glove inspection, this step should be thorough. It could take several days. Information gathered in this step is a major determinant of how valuable and useful the IAQ profile will eventually be. Subparts of this step include:

- Visually identify areas and rooms shown in building drawings.
- Verify duct locations, sizes, outlets, returns, and other components of the air supply system.
- Observe general attitude of occupants through clues such as open windows while the HVAC system is trying to heat or cool the work area, material placed where it will disrupt intended air flow, smoking where not permitted, or dangerous material handling rules being ignored. This information will tell you how much emphasis has to be placed on educating workers.

- Talk with workers in all levels of the organization.
- Look for indicators of IAQ problems. As described in the preliminary walk-through in Chapter 12, some activities reveal that the occupants sense inadequate ventilation or poorly controlled temperature.

Step 2 should have provided file material that is more specific and detailed than that from the first step. This material will likely include:

- Red-lined drawings or sketches showing windows, ducts, openings, and other features that differ from drawings on file.
- Identification of areas of relative high and low pressure, with indications of where those pressures are correct and where they should be adjusted.
- List of responsible staff and/or contractors, with training and qualification descriptions.
- Lists of locations where occupant activities indicate air quality dissatisfaction.
- Lists of other air quality problem indicators.
- Recommendations for monitoring or further checks.

3. Collect detailed information. Information collected here is generally the highest level of detail before writing the profile and report. There will be exceptions. For example, it may appear that you have all the information you need about a certain pollutant pathway, but later find that one or more minor pathways feed into that one, and you will have to obtain more details. The following list is typical of the subjects about which you should gather detailed information in this step:

- HVAC system condition and operation.
- Pollutant pathways.
- Pollutant sources. This subject implies also a list of pollutants, and is a likely candidate for further details.
- Occupants, human and machine, along with locations, functions, inputs and outputs, vulnerability to pollution, history, future, and other facts.

Upon completion of Step 3, you should have the information necessary for making a decision as to whether there is an indoor air quality problem. The following far-from-complete list is included to indicate the type of information you should have:

- Inventory of HVAC system components needing upgrading, repair, adjustment, or replacement.
- Record of control settings, how they were established, who has access to them, and their effects.
- Record of operating schedules.
- Plan showing actual airflow patterns and pressure differentials.
- Information about pollutants, including types, characteristics, sources, and where they travel.
- Set of Material Safety Data Sheets for supplies and hazardous substances that are stored or used in the building.
- Zone/room record.

If, at this time, you identify a problem with the plant's indoor air quality, it is best to postpone further work on the profile until you have acceptable air quality as a base for the report. Continuing to work on the profile while fixing the problem will probably result in duplicate work because some of the information will have to be reassembled after the problem is fixed.

As mentioned at the beginning of this chapter, we now look at the three steps again, going into further detail with each one.

REEXAMINING THE INFORMATION COLLECTION PROCESS

In some ways the information needed for an IAQ profile is similar to that which is collected when solving indoor air quality problems. However, the profile includes the entire building rather than focusing on areas that have a problem or may have caused one. In addition to being a helpful document when responding to future IAQ complaints, the profile's organized body of records will be valuable in such other situations as planning for renovations, and negotiating leases and contracts.

There are three steps involved in collecting inforamtion: collecting and reviewing existing records, the walk-through, and collecting detailed information. It is recommended that budgets and schedules allow your staff to work through at least the first two steps without interruption. The third step is more tolerant of gaps in the schedule, although you must consider that the longer the gap, the greater the possibility that some information could become out of date.

Step 1. More About Collecting and Reviewing Existing Records

You may choose to perform the two versions of the first step just they way they appear in this chapter, or you may prefer going from basic collection right into the detail described here, without recognizing them as distinct phases of the task.

Review construction and operating documents. Collect any available documents that describe the building's construction and operation. Suggestions include: architectural and mechanical plans, specifications, submittals, sheet metal drawings, commissioning reports, adjusting and balancing reports, inspection reports, and operating manuals. Any missing documents will significantly add to the difficulty of your task. For example, if there are no commissioning reports or balancing reports, you have to consider actual ventilation quantities to be unknown; they can be quite different from those indicated on mechanical design drawings. If there are no operating or maintenance manuals for HVAC equipment, you must spend extra time trying to obtain or create new ones, rather than let preventive maintenance programs run without clear direction.

Study the original architectural and mechanical design so that you understand the building's layout and intended functions. Identify and note locations in which changes in equipment or area usage create a potential for IAQ problems. When planning a monitoring program, expect to give extra concern to those locations. Following are typical items of interest, and the questions they could suggest:

- *Commissioning reports*. Was the building properly commissioned before approval for occupancy? Did the commissioning include testing and balancing of the HVAC system?
- *Operating manuals for all building equipment*. If you located or obtained operating manuals, do they provide all the information you need? Do personnel involved completely understand how the equipment is supposed to operate? Are they able to follow a complete preventive maintenance program?
- *Remodeled areas*. When there has been an increase in occupied area, walls added, partitions rearranged, outside walls modified, or other architectural modifications, were HVAC and other systems that affect IAQ reengineered, tested, and balanced? If modifications placed additional load on some equipment, was the preventive maintenance schedule adjusted accordingly?
- *Replaced or rebuilt HVAC equipment*. If part of the original equipment was replaced or substantially rebuilt, does the system still have sufficient capacity? Has new equipment been properly installed and tested? Was the spare parts list updated?

- *Changes in occupancy*. Warehouses have been converted into factories. Sales floors have been remodeled into offices. Rooms that used to have just drafting tables might now be filled with copiers, printers, and other high speed paper handlers that spew dust into the air. Is there documentation to show that consideration was given to ventilation and other air quality provisions whenever changes were made?

Check HVAC current maintenance records against original documents. Compare the latest maintenance and calibration records to construction documents such as equipment lists and mechanical plans. Flag any components that may not be receiving regular attention. Frequently overlooked are components that have been installed in inaccessible or hard-to-reach locations. Modifications are sometimes added in such a way that they have to be removed to allow access to maintenance points. Check especially for items such as filter boxes and small capacity exhaust fans.

Review records of complaints. If there have been IAQ complaints, and you were fortunate enough to have retrieved the complaint file, you have a valuable resource to study. From its contents, you can learn about patterns of pollution, fixes that worked, fixes that didn't work, weaknesses in the system, and other information.

Step 2. More About the Walk-Through

Keeping in mind that the walk-through's purpose is to acquire an overall picture of occupant activities and building functions, and to look for IAQ problem indicators, you have to keep it mostly free-wheeling. Detailed planning would inhibit your ability to follow leads that turn up along the way. If you know of certain places you should not miss, you can prepare a checklist. You should take along a sketch of each floor, showing major features and reference points for location measurements.

The walk-through is not the time for detailed measurements of air quality parameters; do not purchase measuring equipment for this step. If the company already has some instruments, a few direct readings, such as carbon dioxide, can confirm suspicions about under-ventilation in certain areas. You might use chemical smoke in a few places to observe major airflow patterns that you can show on the floor sketch.

Make note of odors, especially in inappropriate places, such as soldering fumes in an office area or tobacco smoke in a nonsmoking area. As you move from one area to another, pay attention to changes in the atmosphere; certain areas may be drafty, stuffy, humid, hot, cold, etc. In some buildings there is even a haze that can be seen across large areas.

Talk with occupants. You will not learn much by standing in a corner with a clipboard. The subject of communication, described in Chapter 10, applies here—starting with telling occupants ahead of time who you are and why you will be walking through their areas. They are the ones who are in the area all day every day and, once they know you are not checking up on them, can be excellent sources of information. You might learn some history, such as how clean the air used to be before a certain machine was installed. Try to determine if conditions you are observing are typical and if all days are alike. For example, the housekeeping crew of one plant sprayed for pest control once a month, and the odor dominated the atmosphere for a couple of days each time.

Be observant for indications of IAQ problems. There are signs you can learn to recognize. Although the walk-through is not the time for detailed follow-up, you should make note of what, where, and when—written, not mental, notes. When the indicator is a stain or other mark, make a special note to have it cleaned after the problem is fixed, so that a repeat of the problem will be noticeable. The following list offers suggestions for items to keep in mind, but no list can include all the indications that could be possible in every plant:

- Fungal growth or moldy odors.
- Other signs of moisture damage, including stains by windows, at columns, on ceilings, and on floors.
- Inappropriate odors.
- Dirty or unsanitary conditions (e.g., dirt, sticky surfaces, or anything that could be nutrients to pests).
- Unsanitary conditions in equipment such as drain pans and cooling towers.
- Leaks in air duct joints—might appear as dust streaks—often indicating excessive back pressure due to clogged filters.
- Smoke marks. If a fire, or even a smoldering situation, involved electrical equipment, make a note to check on whether polychlorinated biphenyls (PCBs) may have been released.
- Presence of hazardous substances.
- Potential for soil gas entry, such as unsealed openings to earth, or wet earth odors.
- Dry drain traps.

- Overcrowding. If any room or area seems to have had an increase in occupants recently, obtain a count. Later you can compare the number to the ventilation system specifications to see if the system was designed to supply that many people.
- Ceiling plenums. Lift a ceiling tile and examine the plenum for potential problems. If this is the building's first inspection, you may find construction debris and damaged or loose material. Make a note to have the plenum cleaned, especially if there is material that could catch dust, particles, and fibers; such build-ups could be released into working areas. You also might find poorly designed full-length partitions that extend through the plenum to the floor above. They can interfere with air movement unless they have been provided with transfer grilles. If fire dampers have been installed to allow air circulation through walls or partitions, confirm that they are open.
- Air intakes. Note any ventilation intakes that are located where they might pick up pollutants and circulate them in the building. The concern is not only pollutants from other sources, but from inside the building as well. For example, air intakes might have been located where they can pick up exhausts from plumbing vents.
- Misused mechanical rooms. Check the place where the HVAC system is housed to see if it is clean and dry. Sometimes the urge to have the HVAC room double as a storage area is irresistible, and you might find cleaning supplies and other items stored there. Unsanitary conditions in the mechanical room are a problem, especially if unducted return air is dumped into and circulated through the room.

Step 3. More on Collecting Detailed Information

If this part of the project is to be completed as time becomes available, be sure to assign high priorities to areas that have been identified as potential IAQ problem areas. By this time you have given the program some direction, and you have a good idea of what to expect as you go along. Checklists were recommended for use during the preliminary walk-through, but this time, you should take forms that ask you to fill in specific types of information. The forms described in the next paragraphs are included in Appendix F.

- *HVAC system*. Supplement your current maintenance records with Figure F-9, HVAC Checklist, Appendix F. For each large air handling unit, make a separate copy of the form.

- *Pollutant sources*. Use the *Pollutant and Source Inventory* form, Figure F-10, Appendix F, to record potential pollutant sources in the building. As you fill out the form, note the locations of major sources, and add them to the floor plan.
- *Building occupancy information*. You can use the *Zone/Room Record* form, Figure F-3, Appendix F, to maintain an up-to-date record of the way each area of the building is used, its source of outdoor air, and whether it is equipped with local exhaust. If you suspect inadequate ventilation, the form will help you estimate ventilation rates in cubic feet per minute per person or per square foot of floor area. You can compare these numbers to guidelines such as design documents, applicable building codes, or the recommendations of ASHRAE 62-1989.

COMPLETING THE PROFILE

With all the information at hand, assembling and writing are the nontechnical tasks remaining. Decide whether you will use a full explanation or an outline format. With either one you can select a system for identifying paragraphs and sections. A highly structured system uses increasing decimal notation for subordinate paragraphs or headings (1, 2, 3, etc. identify major headings; 1.1, 2.1, and 3.1 are the headings under them; 1.1.1, 2.1.1, and 3.1.1 identify the next level of headings, etc.). Some prefer Roman numerals, and others prefer a mixture of capital and lower case letters along with numerals.

Existence of the finished profile should be publicized. The policy on distributing copies is up to management, where the cost of distributing will be a major factor.

In few plants will the profile remain valid over the years. Responsibility for keeping it up to date should be a clear assignment. Small changes can be maintained in a red-line copy, and then the entire profile can be re-issued when there are either several small changes or a major change. The main point to make about updating is that the profile should not remain unchanged until it is needed again, because everything in it would then have to be checked.

CHAPTER 16

Moving Toward Acceptable Indoor Air Quality

Although there are hundreds of approaches for controlling indoor air quality, they are all variations or combinations of four basic strategies that this chapter describes. The next major section looks at a management plan, and the chapter concludes with comments about using outside consultants.

IDENTIFYING BASIC IAQ STRATEGIES

As you plan your indoor air quality program, it helps to identify where your methods place along the range of possible strategies. Opposite ends of the continuum of strategies are broadly described as *source control* and *target control*. The first involves various approaches that range from not producing the pollutant to keeping it contained; the other extreme involves approaches that keep the target from being affected by the pollutants around it. This section looks at these two different strategies as well as those between them.

Stop Producing the Pollutant

Instead of ventilating to eliminate the pollutant, filtering it out, or protecting workers from breathing it, this strategy just stops it from coming into existence in the plant. For example, some manufacturing processes, especially in micro-electronics technology, are based on dangerous chemicals such as gallium arsenide which require careful handling. The process is highly efficient, but some manufacturers have chosen to use other methods to avoid the high degree of care it requires, and the concern over even the tiny possibility of an accident. This approach keeps the pollutant from entering the plant.

Another example of stopping the pollutant from coming into existence is prohibiting smoking in the plant. This approach keeps the pollutant from being created in the plant. Eliminating the very existence of a pollutant offers many advantages, which include:

- It is not necessary to establish an acceptable concentration.
- Even those who are susceptible to concentrations below the acceptable level or who have a special allergy will not be bothered by zero concentration.
- Eliminates the need for ventilation to remove the pollutant.
- Even when total ventilation is sufficient, there is the possibility of pockets and areas that receive inadequate air refreshment.
- When the pollutant is not present at all, there is no risk of it accidentally getting into workers' breathing air.

Not every indoor air quality problem can or should be solved by this method, and it sometimes is not a complete solution. A few of the reasons why a plant manager would not choose to solve an air quality problem by eliminating the presence of the pollutant are:

- Manufacturing processes that do not use the chemical are costly.
- There are no alternative processes. This might especially apply to cleaning fluids, disinfectants, and products that end in *-icide*.
- By their nature, certain actions always affect the indoor air. Grinding, sandblasting, and even breathing produce dust, particles, and carbon dioxide.

Contain the Pollutant

The strategy just described might be thought of as a no-compromise approach; all others involve some degree of compromise. Another approach within the category of source control might be called *containment*. Here the pollutant is created or brought in, but is kept from leaving a certain area and polluting the general breathing air of the building. This is a more popular approach because it allows management to use efficient processes as well as processes for which there are no alternatives. A smoking room is an example of controlling pollution by containment.

General components of this type of arrangement are a closed area, clean air entrance, and exhaust air outlet. A closed area can take several forms, such as:

- An actual room with hard walls and controlled ingress and egress. Workers inside might wear special protective clothing and/or breathing apparatus.
- An enclosure, usually of a few cubic feet, with a door that workers can open to place chemicals and other objects inside. An exhaust removes fumes and other pollutants. The door may not seal tightly, because the exhaust creates a negative pressure inside; any exchange of air will be from (never into) the room, into the enclosure, and out the exhaust.
- An enclosed space that a worker accesses via built-in gloves. Such work stations are typically used for small sand-blasting operations.
- A physically open area with a hood or exhaust tube close to the work It is often used for soldering or small welding operations, with the smoke being drawn away without entering the general room atmosphere.

Special care must be taken to ensure that the outlet for contaminated air is not located in a way that its air can be picked up by a fresh air intake.

Dilute the Pollutant

With this common strategy a pollutant that is not highly toxic is not completely drawn away from its source. By one or more methods it is diluted to an acceptable concentration and then is circulated in the general atmosphere. Methods of dilution include:

- Mixing with fresh or clean air, ventilation
- Filtering
- Impaction
- Diffusion

One of the advantages of using this strategy is that the HVAC system can perform the entire dilution process by drawing air from the work areas, filtering it by one of several methods, mixing it with fresh air, and distributing it back through the work areas. At the same time, the system can adjust and treat the air's temperature, humidity, and other measures of air quality.

One disadvantage with this strategy is that the acceptable pollutant concentration must be defined. Once operational, the system usually operates open-loop—that is, the ordinary HVAC system does not automatically provide more filtering when a sensor detects too high a concentration of a pollutant. It does not remove and clean or replace filters when it senses that the filter needs it. Some systems measure pressure on each side of a filter bank and, by determining differential pressure, trigger an alarm when filters are clogged.

Protect the Target

At the other end of the continuum of strategies for controlling air quality is controlling the air where it is supplied to individuals for breathing. This method generally uses respirators, as described in Chapter 7, and personal protective equipment. Reasons for protecting workers in this way include the following:

- A confined space continuously emits pollutants, as when a storage tank has to be cleaned.
- The work is such that it must be performed by humans, in a dangerous atmosphere.
- An emergency situation has developed.

ESTABLISHING AN IAQ MANAGEMENT PLAN

Indoor air quality is too important to be handled by a see-what-comes-up-today method. A management plan forces you to list all the aspects of your IAQ program, integrate them, and prepare a schedule. Ideally, the plan anticipates whatever will come up, even emergencies, and gives a procedure for it. Once the plan is finished, it gives a free side benefit: the information needed for budgeting the IAQ program.

Several topics in this book have mentioned that the variety of conditions in industrial plants make it impossible for a book to give any detailed procedure that all readers can use. That applies at least equally when it comes to the IAQ management plan. This section makes suggestions, lists typical subjects to include, and gives reminders, but it cannot provide a management plan that would be right for your plant.

Background Information

Preparing the plan will go more smoothly if, before you start writing the plan, or even laying it out, you gather certain key information. This action will make you aware of facts, such as the number of departments involved and the number of possible sources of pollution. The following are notes you might make regarding subjects on which to collect background information, and suggestions for subjects to consider including. This list shows items that might slip your mind when writing the plan—not obvious items such as "perform preventive maintenance." At this point in the process, you might change the responsibilities of some individuals and groups as you realize for the first time exactly how they fit into the plan.

OPERATE AND MAINTAIN HVAC EQUIPMENT

- Keep controls in working order.
- Keep interior of equipment and ductwork clean and dry.

MONITOR ACTIVITIES OF STAFF, TENANTS, CONTRACTORS, AND OTHER BUILDING OCCUPANTS WHO MIGHT AFFECT INDOOR AIR QUALITY

- Smoking
- Housekeeping
- Building maintenance
- Shipping and receiving
- Pest control
- Food preparation and other special uses

MAINTAIN COMMUNICATION WITH OCCUPANTS; KEEP MANAGEMENT INFORMED

- Identify building management and staff with IAQ responsibilities.
- Use health and safety committees.

EDUCATE STAFF, OCCUPANTS, AND CONTRACTORS REGARDING THEIR RESPONSIBILITIES

- Training sessions
- Lease arrangements
- Contracts

IDENTIFY WAYS THAT PROJECTS, FUTURE AND PRESENT, COULD AFFECT IAQ

- New construction

- Redecorating, renovation, remodeling
- Relocation of personnel or functions within building

The Role of the IAQ Manager

Going back again to Management 101, recall that IAQ, like any other function, is better controlled when just one individual has overall responsibility and authority. In a small plant, managing indoor air quality might be just a part of the person's assignment, but the manager in a large plant would probably have a title such as IAQ Manager. Main qualifications for this position should include a good understanding of the building's structure and function, and the ability to communicate with the public, official regulators, facility personnel, occupants, building owners, and others. The following list suggests responsibilities that might be included in the manager's job description.

- Be the primary person responsible for IAQ issues in matters such as OSHA inspections and compliance.
- Review all projects to determine if there are IAQ effects.
- Review all contracts and participate in negotiations, to determine if indoor air quality will be affected.
- Develop the Indoor Air Quality Profile (see Chapter 15).
- Establish and control a system for communicating with appropriate people about IAQ issues.
- Oversee the adoption of new procedures and the operation of existing procedures.
- Coordinate staff efforts regarding indoor air quality.
- Support staff efforts by ensuring that manuals and other information are available, and that they have the authority to carry out their responsibilities.
- Periodically inspect the building, looking especially for situations that have the potential for developing into IAQ problems.
- Manage records of IAQ efforts.
- Respond to complaints, observations, and comments regarding indoor air quality.

Features of the IAQ Management Plan

As you analyze the background material and confer with managers, maintenance workers, and occupants—who can provide information and experience—special items that the management plan should include will surface. The rest of this section looks at typical items that might apply to your plant.

Equipment operating schedules. Confirm that the timing of occupied and unoccupied cycles of equipment, especially the HVAC system, is correct for actual occupied periods. Having the building's air flushed and clean at the start of each occupied period is of special importance. ASHRAE 62-1989 provides guidance on lead and lag times for HVAC equipment. In hot, humid climates, be sure that ventilation times take into consideration the possibility of mold growth during unoccupied periods.

Control of odors and contaminants. Control nonducted airflow by maintaining appropriate pressure relationships between building usage areas. Avoid recirculating air from areas that are sources of strong contaminants, such as smoking lounges and chemical storage areas. Provide adequate local exhaust for activities that produce odors, dust, or contaminants, or confine those activities to locations that are maintained under relative negative pressure. For example, loading docks are a frequent source of vehicle exhaust odors; keep surrounding rooms under positive pressure to prevent them from drawing exhaust fumes into the building. Ensure that paints, solvents, and other chemicals are stored and handled properly, with adequate (direct exhaust) ventilation provided. If local filter traps and adsorbents are used, provide for regular maintenance.

Ventilation quantities. Compare outdoor air quantities to the building design goal and local and state building codes—make adjustments as necessary. It is also informative to see how your ventilation rate compares to ASHRAE 62-1989, because that guideline was developed with the goal of preventing IAQ problems. If more ventilation is necessary, first check your HVAC equipment to determine if it has the capacity to condition the additional air.

HVAC equipment maintenance schedules. Follow the recommended maintenance schedule to ensure that the equipment is in good condition and is operating as designed. Components that get wet, such as drainage pans, coils, cooling towers, and humidifiers, require careful attention to prevent microbiological growth, which can result in microbiologicals or chemicals entering the indoor airstream.

Building maintenance schedules. When an upcoming maintenance activity will interfere with HVAC operation (e.g., painting, roofing operations), try to schedule it for times when the building is unoccupied. Inform occupants when such activities are scheduled and what they can expect (lingering innocuous odors, etc.) when they return. If possible, use local ventilation to ensure that dust and odors are confined to the work area.

Preventive maintenance management. Some preventive maintenance recommendations are based on calendar time, usage time, or other criteria. Other items require a visual inspection to determine if maintenance should be performed, but such equipment is too often installed without thought given to access for inspection. For example, changing ventilation filters on a time basis will sometimes result in throwing away filters with little use on them, and sometimes keeping filters in use that are clogged and preventing adequate ventilation. Difficult access might tempt maintenance personnel to "catch it next time." One solution is to install an inexpensive manometer to monitor the pressure loss across a filter bank, giving a constant indication of the filter's condition.

Housekeeping. As they work throughout a building, cleaning staff or contractors may be the first to recognize potential IAQ problems. IAQ problems can often be prevented if you educate the cleaning staff about topics such as the following:

Cleaning practices. It's a fact that cleaning fluids and compounds emit strong odors; some are innocuous and some could be harmful, especially to allergic individuals. Almost all odors can be worrisome and can trigger air quality complaints. There are three main guidelines for cleaning practices:

- Most importantly, ensure that no cleaning fluids or compounds are harmful to the indoor air quality.
- Schedule work that introduces strong odors at times when the area is not occupied.
- Make sure that fumes from cleaning products are cleared out before air handling systems switch to their "unoccupied" cycles.

Material handling and storage. This is always a major topic when it comes to indoor air quality. The management plan should include provisions for frequent review of critical material usage to ensure proper handling and storage. There should be a procedure for checking new materials to see if they should be included in the critical list.

Trash disposal. Include proper trash disposal procedures in the management plan. If the building includes a cafeteria or restaurant, arrange for daily pick-up of perishable refuse. Ensure that the containers are covered, pest control is effective, and that the trash collection area is cleaned at least daily.

USING CONSULTANT SERVICES

Even large organizations with large IAQ staffs often find it more efficient to buy the services of a consultant for at least some of their functions. A firm that consults on indoor air quality can afford to have someone make a career of studying regulatory agency requirements and keeping up with their changes. This section looks at some of the types of services you can obtain, and ways they can help ensure that the air in your plant is up to standards.

Services By Hardware Providers

Major HVAC system providers usually offer help with planning, operating, or maintaining a system. Their charges range from free all the way to the full cost of their consulting time. You can sometimes learn so much from a sales presentation explaining the advantages and disadvantages of various types of systems that it is hard to tell it from an HVAC class. They may be able to tell you how different equipment fares when it comes to such regulatory agencies as OSHA.

If all your heating, ventilating, and air conditioning functions are not performed by a single system, you might obtain information from the sales department of a specific part of HVAC. For example, the manufacturer of your furnace can probably give expert advice on heating, thermostats, automated adjustments, holiday and daylight saving adjustments, occupied/unoccupied cycles, and other information.

Comprehensive Consulting

When you pay a consultant who does not sell hardware, you probably have more confidence that nothing is being left out. You can have a consultant provide advice only, advice plus measurements, advice plus measurements plus analysis, or the consultant can be a subcontractor, handling purchase and installation of equipment that affects indoor air quality.

One of the motivators for using consultants is the fact that many common building materials and furnishings today emit fumes, odors, and gases. It is the consultant's business to keep abreast of such information. Plywood, particle board, carpet, glue, and organic chemicals can cause many of the symptoms in Table 16-1. Severity or likelihood of the symptom is shown on a scale of one to twelve.

New design versus existing facility. Whether you are designing an entire building or a modification to an existing building, a consultant's advice can help the new facility meet standards when it begins operation. The question with new design is whether the building will function without having indoor air problems. With an existing facility, the question usually is whether the building has an air quality problem now. Sometimes a consultant is called in when management knows there is a problem—then the question is how bad the problem is, and how it can be fixed.

You can expect a consultant to first outline the levels of service available. For new design, it would be best to include the consultant from the early design stages. Basically, for existing facilities that have, or might have, a problem, think of consultant services in increasing order of involvement, as follows:

- A consultant might contract for a visual inspection and report. An experienced person could inspect your plant and point out areas where there is a potential for air pollution, or even a clear indication of pollution. The feeling here is that you will have to clean up those areas, regardless of the findings of a full investigation, so you might as well clean them up before going further. It is even possible that present pollution might mask other problems if it is not cleaned up.
- Whether you used the first step or not, the next level of involvement could be to hire a consultant to take measurements. The measurements could be all-inclusive, or the consultant might recommend stages of measurement so you could review, make a decision, and then measure further.
- To involve the consultant further, you might include analysis and recommendations in the contract.
- After the preceding steps, a consultant who is completely involved would take over any required repair of the air quality, including ordering and installing new equipment. The contract could include follow-up visits, measurements, and corrective action.

Testing Laboratory

Many companies have used the services of a testing laboratory, where typical, special, or chemically pure environments can be created. Different size chambers can accommodate a range of product and activity testing. Small and intermediate

TABLE 16-1
Worker Symptoms that Can Be Due to Building Materials[1]

Symptom	*Contaminant or Condition*				
	Combustion Materials	*Solvents*	*Formaldehyde*	*Ozone*	*Spores, Molds, etc.*
Asthma	2	0	11	11	11
Mucous Membrane Irritation	10	9	11	11	11
Eye Irritation	9	10	11	11	11
Congestion, Runny Nose	9	1	11	11	11
Headache	12	7	11	6	6
Fatigue, Weakness	9	11	11	7	11
Dizziness	7	7	0	4	2
Nausea	10	6	0	1	3
Diarrhea, Constipation	1	0	0	0	0
Rashes	1	11	2	11	11
Sneezing, Cough	10	10	2	11	11
Chest Pain	7	6	3	5	6
Intestinal Cramps	1	0	0	0	0
Discomfort	12	12	12	12	12
Chronic Conditions	6	7	8	6	10

1. Courtesy of Professional Service Industries, Inc., Portland, Oregon.

chambers can be used for testing such products as apparel, carpets, wall coverings, computers, laser printers, and office furniture. Large chambers typically are used to test products requiring human intervention, such as paint, photocopiers, and carpet cleaning fluids.

Full-service laboratories can take measurements in a client's plant, collect samples to take to their testing facilities, and perform evaluations in their test chambers. They also conduct seminars and help building managers learn advanced air quality control techniques.

BUILDING COMMISSIONING[1]

The term *commissioning* refers to the comprehensive verification of systems performance. A process of design review, construction observation, functional performance testing, evaluation, troubleshooting, and problem solving is used to verify that the building systems actually operate to satisfy the owner's and occupants' needs. Electrical, mechanical, and all building systems, including HVAC and everything that affects indoor air quality, are included.

Architects, engineers, contractors, and manufacturers each have a role. They all work as a team in commissioning a building because modern facilities encompass many system control elements. Environmental quality is a result of their interaction. The commissioning agent's purpose is to ensure that the environmental quality of the facility has been tested and verified to be acceptable before occupancy.

The commissioning agent typically supervises and oversees the process, which is a systematic verification to determine that the systems work together as intended by design. If, during the testing process, system deficiencies are identified, discussions with the design team, contractor, and owner will bring awareness of the situation. The result is usually changes to the system(s) involved to obtain the intended level of performance.

Before spending the money for having your building commissioned, you will want to identify benefits whose value justify the cost. Following are some benefits you can evaluate:

- Building systems operate as specified in the contract when the building is turned over by the builder. The owners get the building they contracted and paid for.

1. Courtesy of Engineering Economics, Inc., Golden, Colorado.

- System discrepancies are identified early and resolved quickly while all contractual parties are still on the job. The walls are often still open, and corrections can be easier.
- Systems' performance is verified more thoroughly and under more actual working conditions than are covered by most manufacturers' sign-off procedures.
- A commissioner serves as arbitrator between the design, construction, and manufacturing professionals to resolve conflicting viewpoints, restricted contractual obligations, and/or the lack of specific knowledge of the actual system performance. The commissioner provides information to management with facts and data taken on the job by technicians working side-by-side with the design/construction team.
- The traditional test and balance services normally provided as a subcontract to the HVAC subcontractor are incorporated into the commissioning process, with a broader scope and direct responsibility to the owner. The result is a more comprehensive approach with no conflict of interest on the part of the test/balance contractor and the HVAC contractor.
- Training of operations personnel is especially important in large buildings and facilities that are controlled by sophisticated control systems and equipment. They are often more intricate than can be effectively maintained and adjusted by personnel who receive no more than the standard introductory explanations. The commissioning agent develops an overview of systems via the commissioning process, and the commissioning manual will become, in effect, the building operations reference and training manual.

SUBSCRIPTION AND OTHER SERVICES[2]

Another type of service is one that provides information via reports, newsletters, computer software, seminars, and other media. A few of the titles and descriptions are:

> *Industrial Safety Report*—a monthly publication that describes OSHA regulatory actions and important safety concerns. It frequently has articles about indoor air quality, and describes proposed as well as new regulations.

2. Courtesy of J. J. Keller & Associates, Inc., Neenah, Wisconsin.

OSHA Safety Training Newsletter—a monthly publication that gives training outlines and checklists, regulatory developments and changes, quick tips, information sources and agency contacts, and other information.

Regulatory Update—published twice a month, with latest news on OSHA, EPA, DOT, and other regulatory actions. Descriptions of proposed regulations include comment periods and compliance dates. There are articles on relevant court cases, government reports, policy statements, and private studies that impact regulatory development.

There are also computer programs that allow you to quickly search OSHA regulations, printed OSHA guides with index, compliance manuals, standards manuals, safety manuals, training manuals, hazardous materials manuals, recordkeeping manuals, confined spaces manuals, and others.

GETTING INFORMATION ON THE INTERNET

One more source of information is the Internet, where you can read, download, and/or order a considerable amount of information. One location is FedWorld, located at http://www.fedworld.gov. From there you can hyperlink to several agencies, including OSHA. Another way is to go directly to OSHA's page at http://www.osha.gov/ or NIOSH's page at http://www.cdc.gov/niosh/homepage.html. Figure 16-1 shows the screen that http://www.osha.gov/ brings up, and Figure 16-2 shows the selections available after selecting Standards from the first figure. Programmers who manage the screens change them as the need arises, so your screen might be different from the figure.

FIGURE 16-1
Choices Available from First OSHA Screen

The Assistant Secretary	OSHA Software
Information about OSHA	Frequently Asked Questions
Office Directory	Statistics
What's New	Standards
Media Releases	Other OSHA Documents
Publications	Technical Information
Programs & Services	US Government Internet Sites
Compliance Assistance	Safety & Health Internet
Sites	

[Comments & Info | OSHA Home Page | OSHA-OCIS | US DOL Web Site | Disclaimer]

FIGURE 16-2

Choices Brought up by Selecting Standards in Figure 16-1

Standards & Related Documents

Provided on WWW & Gopher by USDOL OSHA - OCIS

Search of OSHA Standards & Related Documents

OSHA Regulations (Standards - 29 CFR)

OSHA Regulations (Preambles to Final Rules)

Federal Register - OSHA Notices Only

OSH Act of 1970 (Amended 1990)

Standard Interpretations

Standards Development

OSHA Semi-Annual Unified Agenda

Draft of OSHA's Proposed Ergonomics Protection Standard

OSHA's Compliance Assistance Page

[Comments | OSHA-OCIS | OSHA National Office | USDOL Web Site | Disclaimer]

CHAPTER 17

Handling Two Pollutants Found Throughout the Plant

The two pollutant types covered in this chapter are carbon monoxide (CO) and volatile organic compounds (VOCs). Carbon monoxide is a single gas and is the deadlier of the two types. The term *volatile organic compounds* refers to dozens of chemicals that, while not as deadly, can become part of the plant's environment from many more sources.

CARBON MONOXIDE

This gas results from incomplete combustion of materials that contain a significant amount of carbon. Instead of forming carbon dioxide (CO_2), as a complete burn does, carbon monoxide (CO) is missing the second oxygen atom—an important factor in explaining how it kills. (See Health Effects of Exposure to Carbon Monoxide, later in this section.)

One strange reason CO is so dangerous is that people know it is a major component of automobile exhaust. Because of that, it is commonly felt that, if carbon monoxide is present, it will have the familiar odor of automobile exhaust. The fact is, we smell the other components of engine exhaust; carbon monoxide is odorless, colorless, tasteless, and noncorrosive. Carbon monoxide is only slightly lighter than air, so it does not dissipate rapidly.

Source of Carbon Monoxide

Internal combustion engines that use a carbon base fuel are the largest overall source of carbon monoxide. Industrial plants use a variety of internal combustion engines inside and outside the buildings—for transportation, compression, landscaping, cutting, rigging, removing snow, and many other purposes. Carbon monoxide is also a significant component of tobacco smoke.

Heating devices that use a flame can create CO if not adjusted properly. This category includes furnaces for heating work areas, water heaters, stoves, and dryers. Employee lounges sometimes have gas stoves, and lobbies might have fireplaces that burn wood or gas. Charcoal grills are sometimes used in semi-enclosed lunch and picnic areas.

Industrial processes that burn a carbon base fuel can create CO, especially when the flame does not receive sufficient oxygen. Because this source includes the process of incinerating solid waste, plant designers should be sure that output from incinerators cannot get into the plant's air distribution system.

On the subject of outdoor air entering the plant's ventilation intakes, indoor air quality concerns apply to industrial plants that operate in every kind of environment. Therefore, CO as a component of marsh gases, uncontrolled fires such as forest and brush fires, and volcanic gases are a concern to some plants.

How to Protect Workers

The first rule is to keep carbon monoxide sources out of the plant. Do not allow forklifts, transportation vehicles, or material-handling equipment to enter occupied areas, even for a brief pass-through. Gasoline-powered pressure washers, tampers, chain saws, space heaters, and other equipment must not be allowed to operate indoors without worker protection, or in any place where exhausts can get into the plant's ventilation system. When vehicle motors must run indoors, as in maintenance and repair facilities, exhaust removal systems should be used.

Car and Truck Exhaust Removal Systems.[1] An excellent way to exhaust fumes from one or several vehicles is to install a retractable hose reel for each exhaust pipe that will be emitting gases. Each reel can be mounted overhead or on a wall or column, and each can exhaust either directly to the outside, or to a manifold that collects gases from several reels. A fan should draw 235 cubic feet per minute (cfm) per reel for cars, or 600 cfm per reel for trucks.

Some models have dampers that automatically close when the hose is retracted on the reel, to allow all exhaust force to go to reels that are working. Other models automatically turn the fan off to conserve energy when the hose is in its rewound position.

Manual reels are available with centrifugal locking mechanisms that hold the extended hose in convenient positions. The hose can be pulled to a new locked position or allowed to return to the reel. Motor-driven reels are also available, and can be operated from a 24-volt wired switch or from a remote infrared transmitter. Both types control the motor to raise and lower the hose, as well as the damper or fan. The wired switch can be mounted on a wall or hung as a pendant; the infrared remote unit can operate from up to about 65 feet away.

The hose in one system ends in a nozzle that clamps securely to the vehicle tailpipe. Rubber nozzles prevent damage to vehicle bodywork and therefore are suitable for work stations with moving vehicles. Metal nozzles have a simple but heavy duty construction and a spring-loaded flap that helps attach to the vehicle and works as a damper when disconnected. The metal type of nozzle, along with high-temperature hose, is recommended for use where high temperatures will be encountered, but should not be used with moving vehicles because of risks of worker injury and paintwork damage to vehicles.

Emergency Vehicle Exhaust Systems.[2] When vehicles must get underway quickly, special exhaust extraction systems that remain attached, and automatically disconnect and retract when the vehicle leaves, are available. One such system operates as follows:

- When the vehicle is at rest, the nozzle of a flexible, compressed hose fits over the tailpipe. The nozzle stays firmly in place because of an electromagnet on the hose and a disk on the vehicle.

1. From *Exhaust Hose Reel Technical Description*, courtesy of Nederman, Inc., Westland, Michigan.

2. From *Nederman MagnaTrack Exhaust Extraction Systems for Emergency Vehicles;* courtesy of Nederman, Inc., Westland, Michigan.

- As the vehicle begins to move out of the building, the entire nozzle and hose assembly travels with it. This action uses an overhead guide rail, which mounts the horizontal portion of the exhaust hose, to allow a smooth expansion of the compressed hose.
- As the vehicle tailpipe nears the door threshold, a micro-switch contacts a striker plate (bolted on the guide rail), de-energizing the electromagnet and causing it to fall away from the anchor disk. The nozzle comes away from the tailpipe.
- The balancer can now pull the entire vertical section of hose away from the vehicle and store it in a hanging position off the floor, ready to be remounted when the vehicle returns.
- As soon as the vehicle returns, the nozzle fits over the tailpipe and the electromagnet is attached to the anchor disk. The entire assembly travels with the vehicle to the parked position.

Consider that "indoors" is every area that has any amount of restriction to free air flow. Indoors, for protection from carbon monoxide poisoning, is every area that is not clearly outdoors. In 1992 a worker suffered fatigue, dizziness, and confusion after working with a pressure washer inside a building. The machine was powered by a small (four horsepower) engine, but it emitted enough carbon monoxide to override the benefits of exhaust fans and three open doors. A hospital emergency room confirmed that the worker had been poisoned by carbon monoxide and, fortunately, they were able to treat her.

Another pressure washer incident involved someone who had placed the machine in the outside doorway of a building. That worker received a more serious CO poisoning, but also survived. He was in the hospital for seven days, receiving intensive oxygen treatment in the trauma unit.

Early Warnings of Carbon Monoxide

If workers learn to recognize early signs of carbon monoxide's presence, they can avoid the sickness, and even death, that results from continued exposure. Because CO is odorless, the sense of smell is completely unreliable.

Carbon Monoxide Detectors. The most desirable way to learn that CO is entering the internal air is from a commercial carbon monoxide detector. It can quickly detect low concentrations of the gas and sound a warning. Detectors that meet the June 1995 revision of Underwriters Laboratory standard UL 2034 have minimum sensitivity characteristics and good false alarm control.

The type of sensor employed is usually determined by the source of electricity. Detectors that use the plant's regular electric power generally have a solid state sensor that takes in a fresh sample of the air on a regular basis. When the source of power is an internal battery, power consumption is a major design consideration, and these detectors can be expected to have sensors that work on continuous exposure to the air. Their reaction time therefore depends on the concentration of CO and the length of time the gas has been affecting the sensor.

Other differences that relate to the power source are choice of location, ease of maintenance, and reset time. Battery operated models are easy to locate exactly where you want them, but AC models are limited to locations where they can be wired in. Manufacturers usually recommended that detectors be located high on walls or ceilings. A major disadvantage of the battery operated models is, of course, that the batteries have to be checked and replaced. Because AC models clear the sensor and draw in new air frequently, they reset quickly if the carbon monoxide intrusion was transitory. Battery models take longer, depending on how long the intrusion lasted and how strong it was.

Indications that Might Be Sensed by Workers. Sometimes condensation on surfaces where such condensation does not normally form might be a warning. It is an indirect indicator—not due to the carbon monoxide itself, but resulting from a malfunction in certain equipment.

Combustion products from properly designed and operating equipment that burns wood or oil should not leak into the room. Therefore the smell of smoke or other combustion products (you can't smell carbon monoxide) indicates a problem that should be investigated.

Worker Health Reactions. Everyone hopes to detect the presence of carbon monoxide before it gets to this point, but detectors might malfunction and the first indicator could be the effect on workers. Symptoms have been described as "flu-like" and can include nose and throat irritation, burning eyes, dizziness, and headaches. In low concentrations of CO, workers might ignore these symptoms, continue at work, and then see the symptoms stop when they leave work. Carbon monoxide should be suspected if such symptoms appear at work and leave when the workers get out in the air, especially if all or several workers have symptoms that follow the same pattern.

Health Effects of Exposure to Carbon Monoxide

It was mentioned earlier in this chapter that the one atom difference between the carbon monoxide (CO) and carbon dioxide (CO_2) molecules is a factor in making the former a killer. The deadly process begins in the lungs, where hemoglobin (Hb) in the blood normally picks up oxygen for distribution to body cells that need it. However, if air in the lungs contains carbon monoxide, the hemoglobin will pick it up to form carboxyhemoglobin (COHb) much easier than it will pick up oxygen. Hb-CO combinations are strong, so the Hb keeps its CO attachment and does not become available to pick up oxygen on the next passes through the lungs.

Now we come to the missing oxygen atom in carbon monoxide. With CO binding to Hb, there is a tendency to hold on to any oxygen that is being carried, so it is not dropped off at cells that need it. The end result is that cells do not receive the oxygen they need for surviving and performing their functions. For any given amount of CO in the air, each of the following factors can magnify the effect:

- Exercise or active work—the more strenuous, the more the effect.
- Some people are susceptible because they normally carry a higher level of COHb.
- Cigarette smoking—heavy smokers can have as much as 9% COHb.
- The presence of other toxic pollutants.
- Pregnant women are more susceptible.
- Older people are more susceptible.
- Higher altitudes.
- Diabetes.
- Anemia.
- Fever.

Table 17-1 shows the symptoms and effects of various levels of COHb, expressed as percent of COHb. However, the table is mostly for background information because, except in a medical facility, we cannot measure the percentage of COHb in a person's blood. From the table you can get a qualitative view of the progression of harm as the blood becomes less and less effective at distributing oxygen to cells.

Information of more immediate use is given in Table 17-2, which shows symptoms and effects as a function of concentration levels and time of exposure. Concentration levels are given in parts per million (ppm) of CO in the air. Notice how, as the concentration increases, it takes less time to suffer the same effect.

Assisting Victims After Exposure

Experts agree that the rescuers should give top priority to getting victims to fresh air. The other measure nonmedical personnel can take is to keep victims calm and prevent them from any physical exertion. In most situations this treatment will be sufficient, unless the exposure has been extended or severe, or the victim is especially susceptible to carbon monoxide poisoning. Administering oxygen is one of

TABLE 17-1
Symptoms and Effects of Different Levels of COHb

COHb	*Symptoms and Effects*
10%	No symptoms
15%	Mild headache
25%	Nausea and serious headache. Recovery is fairly fast after treatment with oxygen and/or fresh air.
30%	Generally the same symptoms, but intensified. There is a potential for long term effects, especially in infants, children, the elderly, pregnant women, and those with heart disease.
45%	Unconsciousness
50%+	Death

several steps that medically trained people might take to speed up recovery. They will also check for complications.

The length of time for recovery, and the extent of possible after-effects, depend on the percentage of carboxyhemoglobin (COHb) which, in turn, is a function of duration of exposure and concentration level. The half-life of COHb when in an atmosphere free of carbon monoxide is about five hours, meaning that, from any point, the percentage COHb will drop to half of that level in about five hours. In the next five hours, the level will drop to half of that value, and so on. With halves of ever-decreasing amounts, it takes a long time before the blood can be declared free of the effects of carbon monoxide.

VOLATILE ORGANIC COMPOUNDS

Hydrocarbons that are able to react in the atmosphere and are involved in forming acid rain and ground-level ozone are known as Volatile Organic Compounds (VOCs). A loose definition follows from the group's name—*volatile* suggests that members of the group are gaseous or, if liquid, vaporize easily; *organic* suggests that they include carbon in their chemical composition.

Many products give off VOCs as they dry, cure, set, and otherwise age. The smell of paints and thinners drying, floor tiles setting, and upholstery aging are

TABLE 17-2
Symptoms as a Function of CO Concentration and Exposure Time

Concentration (PPM)	*Exposure Duration*	*Symptom or Effect*
35	8 hours	This is the maximum exposure allowed by OSHA in the workplace over an eight hour period
200	2-3 hours	Mild headache, fatigue, nausea, and dizziness
400	1-2 hours	Serious headache; other symptoms intensify
	>3 hours	Life threatening
800	45 minutes	Dizziness, nausea, convulsions
	within 2 hours	Unconsciousness
	within 2-3 hours	Death
1600	20 minutes	Headache, dizziness, and nausea
	within 1 hour	Death
3200	5-10 minutes	Headache, dizziness, and nausea
	within 1 hour	Death
6400	1-2 minutes	Headache, dizziness, and nausea
	25-30 min.	Death
12,800	1-3 minutes	Death

common examples of VOCs. Some odors are from catalysts that were added to a product to control or speed the drying process. Not all VOCs are harmful in normal concentrations, but regulatory agencies have specified exposure limits for many of them.

An IAQ manager involved with something new or different should be alert to the likelihood of VOCs being emitted. Activities that typically result in increased VOC levels include bringing new furniture, fixtures, or floor covering into the area, or moving personnel into a new or newly remodeled or redecorated area.

Because off-gassing of VOCs is most intense when an item is new, some plants take the precaution of placing new furniture in a well-ventilated area for at least several days or weeks before moving it into closed working areas. Climate permitting, outdoors is an excellent area for allowing VOCs to disperse. Plants in other locations can use unoccupied areas that open to the outside, and fans can increase the purging action. Occupants can be kept out of new and remodeled areas while ventilation is increased during the early period of peak off-gassing.

Table 17-3 lists several common VOCs and includes, where applicable, brief information about regulatory exposure limits as well as body organs that the chemicals attack. This information is not complete, and is just for general reference; the agencies change it as studies reveal new data. More complete exposure information, as given in books such as *NIOSH Pocket Guide to Chemical Hazards*, includes ceiling values, peak values, and other numbers. "Ca" by a NIOSH limit means the agency considers the chemical to be a potential occupational carcinogen. Unless otherwise noted, NIOSH exposure limits are time-weighted average (TWA) concentrations for up to a 10-hour workday during a 40-hour workweek. "ST" in that column means the value is a short term exposure limit (STEL), usually a 15-minute TWA exposure that should not be exceeded at any time during the workday. Meanings of abbreviations in the Target Organs column are also given after the table.

TABLE 17-3
Common VOCs

VOC	*Exposure Limits*		*Target Organs*
	NIOSH	OSHA	
Acetone	250 ppm	1000 ppm	Eyes, skin, resp sys, cns
Benzene	Ca, 0.1 ppm	1 ppm	Eyes, skin, resp sys cns, bone marrow
2-Butanone	200 ppm	200 ppm	Eyes, skin, resp sys, cns
2-Butoxyethanol	5 ppm	50 ppm	Eyes, skin, resp sys, cns, hemato sys, blood, kidneys, liver, lymphoid sys
Butyl acetate	150 ppm	150 ppm	Eyes, skin, resp sys, cns
Carbon disulfide	1 ppm	20 ppm	cns, pns, cvs, eyes, kidneys, liver, skin, repro sys

Continued

TABLE 17-3
Continued

VOC	*Exposure Limits*		*Target Organs*
	NIOSH	OSHA	
Carbon tetrachloride	Ca, ST 2 ppm	10 ppm	cns, eyes, lungs, liver, kidneys, skin
Chloroform	Ca, ST 2 ppm	C† 50 ppm	Liver, kidneys, heart, eyes, skin, cns
Cyclohexane	300 ppm	1300 ppm	Eyes, skin, resp sys, cns
n-Decane			
p-Dichlorobenzene	Ca	75 ppm	Liver, resp sys, eyes, kidneys, skin
1,1-Dichloroethene	100 ppm	100 ppm	Skin, liver, kidneys, lungs, cns
1,2-cis-Dichloroethene			
Dichloromethane	Ca	500 ppm	Eyes, skin, cvs, cns
Dimethyl-disulfide			
n-Dodecane			
Ethanol			
Ethylacetate	400 ppm	2000 ppm	Eyes, skin, resp sys
Ethylbenzene	100 ppm	800 ppm	Eyes, skin, resp sys, cns
2-Ethyl-1-hexanol			
Freon 113			
Hexanal			
Limonene			
Methyl, tertiary-butyl ether			
Naphthalene	10 ppm	10 ppm	Eyes, skin, blood, liver, kidneys, cns
n-Nonane	200 ppm	None	Eyes, skin, resp sys, cns
n-Octane	75 ppm	500 ppm	Eyes, skin, resp sys, cns

TABLE 17-3
Continued

VOC	*Exposure Limits*		*Target Organs*
	NIOSH	OSHA	
2-Pentanone	150 ppm	200 ppm	Eyes, skin, resp sys, cns
4-Phenylcyclo-hexene			
α-Pinene			
Styrene	50 ppm	100 ppm	Eyes, skin, resp sys, cns, liver, repro sys
Tetrachloroethylene	Ca*	100 ppm	Eyes, skin, resp sys, liver, kidneys, cns
Toluene	100 ppm	200 ppm	Eyes, skin, resp sys, cns, liver, kidneys
1,1,1-Trichloro-ethane			
Trichloroethylene	Ca	100 ppm	Eyes, skin, resp sys, heart, liver, cns
Trichlorofluoro-methane			
1,2,4-Trimethyl-benzene	25 ppm	None	Eyes, skin, resp sys, cns, blood
Undecane			
Vinyl chloride	Ca	1 ppm	Liver, cns, blood, resp sys, lymphatic sys
m,p-Xylene			

* Because tetrachloroethylene is a possible workplace carcinogen, management should minimize workplace odor, exposure, concentrations, and the number of workers exposed.

† Ceiling value; should not be exceeded at any time.

Abbreviations:

cns	central nervous system	repro	reproductive
cvs	cardiovascular system	resp	respiratory
pns	peripheral nervous system	sys	system

CHAPTER 18

Filtering the Plant's Air

Before the 1970s, oil and other sources of energy were plentiful and relatively inexpensive. High R values of insulation, careful sealing of buildings, and other measures to keep heated or cooled air from exchanging with outdoor air were not considered cost effective. One common method of keeping indoor air fresh was to bring in outside air, heat it (or cool it, depending on the season and location), and exhaust it. The cost of heating or cooling the air for a single use was not important. Many plants that didn't exhaust the conditioned air directly lost a significant amount of it through poor seals, fits, closures, and other leaks. The Arab oil embargo, long lines at gas stations, having to wear heavy sweaters at work, rapidly increasing energy prices, and government incentives changed our outlook.

By the 1990s, it was a rare building that was not sealed and insulated, and did not recycle most of its conditioned air. Couple that fact with an increase in the use of chemicals that, at best, are not as good for us as clean air, and, at worst, are highly lethal, and it is easy to see why we are concerned about indoor air quality. If the recycled air is not cleaned, it will recycle dust, dirt, chemicals, toxins, allergens, bioaerosols, viruses, and other undesirable items to quickly pollute the indoor air quality. Filtering, by methods to be covered in this chapter, is the main method of cleaning the recycled air.

This chapter describes some of the particles that have to be filtered out of the air, and examines the types of filters used for removing particles, gases, odors, and other contaminants.

UNDERSTANDING PARTICLES[1]

To fully understand what removing particles from the air involves, it is necessary to examine the nature of particles. They are measured in microns, which are millionths of a meter. In the English system, there are 25,400 microns in one inch. The smallest mark on a machinist's precision rule is usually 1⁄64 of an inch, which is 397 microns long, so it is clear that a one-micron particle is a very small one.

However, one micron is not the smallest particle that filters must remove. The IAQ community defines particles as having diameters ranging from 0.001 to 100 microns. Those we see on furniture or floating in a ray of sunlight are typically 50 microns or larger. The naked eye cannot see individual particles of five microns or smaller. These particles tend to be perpetually suspended in the air and are generally referred to as "respirable," meaning they are capable of being readily ingested into the lungs, where they tend to remain.

Respirable particles make up more than 90 percent of the total number of particles in the air we breathe. Although five microns is the smallest single particle we can see unaided, we often see small particles when they are grouped. For example, a stream of cigarette smoke comprises 0.01 to 1 micron particles, but we see the stream because of the high concentration of the particles. Once the stream of smoke disperses, we no longer see these respirable particles, but they are still present. The following list gives a few facts related to particle size.

TYPICAL PARTICLES:

- Rosin smoke: 0.01 to 1.0 micron diameter
- Smoke from various oils: 0.03 to 1 micron diameter
- Atmospheric dust: Less than 0.001 to 40 microns diameter
- Carbon black: 0.01 to 0.5 micron diameter
- Zinc oxide fumes: 0.01 to 0.5 micron diameter

1. *Clean Air Fact Guide*, courtesy of United Air Specialists, Inc., Cincinnati, Ohio.

METHODS FOR PARTICLE SIZE ANALYSIS:

- Ultramicroscope: 0.006 to 0.4 micron diameter
- Microscope: 0.4 to 100 microns diameter
- Visible to eye: 50 microns diameter and up
- Light scattering: 0.01 to 10 microns diameter

TYPES OF AIR CLEANING EQUIPMENT:

- Common air filters: 1 to 100 microns diameter
- High efficiency air filters: Less than 0.001 to 8 microns diameter
- Electrical precipator: Less than 0.001 to 90 microns diameter

Airborne particles can be in the form of smoke, mist, fumes, dust, aerosols, and vapors. To remove airborne particles, we rely on air filtration, which involves separation of particles from airstreams.

When very small particles are visible only because they are present in dense concentrations, it is difficult to tell if they are suspended in air (true suspended particles) or diffused throughout it (gas or vapor). The lower boundary where particles act as true particles is about 0.01 micron. The normal theory of separation does not apply to particles below that size; removing them from air requires technologies used for gaseous materials. Particles above 0.01 micron can be filtered or separated.

TYPES OF FILTERS

This section takes a broad look at several basic types of filters. Different manufacturers and organizations choose their own way of breaking down the classifications; the following is one way.

Fiberglass Filter[2]

Off-the-shelf fiberglass filters are made of passive glass fibers, and were originally meant to stop larger particles from damaging heating and air conditioning systems. Their use has extended to filtering breathing air, although they have no effect on small, allergy-causing particulates. Their efficiency ranges from about 5 to 7 percent. Fiberglass filters are generally meant to be discarded after they collect a specified amount of particles.

2. Filter descriptions, with the exception of electronic air cleaners, are courtesy of AirClear Products, Snyder, Texas.

Hog Hair Filter

Made from synthetics, this filter is usually green or blue. It has a rough texture, but its filtering capabilities are about the same as a fiberglass filter. Hog hair filters are washable for reuse.

Cotton/Polyester Filter

Various versions of this filter, such as flat, bag, and pleated, are in common use because they trap about four times the allergy causing particulates that fiberglass filters trap. These filters generally require a monthly change.

Washable Electrostatic Filter

Made of woven polypropelene and a foam or polyester center, these filters will trap about 75 percent of allergy-causing dust. They are washable, and usually require washing at least every three weeks. Efficiencies may decrease with use, because particulates cannot be washed completely out of the filter. Some investigators say this type of filter contributes to mold build up. They have high pressure drops and must be maintained properly to avoid damage to equipment.

Spun Glass Filter

These filters are four inches thick and are for systems that were designed to use them or systems that have been modified to accept them. They will trap about 90 percent of particulates.

Electronic Precipitator

As with the spun glass filters, electronic precipitators are for systems that were designed to use them or systems that have been retrofitted for them. In many situations they are worth the high cost of retrofit, because their efficiency is about 95 percent. They require high maintenance, which includes washing every 15 to 30 days, and they create ozone when running.

Electronic Air Cleaners[3]

Air first passes through a pre-filter for removal of larger particles, and then it passes through an electronic field. The airborne particles, some as small as 0.01 micron,

3. Courtesy of Trion Inc., Sanford, North Carolina.

pass into this strong field (ionizing section), where each particle is given an electrical charge. Charged particles then move into a collector section which consists of a series of equally spaced parallel plates with alternate positive and negative charges. Those plates that have the same charge as the particles repel them; plates with the opposite charge attract and collect them. This process removes up to 99 percent of the impurities.

CHOOSING THE RIGHT AIR CLEANER[4]

The physical nature of your particulate (dry or wet) will determine whether the media, cartridge, or electronic type air cleaner is best for your application. The air cleaner size is generally based on air volume (cfm) needed to effectively entrain and move the particulate to the air cleaner. However, the "dirt-holding capacity" of the air cleaner needs to be considered, particularly when heavy concentrations of particulate are being treated.

Step 1. Determine the physical nature of your particulate. The classifications are usually dry (such as dust and powder) or wet (such as oil mist).

Step 2. Determine how the particulate will be entrained and moved to the air cleaner. The following descriptions will help with this step.

Source capture is the ideal method because the particulate is entrained directly at the point of origin and not allowed to migrate into the workplace environment. Capture hoods, either enclosing or exterior to the emitting process, serve as the point of entry for the particulate. The type and design of hood to be used will depend on the physical characteristics of your process equipment, the contaminant generation mechanism, and the operator/equipment interface. Ducting is then used to connect the hood to a local air cleaner. Your final hood and duct design will determine the air volume needed to effectively entrain and transport particulates to the air cleaner.

Central system uses the ductwork and fan in the general HVAC equipment to entrain fugitive airborne particulates within the recirculating air. The effectiveness

---

4. From *Engineered Solutions for Clean Air,* courtesy of Trion Inc., Sanford, North Carolina.

of this method depends upon how well the general HVAC system recirculates air throughout the workplace environment. The size of the air cleaner is generally based on the HVAC system fan.

Ambient capture is the strategic placement of air cleaners which create effective air patterns to move particulates to the air cleaner. The discharge air of one air cleaner blows particulates toward the intake of another. This method is effective when good airflow patterns can be developed and the volume of air in the workplace environment can be turned over or recirculated through the air cleaners at a frequent enough rate. The frequency, called Air Exchange Rate (AER), is expressed in air changes per hour (ACH). Depending on your airborne particulate concentrations (light to heavy), an AER is selected and total air volume is determined by:

$$\text{cfm} = \frac{VR}{T}$$

where: cfm = air requirement in cubic feet per minute
V = volume of workplace in cubic feet
R = rate of air changes per hour (generally between 4 and 15)
T = time (60 minutes)

Step 3. Estimate the particulate concentration. The following loading guide is usually all you will need for making this estimate, or you can consult an environmental testing company or an air cleaning equipment manufacturer:

Light—dirty city air, designated smoking areas, office environments. This category involves concentrations up to 0.05 grains per cubic foot (100 mg/m^3).

Medium—industrial process exhaust, factory indoor air, commercial kitchen exhaust. This category involves concentrations from 0.05 to 0.275 grains per cubic foot (100 to 600 mg/m^3).

Heavy—metal machining oil mists, grinding dust, heat treating operations, heavy welding. This category involves concentrations from 0.275 to 0.5 grains per cubic foot (600 to 1000 mg/m^3).

Step 4. Locate your plant's position in the Table 18-1. From there you can check with equipment manufacturers to select models that will best meet your specifications.

TABLE 18-1
Step 4 in Choosing the Right Air Cleaner

Particulate Type	***Entrainment and Moving***	***Air Volume in cfm***	***Loading Capacity***	***Filter Type***
Commercial; Dry Particulate Typically Dusts Powders Smoke Pollen Mold Spores Bacteria	Central System, typically HVAC system	800–1600	Medium	Electronic
		1000–2600	Medium	Electronic
		2000–30,000	Heavy	Electronic
		4000–6000	Medium	Electronic
		8250–∞	Heavy	Electronic
	Ambient Capture, typically free hanging, wall mounted, ceiling mounted with established air patterns	140–250	Light	Elect./Media
		160–325	Light	Electronic
		200–750	Medium	Media
		210–450	Medium	Media
		300–600	Medium	Electronic
		330–640	Medium	Electronic
		480–1100	Medium	Electronic
		480–1100	Medium	Media
		600–1025	Medium	Media
		630–1150	Medium	Electronic
		1150–2175	Medium	Media

Continued

TABLE 18-1
Continued

Particulate Type	*Entrainment and Moving*	*Air Volume in cfm*	*Loading Capacity*	*Filter Type*
Industrial; Dry Particulate Typically Dusts Powders Grinding dust Welding fumes Smoke	Central System, typically HVAC system	800–1600	Medium	Electronic
		1000–2600	Medium	Electronic
		2000–30,000	Heavy	Electronic
		4000–6000	Medium	Electronic
		8250–∞	Heavy	Electronic
	Source Capture, typically using hoods, arms, portable units	380–800	Light	Electronic
		380–710	Light	Media
		800–1000	Heavy	Cartridge -p
		800–2000	Medium	Elect./Media
		800–6000	Medium	Electronic
		900–1100	Medium	Media -p
		950–1050	Medium	Electronic -p
		1300–3300	Medium	Media
		2000–2700	Heavy	Cartridge -p
		2000–7000	Heavy	Cartridge
		2800–6000	Medium	Electronic
	Ambient Capture, typically free hanging with established air patterns	800–6000	Medium	Electronic
		1150–2175	Medium	Media
		1300–3300	Medium	Media
		2000–30,000	Heavy	Electronic
		2800–6000	Medium	Electronic

Table 18-1
Continued

Particulate Type	*Entrainment and Moving*	*Air Volume in cfm*	*Loading Capacity*	*Filter Type*
Industrial; Wet Particulate Typically Oil mist Oil smoke Coolants Oily welding fumes	Central System, typically HVAC system	800–1600	Medium	Electronic
		1000–2600	Medium	Electronic
		2000–30,000	Heavy	Electronic
		4000–6000	Medium	Electronic
		8250–∞	Heavy	Electronic
	Source Capture, typically using hoods, arms, portable units	380–800	Light	Electronic
		380–710	Light	Media
		800–2000	Medium	Elect./Media
		800–6000	Medium	Electronic
		1300–3300	Medium	Media
		2800–6000	Medium	Electronic
	Ambient Capture, typically free hanging with established air patterns	380–1800	Light	Electronic
		380–710	Light	Media
		800–6000	Medium	Electronic
		1300–3300	Medium	Media
		2000–30,000	Heavy	Electronic
		2800–6000	Medium	Electronic

CHAPTER 19

Special Issues in Heating, Ventilating, and Air Conditioning

At the center of indoor air quality efforts is the HVAC or heating, ventilating, and air conditioning system. The subject is so big that entire books, even sets of books, are devoted to it, and one chapter's coverage could be no more than cursory. Instead, this chapter looks briefly at major features of HVAC systems, and then concentrates on two HVAC topics that, considering their importance, do not usually receive sufficient attention: technicalities of HVAC damper controls, and ventilating confined spaces.

MAJOR FEATURES OF HVAC SYSTEMS

There are several ways to categorize HVAC systems, but one way that ties in with damper controls (to be covered in the next section) is to divide them according to whether they serve a single zone or multiple zones. With a single zone system, there is one thermostat that controls when warm or cool air is provided, one set of fans that control the air volume and rate, and one exhaust arrangement. If part of the indoor air is recycled, there is one mixing arrangement. The main feature of a single zone system is there is just one feedback loop for control, even if the zone consists of multiple buildings. A single zone system does not imply that all areas receive the same flow of conditioned air, but it means that there are no separate controls in different areas. If one area needs ten times the air flow of another area, that can be part of the basic design, but if the smaller area temporarily has heat-producing equipment in it one day, it will still receive one-tenth the air flow the large area receives. If the thermostat is located in the smaller area, the larger area will be cold that day, and vice versa.

The advantage of a single zone system is its simplicity and low installation cost. Disadvantages include:

- When a thermostat, timer, or other control device signals conditioned air (heated or cooled), it goes to all parts of the zone at a predetermined rate. The air is probably right for the area where the control device is, but could be just the opposite of what another area needs.
- External influences that are not dependable can upset the balance over the zone. For example, a strong sun might add heat to areas on the south and west side of a building in the afternoon, but those areas will receive the same flow of heat then that they receive in mornings and evenings.
- Internal influences, such as people and equipment, might move between areas. Their movement will subtract heat production from one area and add it to another.
- Thermostat placement is critical. Ideally it should be in a location that represents the entire zone. If it is on an outside wall, or a wall of a cavity that has air flowing through it, the thermostat will signal for heat or cooling as needed to keep the wall at the desired temperature.
- Although installation costs are low, operating costs can be high. There are direct costs because occupants in areas where the conditioned air is not at a comfort level will compensate by opening windows or supplementing the HVAC output with individual fans or heaters.

The difference between a single zone system and a multiple zone system is in the ability to control the airflow. There may be a single source of conditioning the air, but a multiple zone system will control the flow of air into each zone according to its needs. Airflow is usually controlled by dampers or fans. Multiple zone systems are generally more expensive to design, install, and balance, but they operate more efficiently. Advantages and disadvantages are basically the reverse of those for single zone systems.

Testing and Balancing

Previous chapters have mentioned testing and balancing of HVAC systems. It involves testing, adjusting, and balancing of HVAC components so that the entire system provides airflows, temperatures, and other characteristics required by the design specifications, under various combinations of specified conditions. Testing typically includes system parameters and components such as the following. Not all of these apply to every installation.

- Airflow rates at all outlets, return ports, exhaust ports, and mixers
- Control settings and responses
- Air temperatures, humidity, pressure, and other characteristics the HVAC system controls
- Fan speeds
- Filter pressure drops and excessive pressure drop alerts
- Overall proper installation, with accessibility to maintenance locations
- Posted maintenance schedule

The test and balance report should provide a complete record of the design, preliminary measurements, final test data, discrepancies between test data and design specifications, and reasons for the discrepancies. Depending on the authority of the person or group writing the report, it might state whether the discrepancies are acceptable. If the discrepancies or other data are not acceptable, the report can include plans and dates for corrections and for retesting. As a baseline for future performance checks and adjustments, the report should list damper positions, equipment capacities, control types and locations, control settings, operating logic, and any other appropriate data.

After there has been a major change to the HVAC system or to anything that might affect its operation, or any time there is reason to believe the system is not functioning as designed, a complete or partial test and balance is in order. The Associated Air Balance Council recommends the following guidelines for determining when to consider testing and balancing:

- When space or occupancy has been significantly changed
- When HVAC equipment has been replaced or modified
- When control settings have been readjusted
- After the air conveyance system has been cleaned
- When accurate records are required to conduct an IAQ investigation
- When the building owner is unable to obtain design documents or appropriate air exchange rates for compliance with IAQ standards or guidelines

TECHNICALITIES OF DAMPER APPLICATION[1]

According to NIOSH, the ventilation system is the cause of almost half of indoor air quality complaints. Although damper control alone is not responsible for IAQ, it has a central role. Therefore, it is important that the dampers and actuators are selected, sized, and operated correctly. Their simplicity—blades that can be turned inside the ducts to open, block, or partially block the airflow—belies their complexity and the effect their operation has on the overall system.

The following symbols and abbreviations are used in this section:

%OA—percent outside air volume
%RA—percent return air volume
EA—exhaust air
MA—mixed air
OA—outside air
RA—return air
SA—supply air
TMA—mixed air temperature
TOA—outside air temperature
TRA—return air temperature

Economizer Systems

If workers occupy a building, outside air must be provided. The method may be as simple as having open windows, or keeping a building from being completely sealed; or, the building may have a complex HVAC system that draws in outside air. HVAC systems themselves have various ways to provide outside air, starting with simply drawing it in, heating or cooling it, distributing it, and exhausting it. However, that method is costly because it throws away air that has been heated or cooled and used only once. An economizer system exhausts some of the air, draws in enough outside air to replace it, and mixes the fresh air with the remaining (filtered) inside air.

The economizer portion of an air handling system is responsible not only for reducing the operating cost of the system but, more importantly, providing a sufficient volume of outside air to ensure good indoor air quality. It also has to maintain the slight positive pressure that prevents infiltration of unconditioned outside air. Incorrect building pressure can increase operating costs as well as cause damages to the structure due to condensation, freezing, and other actions. Figure 19-1 is a schematic of a basic economizer.

---

1. From *Damper Applications Guide 1*, courtesy of Belimo Aircontrols (USA), Inc., Danbury, Connecticut.

FIGURE 19-1

Schematic of a Basic Economizer that Mixes Some Recycled Air with Outside Air

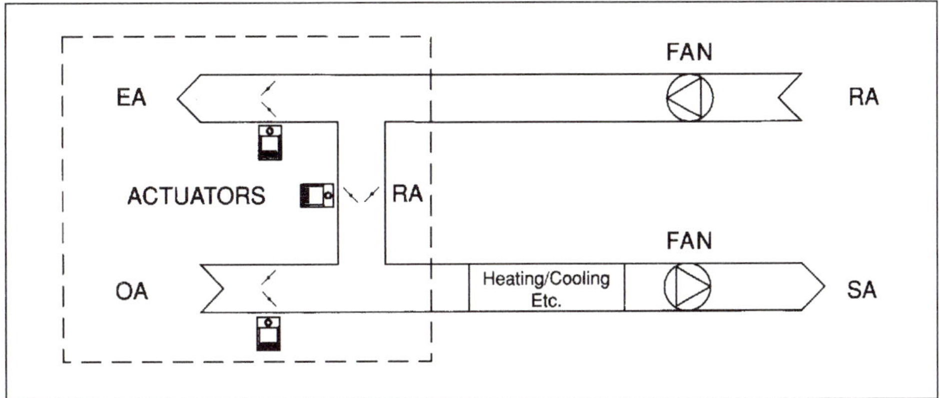

Damper Selection

Two types of rectangular damper arrangement, parallel blade and opposed blade, are used in economizers. Parallel blades move in the same direction and are always parallel to each other. In the opposed blade arrangement, adjacent blades move in opposite directions. Figure 19-2 shows the two arrangements.

The significance of having different arrangements is that they have different inherent flow characteristics, as shown in Figure 19-3. Neither is linear, but the parallel blade damper curve shows less curvature. Its flow increases more rapidly when the damper begins to open, but opposed blade dampers give little change in flow when they begin to open.

In an actual installation, a damper is not the only device that affects airflow. Other influences, such as duct work, filters, and coils, also tend to restrict the flow. Therefore, the damper does not control the flow according to its inherent flow characteristics shown in Figure 19-3. Damper authority is the percentage of total system pressure drop that the damper causes. Figure 19-4 shows that, when the damper is closed, the pressure drop across it will be the same as the total system pressure drop, so the damper authority is 100 percent. However, when the damper is fully open, the pressure drop across it will be small compared to the drop across filters, coils, and other items in the stream. As an example, Figure 19-5 shows an open damper pressure drop that is one tenth the total system drop, making the damper authority ten percent.

FIGURE 19-2
The Two Arrangements of Damper Blades

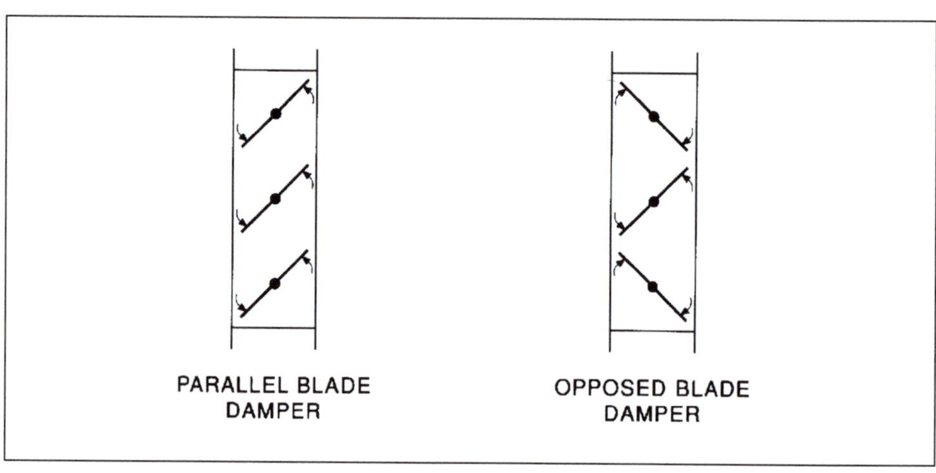

FIGURE 19-3
Inherent Characteristics of Dampers

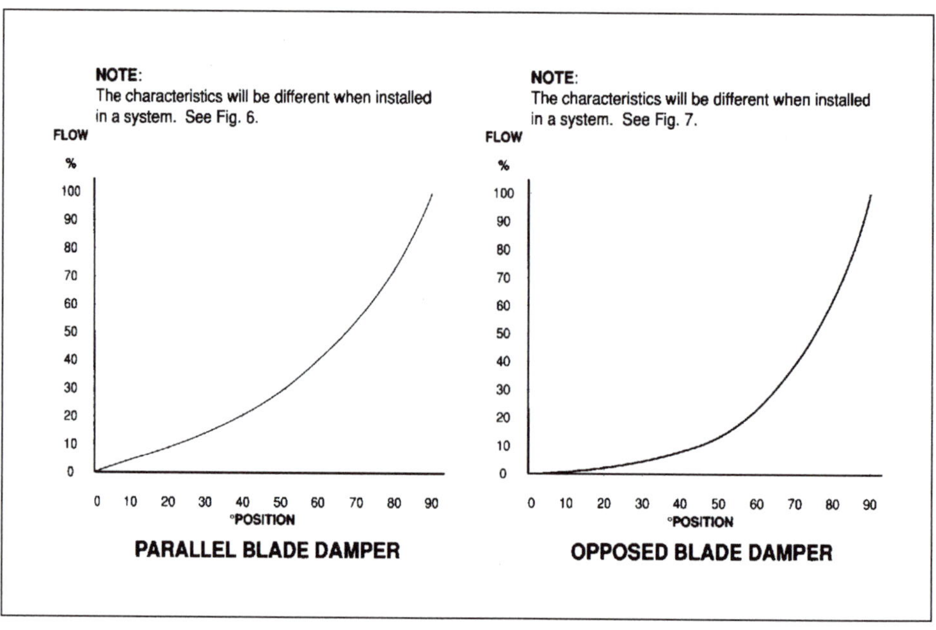

FIGURE 19-4
Damper Authority with Damper Closed

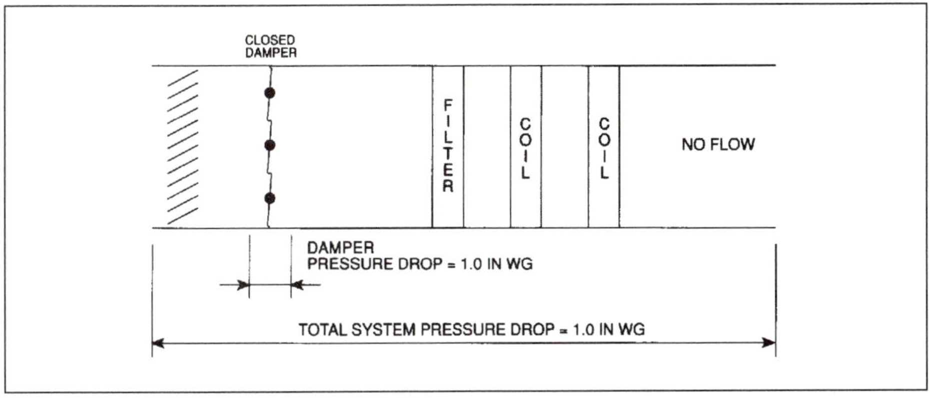

FIGURE 19-5
Damper Authority with Damper Open

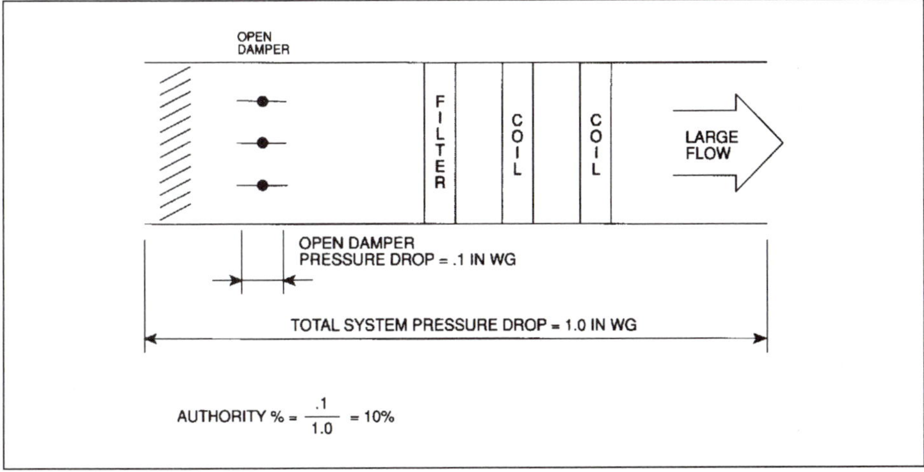

The pressure drop across the damper changes as it operates from closed to open, distorting the flow characteristics curve. When the damper begins to open, the differential pressure is high and the flow increases at a higher rate than the inherent curve suggests. Near the other end of its travel, almost fully open, the damper pressure drop will be small compared to the total system pressure drop.

Figures 19-6 and 19-7 show graphs of actual flow control for parallel blade and opposed blade dampers respectively. Dotted lines in the graphs show inherent characteristics, and solid lines show flow characteristics for a practical range of damper authorities.

It is important to understand that expressions such as "Total System Resistance" and "Total System Pressure Drop" refer to just the part of the system where the flow is controlled by the damper. In Figure 19-8, flow downstream from the mixing plenum (point A) is not controlled by the OA damper; it is constant in the first approximation. The OA damper controls only the flow through weather louver, damper, and OA ductwork. Therefore, the authority of the OA damper is determined by the pressure drop across the OA damper as a percentage of the total differential pressure between OA and point A.

Total system drop for the RA damper is measured between points A and B. System drop for the EA damper is measured between point B and EA. If a single damper is used to control the flow through an air handler and ductwork, then the total system pressure drop is the sum of pressure drops in all parts.

Sizing of the OA, RA, and EA dampers in an economizer is important because it determines the installed damper combined characteristics. If they have been properly sized, their characteristics will complement each other to make the flow constant, as shown in Figure 19-9.

Figure 19-10 uses a mix of parallel and opposing damper arrangements to demonstrate a total flow that does not remain constant as damper settings change. Choosing a mix of parallel and opposing damper arrangements provides a convenient example, but does not imply that such arrangements in one economizer will always give a mismatch. Their inherent characteristics are never linear, and their nonlinearities can combine in various ways to create a match or mismatch.

Most economizer systems have one outside air damper (OA), one return air damper (RA), and one exhaust air damper (EA). They usually operate in unison, so the RA damper opens as the OA and EA dampers close and, if they are sized properly, an increase in the RA flow is matched by an equal decrease in the OA flow, keeping the total flow constant. Resistance to flow will remain constant, as will the pressure in the mixing plenum.

Unfortunately, there are several reasons why it is not easy to achieve this goal of complete balance. There are other considerations, in addition to the previously

FIGURE 19-6
Parallel Blade Installed Flow Characteristics

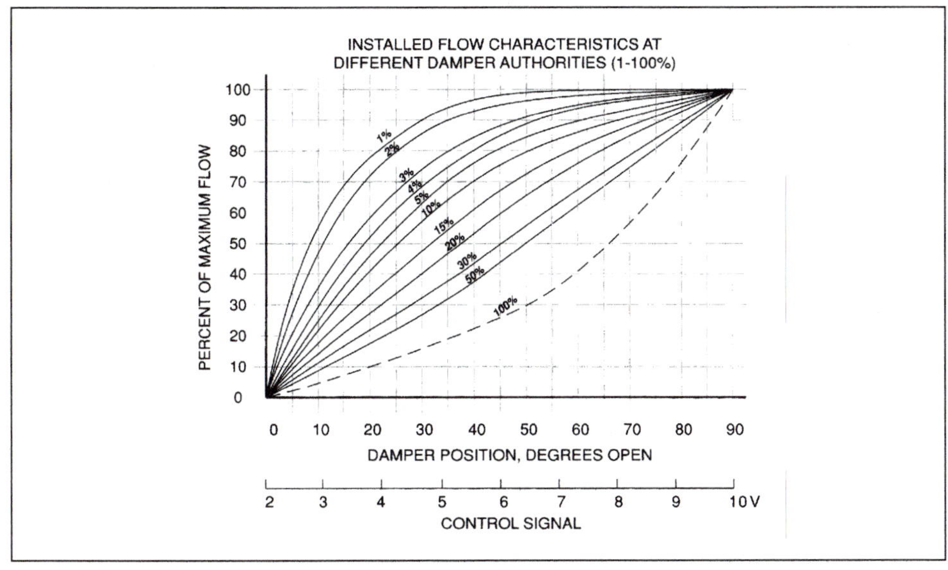

FIGURE 19-7
Opposed Blade Installed Flow Characteristics

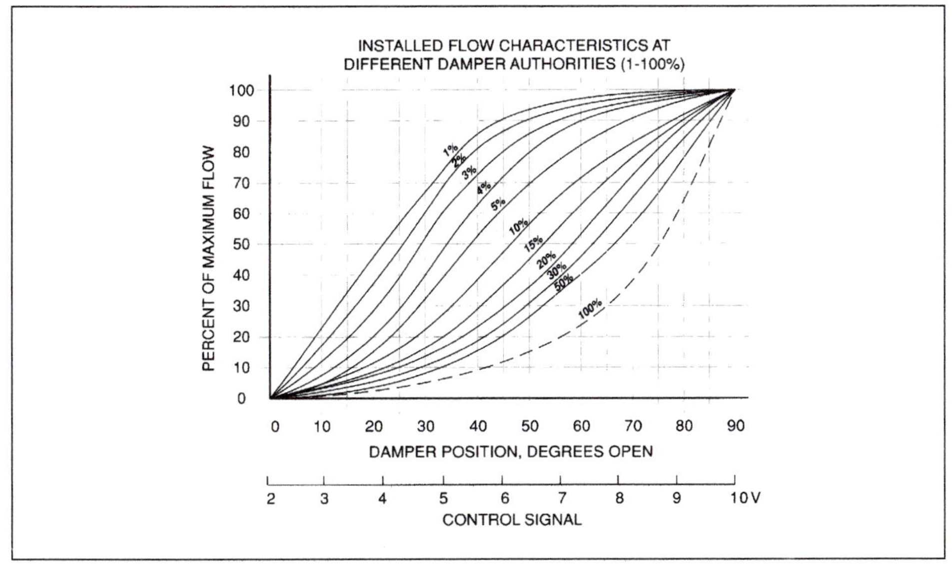

FIGURE 19-8
Separate Flows in an Economizer

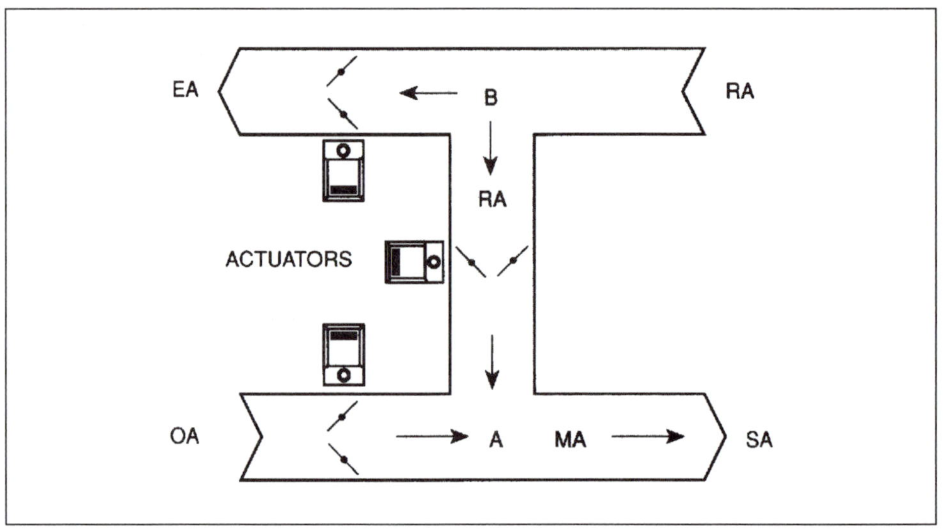

FIGURE 19-9
Ideal Sizing of Dampers Gives Constant Total Flow

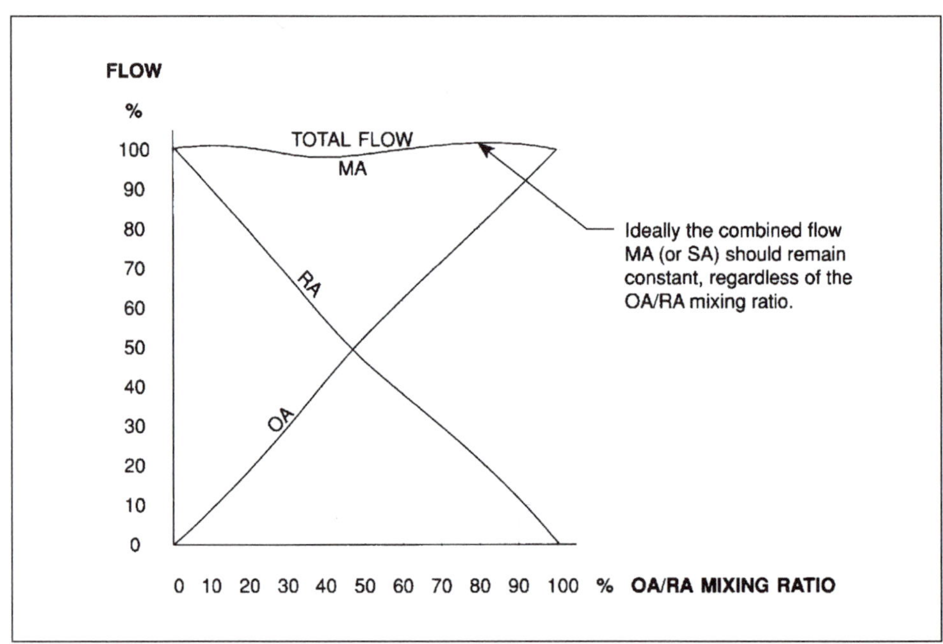

FIGURE 19-10
Mismatched Dampers Cause Uneven Flow

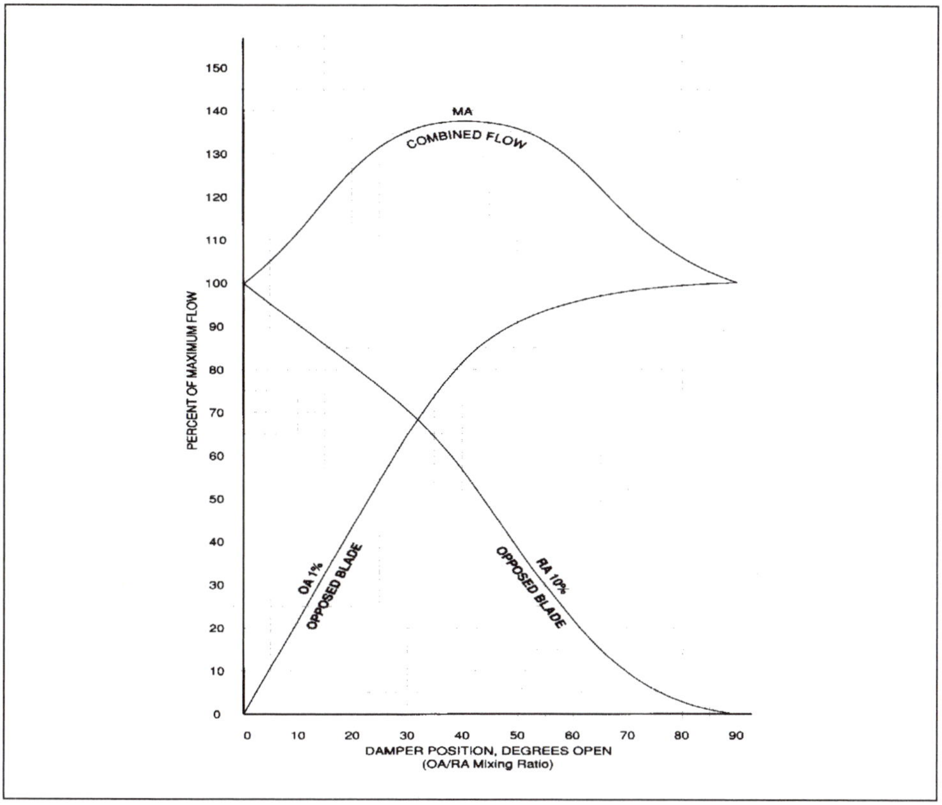

mentioned nonlinear characteristics, especially near the open and closed ends of the curves. The amount of outside air brought into the building is a major determinant of indoor air quality, and the economizer varies the mixing ratio of outside and return air to meet varying conditions. At 100 percent outside air, maximum free cooling is provided and, when the load changes, the mixing ratio must change until outside air volume is reduced to a specified minimum volume that satisfies IAQ requirements. In many systems the economizer is at the minimum position and provides the minimum volume of outside air when there is need for mechanical cooling or heating. At intermediate loads, the mixing ratio must change to maintain a constant mixed air temperature.

Minimum outside air volume can be set in a number of ways, including the following. A pitot tube traverse is usually the most accurate if a completely effective duct length is available to get a velocity pressure profile which is sufficiently uniform for good readings. Measuring outlet volume is possible, but can be in error by 5 to 15 percent because of duct leakage. This formula provides a check on the two methods.

$$TMA = TOA(\%OA) + TRA(\%RA)$$

For example, if %OA is 10 percent, then %RA is 100–10, or 90 percent. If TOA is 60 degrees and TRA is 70 degrees, then TMA will be 69 degrees within the error range, ±2 degrees, of the sensors. Accuracy of the temperature sensors determines overall dependability of this formula.

Some coupled equipment (rooftop units in particular) is difficult to control properly. The lack of a return fan or exhaust often leads to OA entering from the so-called EA damper until the RA damper is nearly closed. The building over-pressurizes during most of the year. Minimum outside air volume is achieved by limiting the modulation of the dampers to a minimum position that provides the desired outside air volume. It is important that this position be controlled accurately because a small positioning error will have a disproportionate effect upon minimum outside air volume.

An alternative solution is to use two OA dampers, one of which is modulated in unison with the RA and EA dampers. The other damper is smaller and is opened when the fan is running, to supply the minimum outside air volume. Figure 19-11 shows this arrangement.

Actuator Selection and Operation

After dampers are selected and matched, and their working positions determined, it is necessary to have actuators that will control the damper positions. When the fans are running, dampers are subject to forces that increase friction which must be overcome by the actuators. There are also dynamic forces that actuators must overcome; forces that act upon the damper blades, and tend to turn them in a direction that depends on the blades' positions.

Actuators usually respond to a variable analog control signal that is either pneumatic (between 3 and 15 psi) or electronic (between 2 and 10 volts). Ideally we would want the dampers to be repositioned in direct proportion to the control signal, and to follow a "nominal Signal/Position Curve." However, hysteresis in the actuators and linkage often precludes this ideal result.

All actuators have some hysteresis, causing there to be one position/signal curve for increasing signals, and a different curve for decreasing signals. A certain amount of the control signal has to first move the actuator through the "dead zone" before there can be any movement of the associated damper. Figure 19-12 shows the curves and gives an example.

FIGURE 19-11
Small Damper Provides for Minimum Outside Air Volume

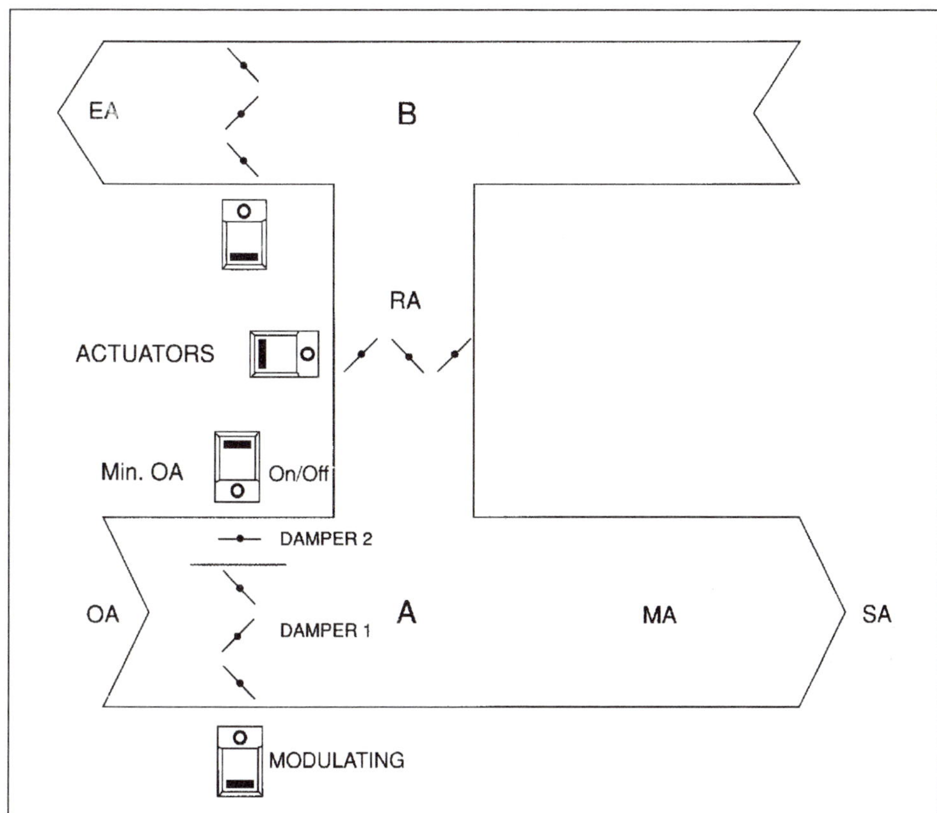

Most actuators are connected to dampers via linkages, which have slack also. Together these hysteresis loops force the actuator to move a significant amount before starting the desired damper movement. Two serious problems result from the combined hysteresis:

- It adds to the instability problems of the mixed air temperature control.
- It makes actuator positioning inaccurate, which is especially serious with respect to the minimum position of the outside air damper. If the control signal is increased from zero (closed OA damper) to the value that represents the minimum OA damper position, a lower than desired minimum position will result. If the control signal is decreased from a high value to the value that represents the minimum position of the OA damper, a higher than desired minimum will result. This difference in the minimum

FIGURE 19-12
The Effect of Hysteresis

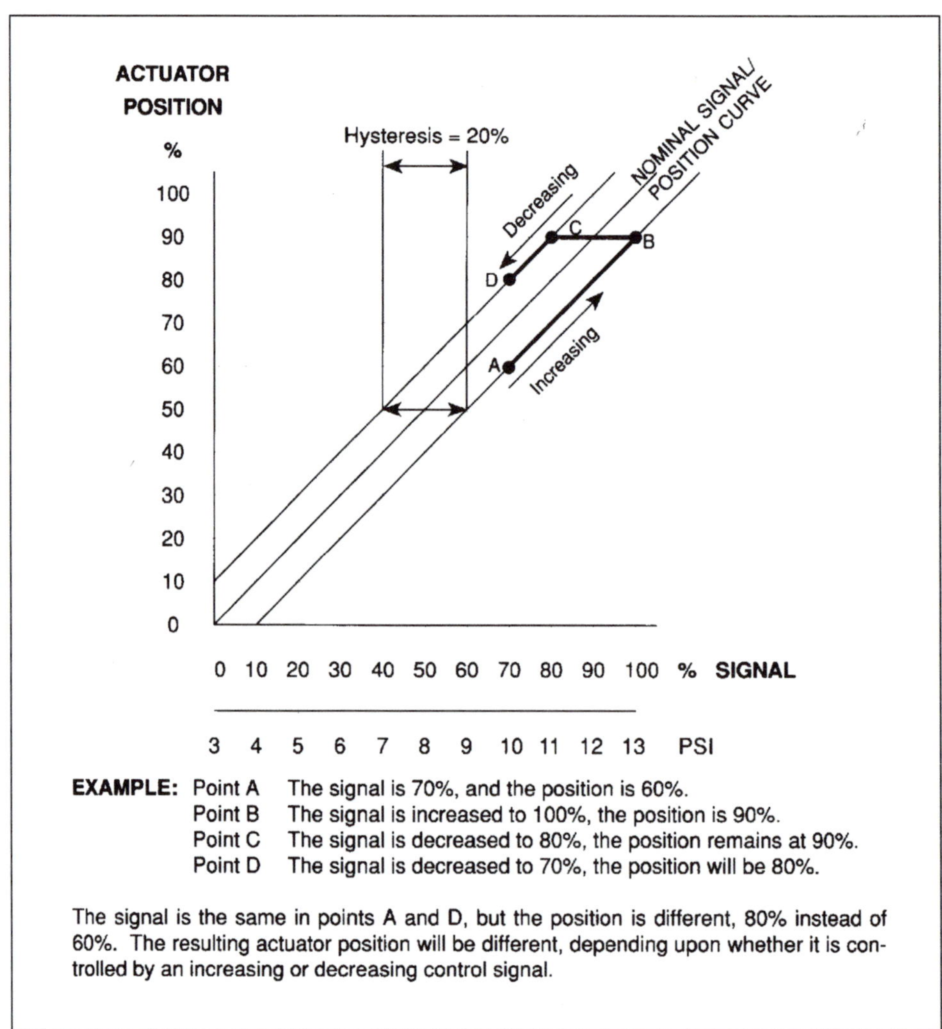

position, depending upon whether it is approached from above or below, has a very large effect upon the minimum outside air volume, as shown in Figure 19-13. A lower than desired outside air volume reduces the indoor air quality, and a higher setting causes needlessly high operating costs.

FIGURE 19-13
Results of Hysteresis on Minimum OA Air Flow

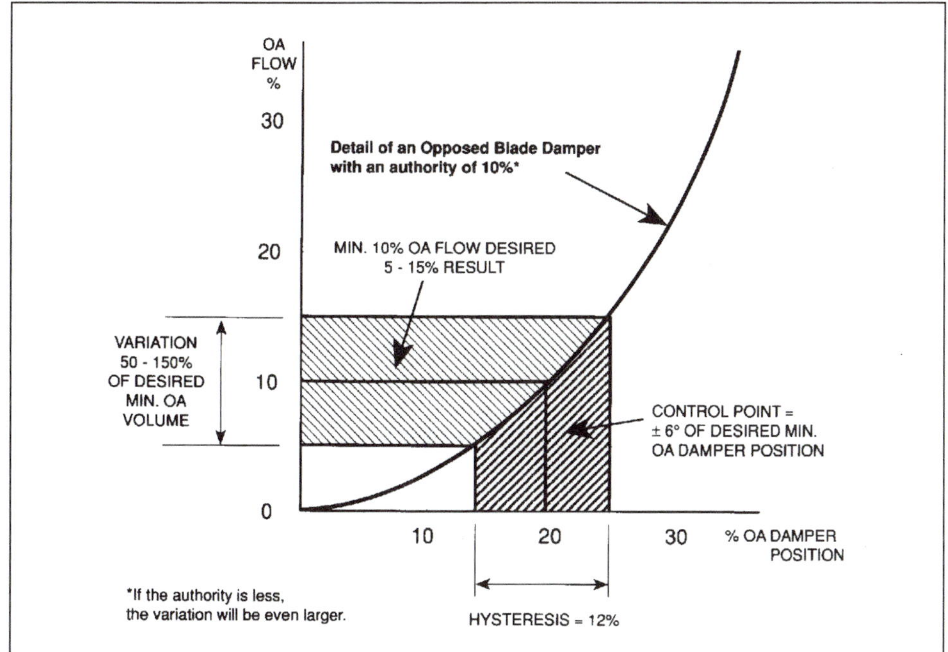

Pneumatic Actuators. Actuators that are controlled by an air signal are in common use because they are reliable, although they operate with high hysteresis. Figure 19-14 shows the main working parts of a pneumatic actuator. Pressure of the control signal acts on the diaphragm and produces a force which pushes the stem out and compresses the spring until the forces are in balance. The shaft's position will change in proportion to the control signal (the nominal signal/position curve).

However, in order to produce a net force at the output shaft, there must be an imbalance between the diaphragm and the spring. The force from the diaphragm must be stronger than the spring in order to produce a pushing force, and the force from the diaphragm must be less than the spring force in order to produce a pulling force. This deviation between the position and the nominal signal/position is called the "spring range shift."

Figure 19-15 shows a pneumatic actuator that is required to produce a clockwise (+) 50 in-lb torque to open a damper and a counterclockwise (–) 50 in-lb

FIGURE 19-14
A Pneumatic Actuator

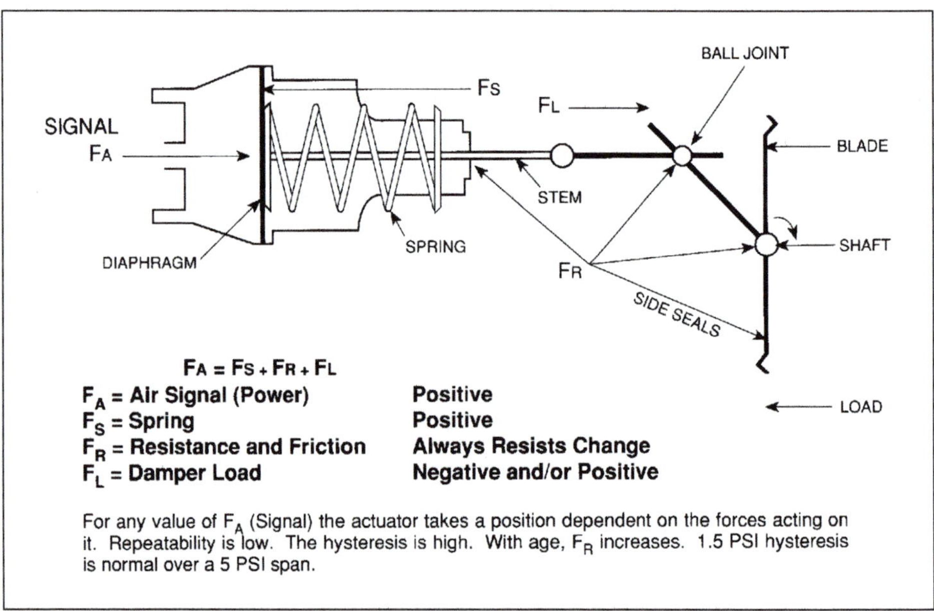

torque to close it. The minimum effective radius of the linkage is two inches, meaning that the pushing or pulling force of the actuator is +25 lb. or −25 lb., according to:

$$\text{Force} = \frac{\text{Torque}}{\text{Effective radius}}$$

The outside diameter of this actuator is 7½ inches, which corresponds to an effective diaphragm (piston) area of approximately 33.3 square inches. When the signal pressure and the position are in balance, no net force is produced at the output shaft. This is represented by the "nominal signal/position" curve. In order to produce a 25-pound force, the signal pressure has to be higher or lower than the nominal signal/position curve. The resultant pressure variation is 25 pounds/33.3 square inches, or 0.75 psi. Therefore, in order to change from +50 in-lb to −50 in-lb torque, the total variation has to be 2(0.75), or 1.5 psi. This is 30 percent of the spring range of a typical pneumatic actuator, which is 8 to 13 psi.

The hysteresis and spring range shift in an actual installation depend on how conservatively the pneumatic actuator is sized in comparison to the required

FIGURE 19-15
Typical Pneumatic Spring Range Shift and Hysteresis

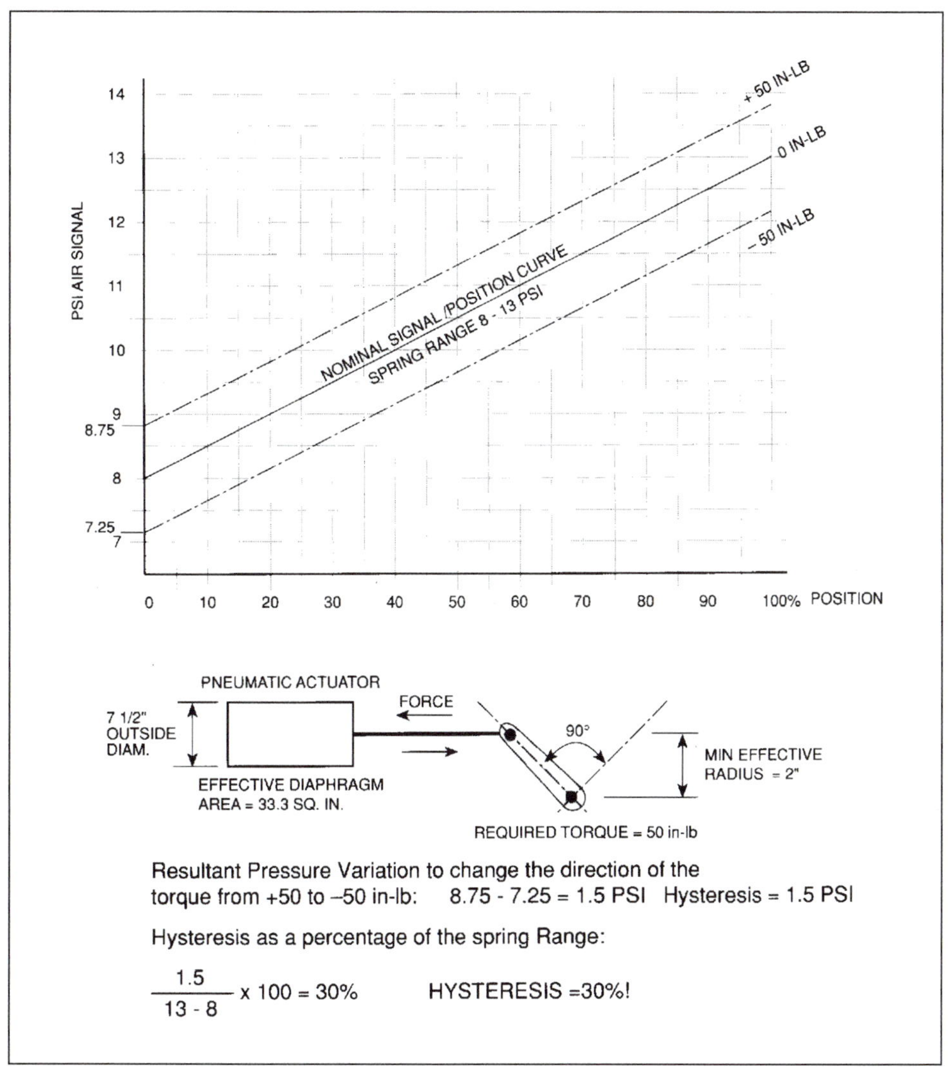

Resultant Pressure Variation to change the direction of the torque from +50 to –50 in-lb: 8.75 - 7.25 = 1.5 PSI Hysteresis = 1.5 PSI

Hysteresis as a percentage of the spring Range:

$$\frac{1.5}{13 - 8} \times 100 = 30\% \qquad \text{HYSTERESIS} = 30\%!$$

torque. Hysteresis will be less with an oversized actuator. Pneumatic actuator manufacturers typically recommend 1¼ to 2 in-lb for each square foot of damper area. Many designers claim that the 12 percent hysteresis this causes is excessive, and they prefer a still larger torque.

One way to improve the accuracy of pneumatic actuators is to provide them with positioning relays, called positioners. A positioner compares the actual position of the actuator with the control signal, and changes the pressure to the diaphragm until the actuator reaches a position that corresponds to the control signal. This arrangement reduces the hysteresis to about ¼ psi, which can be 2½ to 5 percent, depending on the operating range. A disadvantage is that frequent recalibration is needed in order to maintain the accuracy of the positioner.

Linkage between actuator and damper can include a couple of ball joints, pivots, and other elements—each of which can contribute to total hysteresis. Slack in the linkage can easily cause a one to three percent hysteresis when the stress changes between a pushing and a pulling force. The actual amount of hysteresis depends on the linkage's condition and how well it has been adjusted. If the linkage is improperly adjusted and/or the joints are worn, hysteresis can actually be larger than five percent. Taking up this slack requires the actuator to move the same percent before the damper begins to move.

Electronic Actuators

Electronic proportional actuators require both a power source, usually 24 VAC, and a control signal, usually 0 to 10 VDC or 4 to 20 mA. Their operation can be compared to a pneumatic actuator with a positive positioner. However, the electronic actuator requires no field calibration and usually no field maintenance. Position feedback is more precise than with a pneumatic actuator because of a repeatable, geared interface between the actual actuator position and its feedback monitoring system. The feedback signal is also usually available as an output from the actuator to either monitor the actuator position or as a signal for part of a control sequence.

Feedback can be generated by any of several methods. Figure 19-16 shows a typical potentiometer circuit used to measure feedback. In this arrangement, the signal from the potentiometer is fed into a differential amplifier along with the input signal. The amplifier's output, which is the difference between the two signals, moves the actuator until the feedback signal matches the input signal to make the amplifier output zero. Figure 19-17 shows a newer method that uses microprocessor technology. The microprocessor communicates to an application specific integrated circuit (ASIC) which both controls and monitors a brushless DC motor. Pulses generated by the motor allow the microprocessor to determine the actuator's exact position. The microprocessor also allows for special control criteria to be used in either operational characteristics or input signal processing.

FIGURE 19-16
Potentiometer Circuit to Measure Feedback

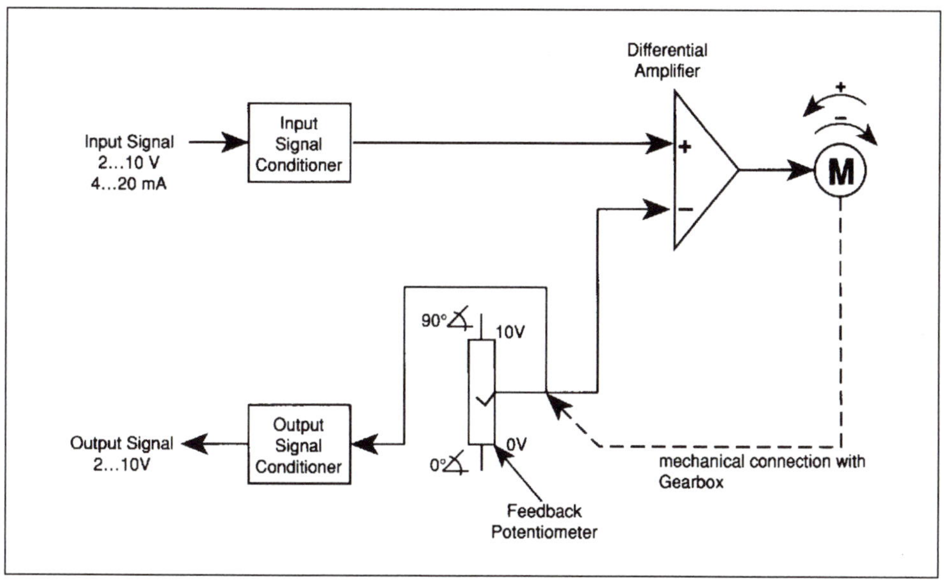

FIGURE 19-17
Microprocessor Feedback Control

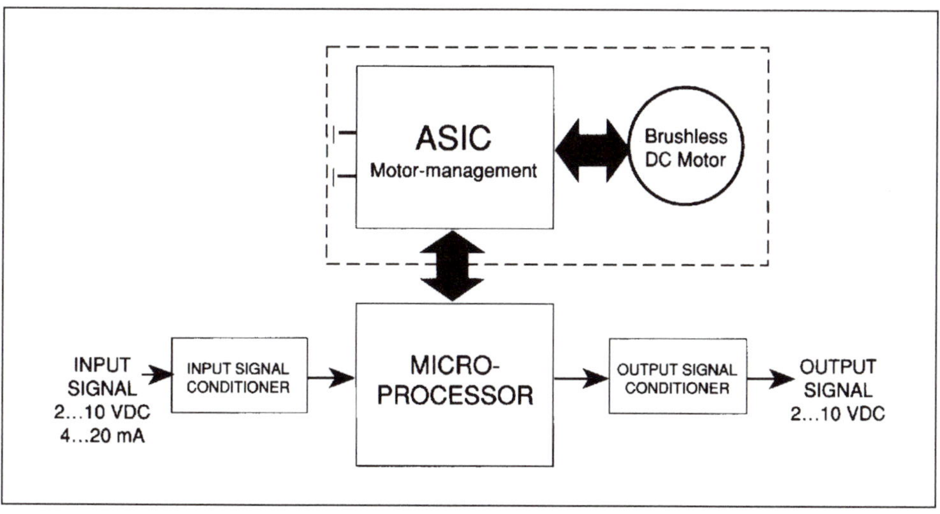

Conventional electronic and electrohydraulic actuators typically have a small hysteresis of about one percent. These actuators are mounted to dampers via linkage that adds one to three percent hysteresis, in a manner similar to the pneumatic actuator mounting. Electrohydraulic actuators are not gear-type actuators; they work similarly to a pneumatic actuator—by building up pressure against a spring—and are therefore subject to spring range shift.

Direct coupling improves damper positioning accuracy. As with crank arm coupling, direct coupling has hysteresis of about one percent. However, without a linkage between actuator and damper, the damper position can be controlled more accurately and the actuator torque is transmitted more efficiently to the damper. In addition, installation time is reduced considerably. Table 19-1 summarizes the accuracies of four arrangements.

HOW TO VENTILATE CONFINED SPACES[2]

The most lethal hazards in confined spaces are usually problems with the atmosphere. Ventilation and related activities are used to correct air quality problems that may exist or could potentially occur within the confined space. The three main types of activities generally considered for correcting confined space air problems are:

- Ventilation
- Purging
- Inerting

Purging

Purging involves the introduction of something other than fresh air to the space as an initial displacement or cleaning agent. Before a confined space is safe to enter, purging must be followed by ventilation. For example, if workers are going to enter an acid tank, it should first be purged with a neutralizing solution to eliminate the acid residue hazard. Then, after the tank has been ventilated, workers can take appropriate precautions and enter the tank.

2. Courtesy of Firecon, East Earl, Pennsylvania.

TABLE 19-1
Summary of Positioning Accuracies

Component	Damper Positioning Accuracy			
	Pneumatic without Positioner	*Pneumatic with Positioner*	*Electronic with Linkage*	*Direct Coupled*
Actuator	12%*	2½–5%†	1%	1%
Linkage	1–3%	1–3%	1–3%	—
E/P Transducer	2%	2%	—	—
Total	15–17%	5½–10%	2–4%	1%

*30% if undersized
†Before calibration drift

Inerting

Inerting involves the introduction of a nonreactive gas such as nitrogen, argon, or carbon dioxide. This method trades one hazard for another, and its main application is where the danger of fire, explosion, or other chemical reaction is the main concern. The inert gas eliminates the possibility of these disasters, but creates an oxygen deficient atmosphere. Workers entering the confined space will be required to wear air supplied respiratory protection.

Ventilation

Ventilation means air flow, and can take place as either forced air or exhaust. Forced air ventilation is the dilution and displacement of a contaminated atmosphere by forcing in fresh air. This method is usually the most effective and efficient for confined spaces.

Exhaust ventilation is usually the accepted method in confined space operations if a point source of contamination exists. For example, if welding fumes are the problem, exhaust ventilation set up to extract the fumes where they are created can be the most expedient method of clearing the air.

One specialized type of forced air ventilation is the use of compressed air. In working environments where moisture of any type could create problems, ventilation may take the form of using dry air from a cylinder. For example, some electrical transformers may be equipped with fittings for this type of ventilation.

The rest of this chapter looks at correct and incorrect methods of ventilating a confined space. Figure 19-18 shows the duct for general space ventilation correctly extending straight down and deep into the confined space, to about two feet from the bottom. Figure 19-19 shows one of the most common mistakes—a setup that does not provide adequate ventilation, and should never be used. It shows clean air being blown into a confined space through a duct that ends just inside the space.

Most manufacturers recommend using no more than 25 feet of duct to maintain adequate airflow. The number and degree of bends should also be minimized because each bend reduces airflow by increasing turbulence in the air stream.

Confined space supervisors should also be aware that, even with the ideal set-up of portable ventilation equipment, complete ventilation is unlikely. Upper corners of the space are the areas most likely to receive the least dilution, as shown by the gray areas in Figure 19-20.

Another potentially dangerous situation is recycling contaminated air that is being exhausted, as shown in Figure 19-21. The ventilation system intake should be at least six to eight feet from the center of the opening if there is no wind or if wind blows exhausted air away from the ventilation system. Always prevent the possibility of wind taking exhausted air to the ventilation system's intake.

Obstructions and barriers to airflow can result in a pocket of undiluted contaminated air remaining in the confined space, as shown in Figure 19-22. Even if workers do not intend to enter that area, it should be ventilated by moving the duct directly into it.

If ventilation is to be shut off after the confined space air reaches an acceptable contaminant concentration, it should first be run for at least 15 minutes. Before testing the confined space atmosphere, allow sufficient time for the inside air to stabilize after turning the ventilation system off.

FIGURE 19-18
Correct Position of Duct at Bottom of Space

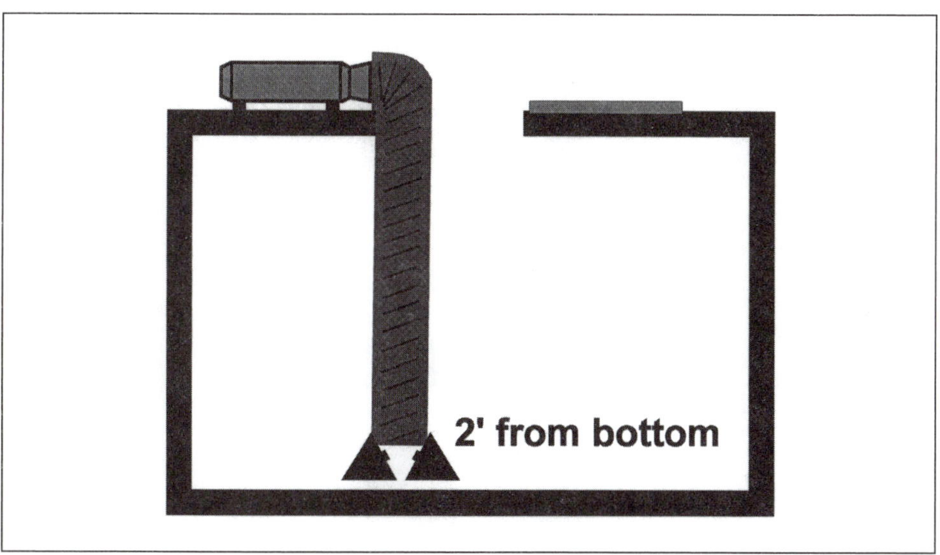

FIGURE 19-19
Inadequate Penetration of Ventilation Air Flow

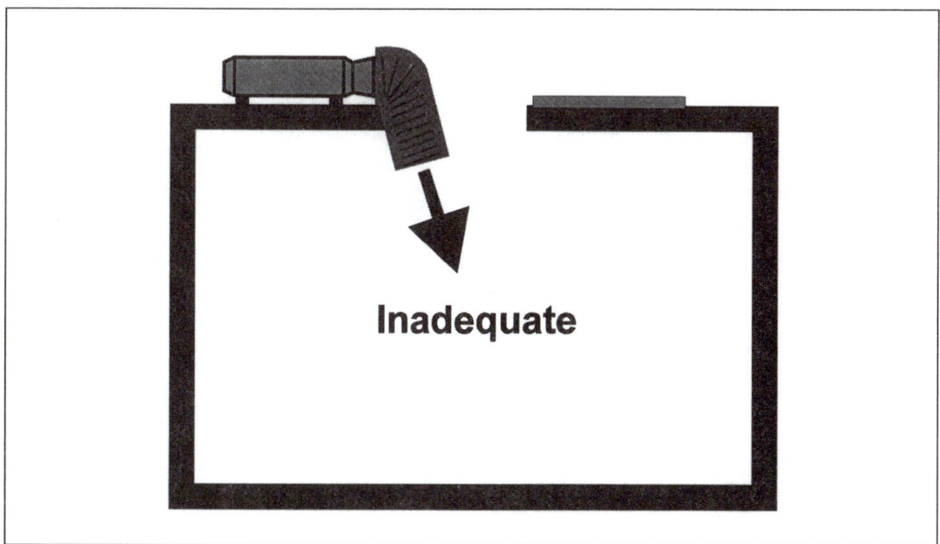

FIGURE 19-20
Undiluted Areas in Upper Corners

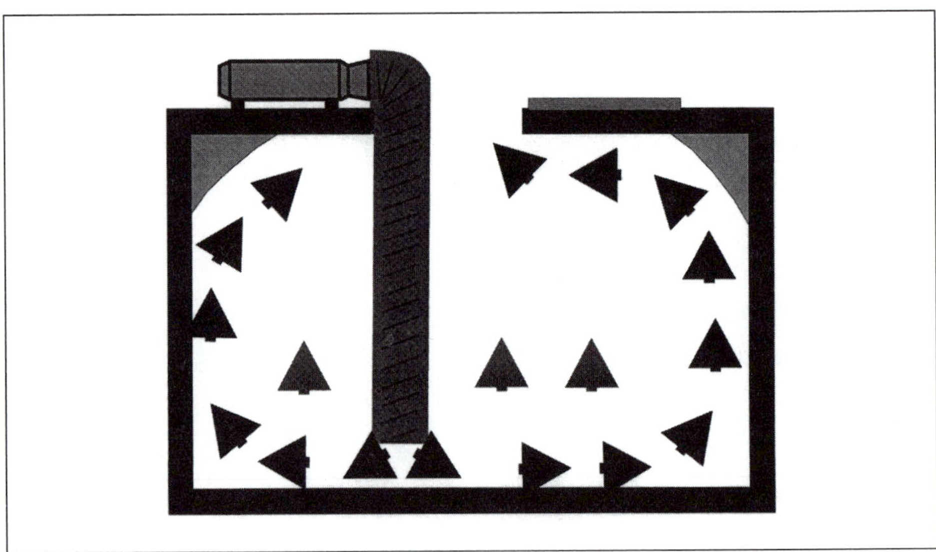

FIGURE 19-21
Exhausted Air Must Not Re-enter the Ventilation System

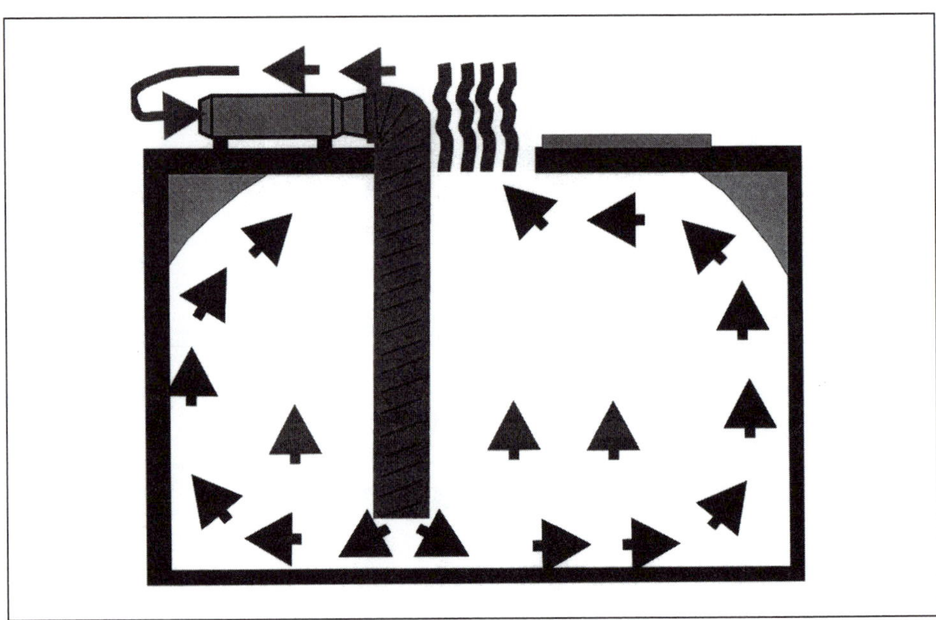

FIGURE 19-22
Obstruction Can Cause a Pocket of Contaminated Air to Remain

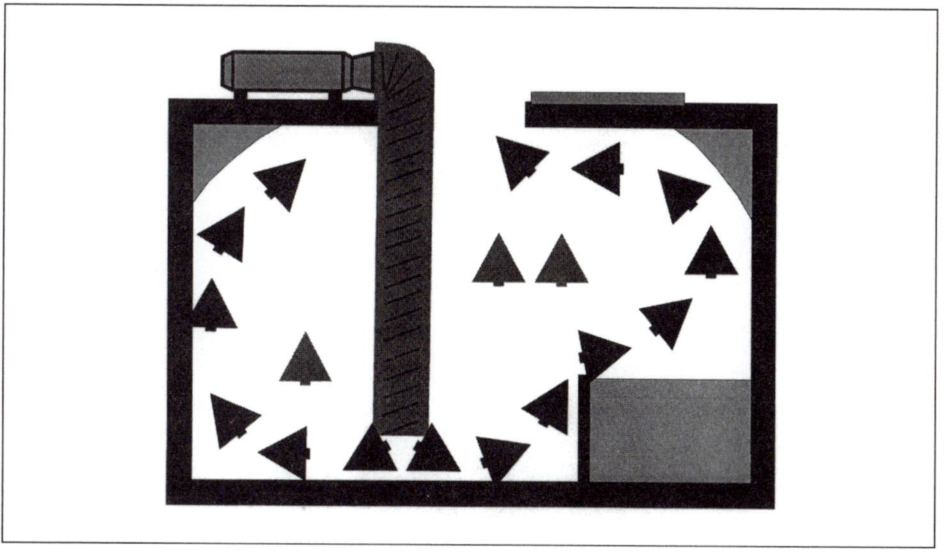

CHAPTER 20

Dealing with Temporary and Local Conditions

Indoor air quality situations discussed in this book have been either relatively permanent (e.g., production processes), or at least continual (e.g., cleaning and maintenance). Their permanence meant the pollution was not going to just go away, and so it forced managers to take corrective action. However, the situations described in this chapter *will* go away, and usually will not return in the near future. The air sometimes clears in the same day, but the potential harm to workers can remain.

DETERMINING THE TEMPORARY POLLUTANT SOURCE

Movable, gasoline-powered equipment is one of the most common temporary pollutant sources. They are often not a part of the company's regular operation. Typically they operate for a short time and then maybe not again for months, or not ever again. Because they are not a constant source of contaminants, it is easy to give low priority to correcting the problem. A few temporary pollutant sources are:

- Pressure washer
- Material-moving equipment
- Pesticide application
- Space heater
- Welder
- Electric generator
- Air compressor

The Main Source of Problems

Most of the air quality problems in the local and temporary category involve gasoline-powered equipment that provides muscle to make work easier and economically efficient. However, gasoline engines consume large amounts of oxygen, and leave in its place large amounts of carbon monoxide. As Chapter 17 pointed out, when carbon monoxide is in a person's lungs, hemoglobin in the blood picks it up much more readily than it picks up oxygen that the body needs. Table 17-2 showed that symptoms and effects (which range from mild headache to death) from even small concentrations of carbon monoxide become more serious as either the concentration or the person's time of exposure increases. NIOSH places a ceiling limit of 200 parts carbon monoxide per million parts of air, and a time weighted average exposure limit of 35 ppm for up to a 10-hour work-day during a 40-hour workweek. The ceiling is a limit that should not be exceeded, even for a short period of time. OSHA sets a permissible exposure limit at 50 ppm.

Pressure Washers

Pressure washers have proven to be efficient devices for cleaning many different types of surfaces as well as places that would be inaccessible with other methods. Therefore, the goal here is not to exclude them, but to learn to eliminate the indoor air quality problems they can cause.

Chapter 17 gave examples of workers who were victims of carbon monoxide poisoning from the use of pressure washers. Some of them even realized the potential danger and took steps to provide ventilation. One placed the gasoline-driven pressure unit outside the door, but suffered from fumes that blew in.

Two problems inherent in the use of pressure washers are release of carbon monoxide and products of the cleaning process. These problems require more attention than they sometimes receive.

Carbon Monoxide. All suggestions given later in this section apply to pressure washers. In addition, extra lengths of high pressure water hose can allow the pressure unit to be located outside, where its carbon monoxide cannot enter the building. The amount of hose is limited only by the length which will cause internal friction to affect pressure at the nozzle.

Products Released by Pressure Washing. As the word implies, "washing" removes soil and contamination from surfaces. Of course, the materials removed are wet and do not fly in the air to the extent they would from sand-blasting; but the mist does fly, and when it settles again and evaporates, it releases the contamination it carried. Forced ventilation of the area is not always sufficient to eliminate this problem. Following the wash with a flush, where the area's physical design permits it, solves any resettlement problem. When mats and other removable items are to be washed, it is best to take them outside for pressure washing.

Some pressure washers heat the water, and some even make steam for washing. Either can raise the area's relative humidity to the point of making it unlivable for others in the area; pressure washing and other activities should be coordinated so the area is unoccupied during washing.

When the waste water contains oil, grease, or other substances, it is considered hazardous waste and requires special handling. Oil and grease in waste water mist are probably not serious air quality concerns, but it would be advisable to determine before the washing what hazardous chemicals might be on the surfaces to be washed, and what precautions should be taken.

Pressure washing the building's outside walls does not usually create IAQ problems, except when removing fungus, mildew, moss, and loose materials from areas near the ventilation system intake. If possible, turn off the system that draws in air, and cover the intake while working in the area.

Material-Moving Equipment

A gasoline-powered forklift drives through an occupied area. Its presence might even be extended if it stops to deliver material to the area, or extended still further if it makes several trips, and sometimes stays in a factory to stack pallets of material. Another tool of convenience is the walk-along powered pallet handler. While these work savers make life easier for material handlers, they can also make life shorter by putting carbon monoxide and other combustion products into the plant's air.

Many plants have kept the efficiencies of powered material movers while avoiding the pollutants of gasoline engines by using electric-powered forklifts and pallet handlers. Little or no handling capacity is given up, because the following features are currently available in electric-powered units.

- Forklift load capacities of at least three tons
- Walk-along pallet truck capacities of at least four tons
- Speeds of up to 625 feet per minute with full load
- Full-load lifting speeds over 60 feet per minute
- Power steering

In addition to not contaminating the indoor air, electric versions of these popular labor-savers offer the extra benefit of reducing noise pollution.

Mitigating Pollution from Temporary Sources

Specific solutions were mentioned as they applied to the preceding examples of temporary pollution sources. The following solutions apply to all temporary pollution sources in all plants. Several suggestions are given because some are not feasible in certain plants, and some that are best in certain plants are less helpful in others. While some of the suggestions are laughable in a few situations, they can be exactly the solution that will solve the problem for another plant. Some of the solutions are admittedly drastic, but contaminated air, even on a temporary basis, can be a serious problem.

- Evacuate the area while using the equipment, and have the operator wear personal protective equipment. The increase in worker safety that results from this plan is unquestionable, but the cost/risk ratio is certainly high for some companies. High costs in the numerator of this ratio, and low risk in the denominator can make the ratio here higher than it is for other solutions.
- An alternative to evacuating the area is to do the task on a weekend or at another time when the area is not occupied. There may be a premium charge for weekend work, but the total cost may be less than it would be for evacuating the area—or for ignoring the problem.
- Don't do the polluting operation at all, or devise an alternative method. This solution can present management with a challenge in creativity, but asking the traditional problem-solving question, "Exactly what is it we want to do?" can bring about significant improvements in the way some tasks are handled. For example, one manufacturer felt that the efficiency of having a gasoline-driven forklift take 1000 subassemblies at a time, once

every two weeks, to the main assembly area was the only way to keep the product's price at a competitive level. Hand-carrying them in, or even transporting them by carts, would cause sales to go to competitors. They already had a small fleet of forklifts, some of which were often idle, and could not justify the expense of buying an electric-powered unit for the twice a month application. The answer—a small roller-type, gravity-powered conveyor—came during a brainstorming session.

- Ventilate heavily, and monitor. This chapter pointed out that a person's perception of adequate ventilation can be in error, and Chapter 17 gave some tragic examples. A consultant should be able to calculate what ventilation will be sufficient, or an expert on staff might plan the ventilation. It is seldom possible to determine a fan's rate of air displacement, but it is always possible to make a conservative estimate and then assume that the true displacement will actually be a small percentage of that estimate. Next, add a generous safety factor by assuming that, although all doors, windows, and other outside openings are unrestricted, there is significant resistance to the fans' displacement (be sure not to allow a pressure differential that will force polluted air to other parts of the building). Finally, after designing ventilation that will give more than adequate air flushing of, say, 10 air changes per hour, back it up with monitoring and alarms. Monitor several places in the area, especially corners that might not receive full benefit of the ventilation.

MANAGING LOCAL CONTAMINATION

Whether the contamination from an operation is local or general depends more on the removal procedure than on the operation itself. Contamination that goes into the area's atmosphere, and is drawn in by the ventilation system along with the rest of the area air is general contamination. However, when a hood or other ventilation intake is close to the operation, and picks up the contamination before it can get into the area's atmosphere, this arrangement is local contamination. For example, the pressure washers' carbon monoxide in the preceding section was general contamination. If a company produced miniature engines for model airplanes, and tested them under a hood that captured all the carbon monoxide, the operation would produce local contamination.

The definition of local contamination highlights its main benefit: When pollutants are drawn away directly, they do not linger in the area, even in diluted form. There can be no dangerous accumulation, and contaminants are removed before they enter the workers' breathing zone. In many situations it makes little difference whether contaminants are removed by general or local ventilation, but some situations require local ventilation. For example, Chapter 21, which explains regulations that cover working with asbestos, tells you to use local systems and be guided by American National Standard's ANSI Z9.2-1979. The chapter also tells you to provide a local exhaust ventilation system for all hand- and power-operated tools that produce or release asbestos fibers.

Ducted Local Exhaust[1]

Shields of industrial power tools such as grinders and saws are generally equipped with a fitting for convenient connection to a local exhaust duct. The duct can take dust and other machine output to the central ventilation ducts, to a self-contained dust collector unit for source capture control of dust and particulates, or to an outside air discharge. Other local collectors and systems are:

- Two-stage electrostatic precipitator to capture smoke at the point of generation from a small process. It can be mounted on an outside wall and is supplied by flexible ducts with hoods that can be placed directly at the work.
- Self-contained cartridge dust collector for general local capture. Ducts with flexible joints can be moved to be close to the work that generates contaminants.
- Portable source capture air cleaning unit. One model has an 8-inch diameter by 10-foot long swing arm ending in a hood that picks up smoke, dust, and other particulate matter within up to 14 inches. It removes 99.9 percent of particulate matter and discharges cleaned air back into the workspace.

Small Enclosed Areas

For small operations that generate contaminants, there are enclosures, usually about 3 to 10 cubic feet in volume. Most of them have exhaust ducts permanently attached, and they often have electrical pass-throughs and mechanical connections for remote control. In a typical application, a person sets the work up inside

---

1. Courtesy of United Air Specialists, Inc., Cincinnati, Ohio.

the enclosure, closes its door, and initiates the operation with a switch or mechanical control. The door, and sometimes the entire enclosure, are usually made of glass or clear plastic, allowing the operator to observe the work as it progresses. Except for specialized small units, this procedure is seldom used in manufacturing; it is more suited to laboratory and test work.

Some enclosures connect directly to a central exhaust system, and some have a duct that takes their contaminated air outside or to a special filtering or air-cleaning unit. With either system it is important to ensure that the exhaust system is functioning and that it maintains a negative pressure inside when the door is closed. Because some enclosures are not designed to close and seal tightly, they depend on negative pressure to prevent contaminated air from leaking into the workers' general atmosphere.

CHAPTER 21

Handling Intractable and Controversial Problems

Some claim that labeling asbestos as dangerous is incorrect and that, whether it is a hazard or not, it is no more of a hazard than the materials that are recommended to replace it. A court in Canada, after hearing nearly two years of testimony, dismissed a multi-million dollar claim, with the judge saying, "Based on the medical evidence that I have accepted, I find that exposure to the low levels of asbestos fibres typically found in the air circulating in public buildings poses no appreciable risk to the health of building occupants." However, OSHA and NIOSH have categorically identified asbestos as a carcinogen, and have promulgated regulations that building owners and facility managers must meet.

Radon is more prevalent in some parts of the world than in other parts, and sometimes requires definite measures to keep it from seeping from the ground into buildings. There are various methods for detecting it, and measures that can be taken to mitigate the risks involved when it is present.

For hundreds of years tobacco smoke has been a part of society, although it was generally accepted as "bad for you." It has been only during the past couple of decades that serious studies have examined and quantified the effects of smoking on smokers' health, and those results led to studies of the effects of tobacco smoke from other people on nonsmokers. Again, reactions to the studies' findings range from complete rejection, to claims that the hazards are overstated, to grateful acceptance. This chapter looks at the evidence and at the regulations that OSHA enforces.

ASBESTOS

One reason asbestos has become such a big issue is that it is everywhere. It is such a convenient and useful material because it withstands high heat without burning, is an efficient insulator, and is durable. Asbestos can be formed and molded into a variety of products that have been used in industrial and residential buildings, inside and out, basement to roof.

However, in spite of its strength and durability, the products made from asbestos occasionally break down and release dangerous particles. Sometimes normal aging of the products weakens the binding agent and releases asbestos fibers. Other times cleaning, abrading, drilling, cutting, damaging, moving, normal use, and other actions cause the product to crumble and release fibers. The fibers are small and light; they travel in the air, settle on surfaces, and cling to clothes.

Once airborne, the fibers are easily inhaled and can cause cancer of the lungs and stomach. We do not see the fibers because they are smaller than some dust; most are less than 1/1000 the thickness of a human hair. They accumulate in the lungs with repeated exposures. It is difficult to quantify the damage they cause; symptoms of asbestosis, cancer, and mesothelioma may not appear for 10 to 40 years after the exposure. Although the hazard of low level exposure is not defined, experts who study the material's effects cannot identify a low level that is completely safe.

One activity that clearly exacerbates the effect of inhaling asbestos fibers is smoking. A study over 10 years from January 1, 1967 to December 31, 1976 divided 81,983 men (an unusually large sample) into four categories: asbestos workers who smoked and didn't smoke, and nonasbestos workers who smoked and didn't smoke. Those who did not work with asbestos formed the two control groups. Table 21-1 shows the death rates and mortality ratios, with the nonsmoking control group defined as a mortality ratio of unity.

When nonsmokers were compared to nonsmokers, and when smokers were compared to smokers (eliminating smoking as a factor), the table shows that an asbestos worker is about five times more likely to die of lung cancer than someone who is not exposed to asbestos. However, an asbestos worker who smokes approximately doubles the risk from his asbestos exposure.

TABLE 21-1
Smoking Exacerbates the Effects of Asbestos Exposure

Groups	*Exposure to Asbestos*	*History of Cigarette Smoking*	*Death Rate**	*Mortality Rate*
Non-smokers				
Control	No	No	11.3	1.00
Asbestos Workers	Yes	No	58.4	5.17
Smokers				
Control	No	Yes	122.6	10.85
Asbestos Workers	Yes	Yes	601.6	53.24

* Rates per 100,000 man-years standardized for age on the distribution of the man-years of all the asbestos workers. Number of lung cancer deaths based on death certificate information.

Testing for Asbestos

The two tests outlined here should be conducted by qualified personnel with adequate instruments. First is the *bulk sampling* test, which involves having a certified person wet a surface and then take a small core or scraping. He or she sends the sample to an approved laboratory for analysis. If the laboratory uses Polarized Light Microscopy (PLM), it can determine the type of asbestos as well as the percentage. The EPA defines asbestos-containing material (ACM) as containing more than one percent asbestos, but other organizations use tighter or looser definitions.

The second technique, *air sampling*, takes its sample from the air instead of examining material that settles out of the air. A major advantage to this technique is that it directly checks the air that workers breathe, so it is preferred as proof that a site has been completely cleaned. Air sampling uses one or both of two types of microscopy: Phase Contrast Microscopy (PMC) and Transmission Electron Microscopy (TEM). The first costs less, but could miss smaller fibers.

Resolving an Asbestos Problem

The least expensive way to resolve an asbestos problem is to simply eliminate access to the area. Of course, this method can be considered only if the problem is limited to a small area whose access is not important for the building's functions.

Some buildings already have areas that are seldom entered, such as crawl spaces and pipe shafts. However, if they are to be closed off as a means of resolving an asbestos problem, they must be sealed and labeled, so that if entry becomes necessary, the person will take proper precautions. Another method that often receives approval is to encapsulate the asbestos to prevent fibers from breaking loose. Some encapsulants can be applied as easily as paint.

If it would not be practicable to seal off the area (sometimes the entire building is involved), and encapsulating would not prevent further damage that would release fibers, it will be necessary to remove the asbestos. The procedure is expensive and must be performed by certified professionals who have the proper equipment. They close off an area and keep it ventilated, with exhaust air filtered to prevent the escape of any fibers. Ventilation also keeps the area under negative pressure so that any air leaks bring air in, not out to areas that are not under control. Upon completion of the removal, the area must tested and certified as clean and safe to occupy.

Regulations and Requirements

Because asbestos fibers are recognized as carcinogenic (cancer-causing), a number of regulations have been promulgated to protect workers, occupants, tenants, children, and others. The federal government has initiated some regulations; state and local governments frequently write codes and other constraints that are stricter. A few of the main regulators are:

- OSHA—aimed mostly at protecting workers
- AHERA—aimed at making schools free of asbestos danger
- ASHARA—applies to public buildings
- NESHAPS—regulates emissions into open air

The rest of this section summarizes features of OSHA's *Asbestos Standard for General Industry*. It is included for general background information and planning, not for control or final action. Be sure to base decisions on a full copy of the latest regulations and, preferably, the experience of an expert. The full standard includes more details and explains consultations and other assistance that OSHA offers.

Scope and application. OSHA's revised standards were developed in recognition of vastly different conditions prevailing in workplaces for general industry (29 Code of Federal Regulations [CFR] 1910.1001), for the shipyard industry (29 CFR 1915), and the construction industry (29 CFR 1926.1101). Information in this section applies to all occupational exposure to asbestos in general industry. With more than 685,000 workers affected by the new standard, OSHA's estimate of avoiding 42 additional cancer deaths per year is considered conservative. This estimate is in addition to the lives saved of those peripherally exposed to asbestos and the lives saved by earlier OSHA standards.

Provisions of the standard. OSHA sets out several provisions employers must follow to comply with the asbestos standard. The agency has established strict exposure limits and guidelines for exposure monitoring, medical surveillance, recordkeeping, regulated areas, and communication of hazards.

Permissible Exposure Limits (PELs). Employers must ensure that no employee is exposed to an airborne concentration of asbestos in excess of 0.1 fiber per cubic centimeter of air (0.1 f/cc) as averaged over an 8-hour time weighted average (TWA) day.

Excursion Limit (EL). Employers must ensure that no employee is exposed to an airborne concentration of asbestos in excess of 1.0 f/cc as averaged over a sampling period of 30 minutes. OSHA adopted the term "excursion limit" for referring to the short-term permissible exposure limit to be consistent with the terminology used by the American Conference of Governmental Industrial Hygienists (ACGIH).

Exposure monitoring. Except for clutch and brake repair facilities where a "preferred" control method is used, each employer who has a workplace or work operation covered by this standard must assess all asbestos operations for their potential to generate airborne fibers. Where exposure may exceed the PEL, employee exposure measurements must be made from breathing zone air samples representing the 8-hour TWA and 30-minute EL for each employee.

Employers must reinitiate monitoring whenever there has been a change in the production, process, control equipment, personnel, or work practices that may result in new or additional exposures to asbestos above the PEL and/or EL, or when the employer has reason to suspect that a change may result in new or additional exposures above the PEL and/or EL.

Medical surveillance. Employers must institute a medical surveillance program for all employees who are or will be exposed to airborne concentrations of asbestos at or above the PEL and/or EL. All medical examinations and procedures must be performed by or under the supervision of a licensed physician. Such examinations must occur at a reasonable time and place and be provided at no cost to the employee. At a minimum such examinations must include a medical and work history; a complete physical examination with emphasis on the respiratory system, the cardiovascular system, and the digestive tract; a chest x-ray; pulmonary function tests; respiratory disease standardized questionnaire as set forth in 29 CFR 1910.1001 Appendix D, Part 1 of the standard; and any additional tests deemed appropriate by the examining physician. These examinations must be made available annually. Chest roentgenogram must be conducted in accordance with Table 21-2.

An abbreviated standardized questionnaire (see CFR 1910.1001 Appendix D Part 2 of the standard) must be administered to employees. When an employee terminates, the employer must provide a termination of employment medical examination to the employee within 30 calendar days before or after the date of termination. If adequate records exist that show the employee has been examined in accordance with the standard within the past year, no additional medical examination is required. A preemployment medical examination may not be used unless the employer pays for it.

Employers must provide the examining physician with a copy of the standard and Appendices D and E. The physician must also be given the following for each employee to be examined:

- Description of duties as they relate to his or her asbestos exposure
- Employee's actual or anticipated exposure level
- Description of any personal protective and respiratory equipment used or to be used
- Information from previous medical examinations

Upon completion of the examination, the employer must obtain a written signed opinion from the physician. It must include at least the following:

- Results of the medical examination
- Physician's opinion as to whether the employee has any detailed medical conditions that would place him or her at an increased risk from exposure to asbestos. The written opinion is not to reveal specific findings or diagnoses unrelated to occupational exposure to asbestos.
- Any recommended limitations on the employee or on the use of personal protective equipment
- A statement that the employee has been informed by the physician of the results of the examination
- A statement that the employee has been informed by the physician of the

TABLE 21-2
Frequency of Chest Roentgenogram

Years Since First Exposure	*Age of Employee*		
	15 to 35	***35+ to 45***	***45+***
0 to 10	Every 5 years	Every 5 years	Every 5 years
10+	Every 5 years	Every 2 years	Every 1 year

increased risk of lung cancer attributable to the combined effect of smoking and asbestos exposure

Regulated areas. Employers are required to establish and set apart a regulated area wherever airborne concentrations of asbestos and/or presumed asbestos-containing material exceed the PEL and/or EL. Only authorized personnel may enter regulated areas. All persons entering a regulated area must be supplied with and be required to use an appropriate respirator. Note that employers cannot leave it up to employees to wear the equipment or not. Implied in that statement (and in other regulations) is a requirement that employers train employees in respirator use. No smoking, eating, drinking, chewing tobacco, or applying of cosmetics is to be allowed in regulated areas.

Warning signs must be provided and displayed at each regulated area and must be posted at all approaches to regulated areas. Where necessary, signs must bear pictures or graphics, or be written in appropriate language so that all employees understand them. The signs must bear the following information:

Danger
Asbestos
Cancer and Lung Disease Hazard
Authorized Personnel Only
Respirators and Protective Clothing Required in this Area

In addition, warning labels must be affixed to all asbestos products (raw material, mixtures, scrap) and to all containers of asbestos products, including waste containers, that may be in the workplace. The labels must comply with the requirements of 29 CFR 1910.1200(f) of OSHA's Hazard Communication standard and must include the following information:

Danger
Contains Asbestos Fibers
Avoid Creating Dust
Cancer and Lung Disease Hazard

Labels or Material Safety Data Sheets (MSDSs) are not required where asbestos fibers have been modified by a bonding agent, coating, binder, or other materials, if the manufacturer can demonstrate that during handling, storage, disposing, processing, or transporting no airborne concentrations of asbestos fibers in excess of PEL and/or EL will be released when asbestos is present in a product at a concentration of less than 2.0 percent.

Information and training. Employers must develop a training program for all employees who are exposed to airborne concentrations of asbestos at or above the PEL and/or EL. Training must be provided prior to or at the time of initial assignment and at least yearly after that. Training programs must inform employees about ways in which they can safeguard their health.

In addition, employers must provide awareness training for employees who do housekeeping operations in facilities where materials that contain or are presumed to contain asbestos are present. The elements of the training must include health effects of asbestos; locations, signs of damage and deterioration of products; proper response to fiber release episodes; and where the housekeeping requirements are found in the standard. This training must be held annually and conducted so that all employees understand it.

All training materials must be available to employees without cost. When requested, the materials must be provided to the Assistant Secretary for OSHA and/or the Director of NIOSH.

Methods of meeting requirements. To the extent feasible, engineering and work practice controls must be used to reduce and maintain employee exposure at or below the PEL and/or EL. The standard, therefore, requires employers to institute the following measures:

- Design, construct, install, operate, and maintain local exhaust ventilation and dust collection systems according to the American National Standard Fundamentals Governing the Design and Operation of Local Exhaust Systems, ANSI Z9.2-1979.

- Provide a local exhaust ventilation system for all hand- and power-operated tools such as saws, scorers, abrasive wheels, and drills that produce or release asbestos fibers.
- Handle, mix, apply, remove, cut, score, or work asbestos in a wet state to prevent employee exposure.
- Do not remove cement, mortar, coating, grout, plaster, or similar material containing asbestos from containers that are being shipped without wetting, enclosing, or ventilating them.
- Do not sand floors containing asbestos.
- Do not use compressed air to remove asbestos or materials containing asbestos unless the compressed air is used in conjunction with a ventilation system designed to capture all dust moved by the compressed air.

Where engineering and work practice controls have been instituted but cannot reduce exposure to the required level, employers must supplement them by using respiratory protection. General use of respirators was covered in Chapter 7. Employee rotation cannot be used as a means of compliance with the PEL and/or EL.

When the PEL and/or EL is exceeded, employers must establish and implement a written program to reduce employee exposure to or below the PEL and EL by means of engineering and work practice controls and by the use of respirators where required and permitted.

Every employee who is exposed to airborne concentrations of asbestos that exceed the limits, must be supplied with free protective clothing such as coveralls or similar full-body clothing, head coverings, gloves, and foot coverings. Employers are responsible for ensuring that employees use protective clothing and other equipment. Wherever the possibility of eye irritation exists, face shields, vented goggles, or other appropriate protective equipment must be provided and worn.

Asbestos-contaminated work clothing must be removed in designated changing rooms and stored in closed, labeled containers that prevent dispersion of the fibers into the ambient environment. Protective clothing and equipment must be cleaned, laundered, repaired, or replaced to maintain effectiveness.

Employees who are required to work in regulated areas must be provided with clean changing rooms, shower facilities, and lunchrooms. Changing rooms must have two separate lockers or storage facilities—one for contaminated clothing, the other for street clothing. They must have sufficient separation to prevent accidental contamination of street clothing. Employees must shower at the end of the shift and cannot leave the workplace wearing any clothing or equipment worn during the work shift. Lunchroom facilities must have a positive pressure-filtered air supply and must be readily accessible to employees.

Employers are responsible for ensuring that no employee enters lunchroom facilities with protective work clothing or equipment unless surface asbestos fibers have been removed by vacuuming or other method that removes dust without causing the asbestos to become airborne. Employers must also ensure that employees wash their hands prior to eating, drinking, or smoking.

RADON

Radon is a radioactive gas that occurs in nature. It comes from the natural breakdown (radioactive decay) of uranium, and can be found in high concentrations in soils and rocks containing uranium, granite, shale, phosphate, and pitchblende. Radon is sometimes found in soils contaminated with certain types of industrial wastes, such as the byproducts from uranium or phosphate mining. In outdoor air, radon is diluted to such low concentrations that it is usually harmless. However, when radon leaks into an enclosed space, it can accumulate to unhealthy levels, depending on its concentration in the underlying soil and on the building's construction.

The only known health effect associated with exposure to elevated levels of radon is an increased risk of developing lung cancer. Scientists estimate that between 5000 and 20,000 lung cancer deaths a year in the United States may be attributed to radon. The probability of developing lung cancer from exposure to radon depends on the concentration and on the length of time exposed, but the relationship is not a linear one (i.e., doubling the time exposed and halving the concentration do not result in the same risk). In general, increasing either the concentration or the exposure time increases the risk, but exposure to a slightly elevated level for a longer time seems to present a higher risk than exposure to a significantly higher level for a shorter time.

Table 21-3 shows eight ranges of radon concentration, in picocuries per liter of air (pCi/l), which correspond to levels from average outdoor concentration to 1000 times average outdoor concentration. For each range, the table gives a comparative multiple of average outdoor concentration. The last column puts the risk in perspective by giving a comparable risk from smoking or x-rays. EPA estimates that, at any radon risk level for never-smokers, the risk for smokers is 15 to 20 times higher; for former smokers the risk may be eight times higher.

Most of the literature today describes radon in homes, but the facts apply also to industrial plants, especially low ones built directly on the ground. Some investigators claim that the effects diminish rapidly above the third floor of a building,

TABLE 21-3
Radon Risk Comparisons

Radon Concentration (pCi/l)	*Comparable Exposure Level*	*Comparable Risk*
0.2 to 1.0	Average outdoor level near the low end of this range; average indoor level near the top	20 chest x-rays per year falls in the lower half of this range
1 to 2	About 5 to 9 times average outdoor level	Nonsmoker risk of dying from lung cancer about in the middle of this range
2 to 4	Ten times average outdoor level falls near the low end of this range	200 chest x-rays per year falls near the middle of this range
4 to 10	Ten times average indoor level falls near the top of this range	Five times a nonsmoker's risk falls about the middle of this range
10 to 20	100 times average outdoor level falls near the top of this range	One pack a day smoker's risk falls about the middle of this range
20 to 40	200 times average outdoor level falls near the top of this range	Two packs a day smoker's risk falls in the upper half of this range
40 to 100	100 times average indoor level falls in the upper half of this range	20,000 chest x-rays per year would be in about the middle of this range
100 to 200	1000 times average outdoor level falls at about the upper end of this range	Four packs a day smoker near the low end; more than 60 times a nonsmoker's risk near the upper end

although that conclusion could be negated by a ventilation system that picks up radon at lower floors and distributes it to higher floors.

Detecting the Presence of Radon

Because we cannot see or smell radon, special equipment is needed to detect it. The two most popular commercially available detectors are the charcoal canister and the alpha track detector. Both are used by exposing them to air for a specified period and then sending them to a laboratory for analysis. Test periods are three to seven days for the charcoal canister, and two to four weeks or longer for the alpha track detector. Other detection methods require trained professionals, and are considerably more expensive.

Radon Mitigation Methods

Because earth and rocks are the source of radon, mitigation methods are usually designated for a building's ground-level floor or for the lowest level. The basics on which mitigation methods are designed include:

- Radon typically seeps from the ground into a building through cracks and other leaks; therefore sealing known and possible leaks is an important measure.
- Allowing critical areas of a building to be at negative pressure with respect to the adjacent soil will make it more likely that radon will be drawn into the building.
- Radon is a gas that is most harmful when allowed to accumulate; therefore ventilation is an effective method for reducing radon that cannot be prevented from entering the building.

Sealing cracks and openings. Keeping radon out in the first place is always a good starting place. Places to check include cracks in continuous concrete surfaces (walls and floors), spaces that open in block walls due to mortar shrinkage, joints between walls and floors, and openings around pipes, flues, and other objects that pass through walls and floors.

Natural ventilation. Opening windows, doors, louvers, and other devices that can be controlled often provides sufficient exchange of outdoor air for radon-laden indoor air. If the lowest level is a crawl space, this mitigation method means ensuring that ventilation openings are not blocked. Whenever possible, provide openings on all sides of the structure, or at least on opposite sides, to achieve good air movement. Drawing air out and not replacing it with an equal amount of outside air, creates a negative pressure that can cause additional radon to enter; opening windows on just the building's lee side can bring about this situation.

If the windows or other openings already exist, there is no equipment or operating cost for mitigating radon with natural ventilation. However, when the area to be ventilated uses conditioned air (heated or cooled), the cost of replacing it could be significant.

Forced ventilation. Instead of allowing naturally circulating air to perform the exchange, this method uses a fan, blower, or other device to move the air. The main advantages over natural ventilation are reliability and control. As long as there is electric power, air will exchange at a given rate which does not depend on outside weather or wind shifts. The disadvantage is cost—of original equipment, installation, and operation. Natural ventilation and forced ventilation both involve the cost of wasting conditioned air. Although it would appear that having the fan exhaust the air results in the same amount of air exchange as having the fan bring air in, it is best to locate the fan at the intake. Forcing exhaust air creates a partial vacuum in the building, which could cause additional radon to be drawn in.

Heat recovery ventilation. In a variation of the forced ventilation method, some radon mitigation methods pass the exhaust air through a heat recovery ventilator (often called an air-to-air heat exchanger). In the exchanger, incoming air draws heat (or cold) from the outgoing air, thus reducing the loss of conditioning that has been applied to the air. A major difference from previous methods is that windows, doors, and other openings are kept closed with heat recovery ventilation. This method provides more control over ventilation.

The original cost of equipment and installation is higher than with previous methods, and operating cost will be higher because a larger blower is necessary for overcoming the airflow resistance of the heat exchanger.

Air supply. Furnaces and other equipment that draw air from the area in which they operate can be provided with intakes to take air directly from outdoors. This method prevents the reduced pressure that results when equipment draws air from indoors, and therefore reduces the tendency to draw radon through small openings in floors and walls.

Covering exposed earth. The concrete floors of some buildings, especially the basements of older buildings, are incomplete or have large openings for access to utilities, plumbing, or sumps. Radon entry to the area can be reduced by covering the areas, usually with sheet metal, and running a pipe through the covering to the outside. A fan will ensure that the pipe draws off radon. It is important to seal all joints such as where sheet metal joins concrete, where the pipe penetrates the sheet metal, and where the pipe goes through to the outside.

Drain-tile suction. Some foundations have water drained away by perforated pipes called drain tiles. If the tiles form a continuous loop around the foundation, they may be used to pull radon from the surrounding soil and vent it away from the building. With this method, a pipe connected to the drain tile loop carries water and radon-laden air away from the building, to the center of a tee connection. A pipe goes up from one side of the tee, to a fitting with a fan that draws radon from the drain tiles. From the other side of the tee, water from the drain tiles goes to a trap, and overflow goes out a drain exit. When water is not flowing out the drain exit, the trap keeps air from coming in the drain exit and slowing the flow of radon-laden air through the fan.

Block-wall ventilation. When a foundation is made of hollow blocks, radon can fill the spaces and then leak into other parts of the building. One method of radon reduction, called "wall suction," is to tap a pipe into the hollow space, and run the pipe to an outside fan that will draw radon away. Another method, called "wall pressurization," is to blow air into the hollow space to maintain a pressure that will keep radon from entering the space.

Sub-slab suction. When the lowest floor of a building is a concrete slab over soil or on top of aggregate, radon can accumulate under the slab, building a pressure that will force it into the building. One mitigating technique is to drill holes through the slab and install pipes to penetrate into the soil or aggregate. The pipes are connected to a main line that passes through a wall to the outside, where a fan draws off the radon. It is important to provide a good seal where the pipes pass through the slab.

Comparison of methods. Table 21-4 summarizes costs and other features of radon mitigation methods. The expected maximum reductions given represent the best that a single method can accomplish. Actual maximums achieved might be higher or lower, depending on unique characteristics of the building. Several methods may have to be combined to achieve acceptable results, especially when initial radon levels are high.

TABLE 21-4

Summary of Radon Mitigation Methods

Method	***Installation Cost***	***Operating Cost***	***Expected Maximum Reduction***	***Comment***
Natural ventilation	Minimal	Very high to moderate*	90%	Useful, intermediate step to reduce high radon levels
Forced ventilation	Low	Very high to moderate*	90%	More controlled than natural ventilation
Air supply	Low to moderate	Low	Site specific	May be required to make other methods work
Heat recovery ventilation	Moderate to high	Moderate	90+%	Air intake and exhaust must be equal
Covering exposed earth	Moderate	Low	Site specific	Required to make most other methods work
Sealing cracks and spaces	Minimal to moderate	None	Site specific	Required to make most other methods work
Drain-tile suction	Moderate to high	Low	97+%	Works best when drain tiles are continuous, unblocked loop
Block-wall ventilation	High to very high	Low	97+%	Applies to block wall basements. Sub-slab suction may be needed to supplement
Sub-slab suction	High to very high	Low	97+%	Important to have good aggregate or highly permeable soil under slab

* Cost of operating natural ventilation can be moderate for crawl space; can be very high for lowest floor when conditioned air is wasted.

Assistance from State Agencies. The following list gives telephone numbers of agencies in each state that offer assistance and information about radon and mitigation techniques.

Alabama 800-582-1866
Alaska 800-478-8324
Arizona 602-255-4845
Arkanses 501-661-2301
California 800-745-7236
Colorado 800-846-3986
Connecticut 203-566-3122
Delaware 800-554-4636
D.C. 202-727-5728
Florida 800-543-8279
Georgia 800-745-0037
Hawaii 808-586-4700
Idaho 800-445-8647
Illinois 800-325-1245
Indiana 800-272-9723
Iowa 800-383-5992
Kansas 913-296-6183
Kentucky 502-564-3700
Lousiana 800-256-2494
Maine 800-232-0842
Maryland 800-872-3666
Massachusetts 413-586-7525
Michigan 800-723-6642
Minnesota 800-798-9050
Mississippi 800-626-7739
Missouri 800-669-7236
Montana 406-444-3671
Nebraska 800-334-9491
Nevada 702-687-5394
New Hampshire 800-852-3345 X4674
New Jersey 800-648-0394
New Mexico 505-827-4300
New York 800-458-1158
North Carolina 919-571-4141
North Dakota 701-221-5188
Ohio 800-523-4439
Oklahoma 405-271-1902
Oregon 503-731-4014
Pennsylvania 800-237-2366
Puerto Rico 809-767-3563
Rhode Island 401-277-2438
South Carolina 800-768-0362
South Dakota 800-438-3367
Tennessee 800-232-1139
Texas 512-834-6688
Utah 800-536-4250
Vermont 800-640-0601
Virginia 800-468-0138
West Virginia 800-922-1255
Wisconsin 800-267-4795
Wyoming 800-458-5847

Radon Mitigation Standards

The 1988 Indoor Radon Abatement Act (IRAA) required the EPA to develop a voluntary program to evaluate and provide information on contractors who offer radon control services. The Radon Contractor Proficiency (RCP) Program was established to fulfill this portion of the IRAA, initially to homeowners. Since then there have been interim standards and improved techniques which led to *Radon Mitigation Standards* by the EPA in 1993, which was revised in 1994. Officially listed as EPA 402-R-93-078, it is a good document for anyone, homeowners and businesses, who plans to have a contractor perform radon mitigation work.

The following table of contents shows the document's coverage:

TOBACCO SMOKE

As mentioned, asbestos, radon, and tobacco smoke all cause lung cancer, and combining tobacco exposure with either of the others greatly multiplies the risk. Since 1964, when the Surgeon General issued the first report on smoking and health—which concluded that cigarette smoking causes lung cancer—employers, employees, and unions have become increasingly aware and convinced of the harm being caused to health and longevity. Continuing research and studies on the toxicity and carcinogenicity of tobacco smoke are amassing proof that the health risk is not limited to the smoker, but includes those who breathe environmental tobacco smoke (ETS), also known as second hand smoke. Recent epidemiologic studies support and reinforce earlier published reviews by the Surgeon General and the National Research Council demonstrating that exposure to ETS can cause lung cancer in nonsmokers. In addition, studies now show a possible association between exposure to ETS and increased risk of heart disease in nonsmokers.

Because of the collective weight of evidence, NIOSH considers ETS to be a potential occupational carcinogen and recommends that employers take steps to reduce exposures to the lowest feasible concentration. All available preventive measures should be used to minimize occupational exposure to ETS. In 1993 the EPA classified ETS as a serious cancer threat and issued guidelines urging every company to have a policy protecting nonsmokers from involuntary exposure. The remainder of this chapter looks at the dangers and at the recommendations to employers.

Components of Tobacco Smoke

To understand the results of research studies on tobacco smoke, you must recognize the three categories of smoke that investigators use.

- *Mainstream smoke (MS)*—Smoke that is drawn through the remaining tobacco (and filter, if present) and into the smoker's mouth.
- *Sidestream smoke (SS)*—Smoke from a combination of two conditions: 1) smoldering tobacco between puffs, and 2) smoke diffusing through the cigarette paper plus smoke escaping from the burning cone while the smoker draws a puff.
- *Environmental tobacco smoke (ETS)*—Smoke in the ambient atmosphere composed of sidestream smoke and exhaled mainstream smoke.

ETS contains many of the toxic agents and carcinogens that are present in MS, but generally in diluted form. The major source of ETS is SS, which contains higher amounts of some toxic and carcinogenic agents than MS when it is obtained in its undiluted form under laboratory conditions. For example, the release of volatile N-nitrosamines and aromatic amines is higher in SS than in MS.

A major reason that undiluted SS and MS have different concentrations of toxic and carcinogenic agents is that peak temperatures in the burning cone of a cigarette reach 800° to 900°C when the smoker draws a puff, but only 600°C between puffs. The result is less complete combustion of tobacco during generation of SS. In addition, most of the burning cone is oxygen deficient during smoldering and produces a strongly reducing environment. Table 21-5 lists 26 toxic and carcinogenic agents identified in SS and MS.

TABLE 21-5
Toxic and Carcinogenic Agents in Undiluted Cigarette SS

Compound	Type of Toxicity	Amount in SS*	Ratio SS/MS
Vapor phase:			
Carbon monoxide	Toxic	26.8 - 61 mg	2.5 - 14.9
Carbonyl sulfide	Toxic	2 - 3 mg	0.03 - 0.13
Benzene	Carcinogenic	400 - 500 μg	8 - 10
Formaldehyde	Carcinogenic	1,500 μg	50
3-Vinylpyridine	Suspected Carcinogen	300 - 450 μg	24 - 34
Hydrogen cyanide	Toxic	14 - 110 μg	0.06 - 0.4
Hydrazine	Carcinogenic	90 ng	3
Nitrogen oxides (No_x)	Toxic	500 - 2,000 μg	3.7 - 12.8
N-nitrosodimethylamine	Carcinogenic	200 - 1,040 ng	20-130
N-nitrosopyrrolidine	Carcinogenic	30 - 390 ng	6 - 120
Particulate phase:			
Tar	Carcinogenic	14 - 30 mg	1.1 - 15.7
Nicotine	Toxic	2.1 - 46 mg	1.3 - 21
Phenol	Tumor promoter	70-250 μg	1.3 - 3
Catechol	Cocarcinogenic	58 - 290 μg	0.67 - 12.8
o-Toluidine	Carcinogenic	3 μg	18.7
2-Napthylamine	Carcinogenic	70 ng	39
4-Aminobiphenyl	Carcinogenic	140 ng	31
Benz(a)anthracene	Carcinogenic	40-200 ng	2 - 4
Benzo(a)pyrene	Carcinogenic	40 - 70 ng	2.5 - 20
Quinoline	Carcinogenic	15 - 20 μg	8 - 11
N'-nitrosonornicotine	Carcinogenic	0.15 - 1.7 μg	0.5 - 5.0
NNK	Carcinogenic	0.2 - 1.4 μg	1 - 22

Continued

TABLE 21-5
Continued

Compound	*Type of Toxicity*	*Amount in SS**	*Ratio SS/MS*
Particulate phase:			
N-nitrosodiethanolamine	Carcinogenic	43 ng	1.2
Cadmium	Carcinogenic	0.72 μg	7.2
Nickel	Carcinogenic	0.2 - 2.5 μg	13 - 30
Polonium-210	Carcinogenic	0.5 - 1.6 pCi	1.06 - 3.7

* Amount in SS per cigarette

Although ETS is diluted before being breathed by a person, the exposure is continuous during the time he or she spends in the ETS-polluted environment. The smoker is only exposed to MS during the time it takes to smoke each cigarette.

NIOSH Conclusion

Workers should not be involuntarily exposed to tobacco smoke and therefore to the risk of developing cancer. The best method for controlling worker exposure to ETS is with a two-part program. Part 1 is to eliminate tobacco use from the workplace while implementing a smoking cessation program for workers. Until tobacco use can be eliminated, part 2 is to protect nonsmokers from ETS by isolating smokers. These two parts are described in the following paragraphs.

Eliminating tobacco use. It is best if management, workers, and workers' representatives work together to develop appropriate policies. Suggestions include:

- Prohibit smoking at the workplace and provide sufficient disincentives for those who do not comply.
- Distribute information about health promotion and the harmful effects of smoking.
- Offer smoking-cessation classes to all workers.
- Establish incentives to encourage workers to stop smoking.

Isolating smokers. Because it is important to achieve cooperation for eliminating smoking in the workplace, the preceding steps will take time. During that time, employers should take action to reduce exposure of nonsmokers to the lowest feasible concentration. The 1986 Surgeon General's report on involuntary smoking concluded that, "the simple separation of smokers and nonsmokers within the same airspace may reduce, but does not eliminate, the exposure of nonsmokers to ETS." In indoor workplaces where smoking is permitted, ETS can spread throughout the airspace of all workers. Therefore it is important for employers to:

- Provide separate, enclosed areas for smoking, with separate ventilation.
- Exhaust the air from this area directly outside and do not recirculate it within the building or mix it with general dilution ventilation for the building.
- Ensure that ventilation of the smoking area meets general ventilation standards and that the area has a slight negative pressure relative to adjacent nonsmoking areas.
- Post warning signs at entrances to the workplace stating that smoking is prohibited except for designated smoking areas.
- Clearly identify the designated smoking areas.

Smoking should not be permitted right outside doors or near building ventilation intakes where the smoke can enter the building. The designated smoking area must not be a work area where any workers are required to be, and it must not be a place that nonsmokers have to pass through in the course of their work, or to enter or exit the building.

Provide indoor smoking areas with at least 60 cubic feet per minute of supply air per smoker. Many buildings supply this air from other parts of the building, such as corridors. The smoking area exhaust port must not be located in a way that will allow its output to be taken into the inlet for general building ventilation.

CHAPTER 22

IAQ Standard for New Buildings and Major Renovations

The last few years have seen an explosion in IAQ interest and concern, and one result has been rapid changes in associated technology. Anyone who is planning a new building or a major renovation to an existing building will find significant changes in building codes and in HVAC and other equipment.

ASHRAE Standard 62-1989, by the American Society of Heating, Refrigerating and Air-Conditioning Engineers, has been adopted by building codes and is presently applied to all types of indoor environments. It has been updated several times to reflect new information about occupant tolerance, industry knowledge of indoor air contaminants, HVAC system operating costs, building envelope permeability, and the growing number and types of contaminants generated within buildings.

If a building code has not adopted ASHRAE Standard 62-1989, new and retrofitted air handling systems should be designed in accordance with whichever is more stringent. Air handling systems designed or retrofitted in the general period from about 1972 to 1989 should be re-evaluated and considered for upgrade to comply with this standard. This chapter presents commentary and interpretation on major features of the standard.

The industry often uses the terms "outdoor air" and "ventilation air" interchangeably; it is important to understand the difference between them. Outdoor air is air from outside the building, while ventilation air is that portion of supply air whose purpose is maintaining acceptable indoor air quality. When a single space air handler serves a single space, ventilation air consists entirely of outdoor air. However, when a single air handler serves multiple spaces, ventilation air may be a mixture of outdoor air and treated recirculated air.

ASHRAE STANDARD 62-1989 COMMENTARY[1]

A building designer whose work does not conform to the standard, in addition to local building codes, is taking a chance with strict product liability claims. However, because even the standard can be open to interpretation, the commentaries to follow represent the best judgment of the Trane Company (an American Standard company in LaCrosse, Wisconsin) of the meaning and purpose of the various provisions. The ultimate responsibility for interpretation and compliance rests with individual designers and installers. Key points from the section are arranged by topic and discussed in question (Q), answer (A), and commentary (C) format.

Throughout the standard, both *suggestions* (signaled by words "should" and "may") and *requirements* ("shall" and "must") are presented. Requirements must be met to claim conformance with the standard and, on the surface, it seems that suggestions need not be met for compliance. However, ignoring these suggestions may not be prudent in the context of potential litigation, since the suggestions reflect the consensus of experts in the HVAC industry.

ASHRAE Standard 62-1989 makes several references to the "occupied zone," which is a defined zone within the occupied space, ". . . between planes 3 and 72 inches above the floor and more than 2 feet from the walls." Occupied space is assumed to refer to all inhabited areas—usually floor to ceiling and wall to wall of entire rooms within the building.

Analysis of Section 5.0: Systems and Equipment

Section 5.0 of the standard presents general suggestions and requirements for ventilation systems and equipment to ensure acceptable indoor air quality.

1. Thanks to the Trane Company, LaCrosse, Wisconsin, an American Standard company, for permission to use this material from their *Engineers Newsletters*, © American Standard, Inc.

Airflow Measurement

Q: Is outdoor airflow measurement required?

A: No. Provision for measurement is suggested by Section 5.1: *When mechanical ventilation is used, provision for airflow measurement should be included.*

C: This phrase seems to suggest continual measurement of outdoor airflow entering the building at the air handler, presumably so that outdoor airflow rates can be monitored, recorded, and used at a future date to prove that the building was ventilated properly. On the other hand, perhaps this phrase suggests only that the outdoor airflow stream be accessible for periodic flow measurements. In any event litigation-conscious designers tend to read this suggestion as a requirement for continuous outdoor airflow monitoring at the air handler. In addition, continuous measurement of outdoor airflow may be required for its control, regardless of the need for monitoring airflow.

Q: Is mechanical ventilation required?

A: Yes. Section 5.1 adds: *When infiltration and natural ventilation are insufficient to meet ventilation requirements, mechanical ventilation shall be provided.*

C: If natural ventilation (windows and cracks) is insufficient to meet ventilation requirements presented later in the standard, a fan, at least, is required to draw outdoor air into the building.

Energy Recovery

Q: Does the standard require the use of energy recovery systems?

A: No. Energy recovery systems are suggested by Section 5.1: *The use of energy recovery ventilation systems should be considered for energy conservation purposes in meeting ventilation requirements.*

C: More outdoor airflow results in more exhaust airflow. Since energy is invested to condition the air that enters the occupied space and is then exhausted, it seems logical to try to recover some of this invested energy from the exhaust airstream. Energy recovery systems include sensible heat recovery systems (coil loops and heat pipes), as well as enthalpy recovery systems (heat wheels). In practice, however, the installation and operating costs of an exhaust air energy recovery system in a commercial building may not result in a reasonable payback. Furthermore, cross contamination potential restricts energy recovery choices. Therefore, this suggestion is seldom pursued by designers.

Room Air Distribution

Q: Should ventilation air supplied to the occupied space reach the occupants?

A: Definitely. This is a requirement according to Section 5.2: *Ventilating systems shall be designed and installed so that the ventilation air is supplied throughout the occupied zone*.

C: Air supplied to the occupied space (the room) must be effectively distributed throughout the occupied zone. Room air distribution is directly related to room geometry; air temperature; diffuser type, size, and location; air velocity; and return air type and location. The building designer is responsible for a properly designed room air distribution system, and proper installation is required. But does the standard require the designer to be responsible for the installer's work? On the other hand, who is responsible if the design is weak? The standard is not clear on this point, which could open the door to finger pointing.

To help building designers properly design room air distribution systems, manufacturers of variable air volume terminal units, room air diffusers, zone-mounted terminal units (such as fan coils and unit ventilators) and unitary air conditioners should provide pertinent air distribution data, such as throw patterns and distances.

Q: Can the occupied zone be uncomfortable, in terms of air movement?

A: No. The standard requires that no conditions in the occupied zone conflict with the human comfort factors presented in ASHRAE Standard 55-81. Section 5.3 includes: *Ventilating systems shall be designed and installed so that they do not cause conditions that conflict with ASHRAE Standard 55-1981*.

C: In a properly designed room air distribution system, occupied zone air velocities do not exceed 30 fpm in the winter or 50 fpm in the summer, and the temperature gradient does not exceed 5°F. Designers must be aware of the factors that affect human comfort, and design the room air distribution system to maintain these factors within acceptable ranges. In general, the diffusers must not dump cold air at high velocities into the occupied zone at any load condition, and the occupied zone temperature must not be allowed to stratify. The standard requires proper installation of the ventilating system, but does not clearly assign this responsibility to the building designer or the installer.

Q: Can room air distribution requirements be relaxed at part-load conditions?

A: No. Acceptable indoor air quality in the occupied zone is required at all load conditions, according to Section 5.4: *When the supply of air is reduced during times the space is occupied (e.g., variable air volume systems), provision shall be made to maintain acceptable indoor air quality throughout the occupied zone.*

C: One provision often incorporated in variable air volume (VAV) systems to assure proper room air distribution at part-load conditions (reduced primary airflow) is the fan-powered or induction VAV terminal unit. In effect, the occupied space is actually supplied with variable temperature air at a constant diffuse airflow. The room air distribution system is designed for constant air volume, so room air distribution always operates as designed.

An alternative provision, which also fulfills this requirement, is the use of minimum flow settings at the shutoff VAV terminal unit in conjunction with properly selected and located diffusers (that is, selected and located to operate properly over the full range of flows). To assure proper low-flow operation, the flow through the VAV terminal unit is not allowed to drop below the minimum flow required by the diffusers. Typically, two minimum flow settings are needed: one for the delivery of cool air, and another higher setting for delivering warm air. Experience shows that linear slot diffusers operate over the widest range of flows and are, therefore, commonly specified in VAV systems.

Design Documentation

Q: Should building designers document assumptions and design calculations?

A: Yes. This is a requirement according to Section 5.2: *The design documentation shall state assumptions that were made in the design with respect to ventilation rates and air distribution.*

C: Although not stated explicitly, the standard implies that ventilation system design documentation is a required part of the overall building design documentation package. This documentation is necessary for checking the design (part of the commissioning process) as well as for operating, maintaining and altering the system in the future. Building designers are responsible not only for building plans, specifications, and documents such as installation, operation, and maintenance manuals, but also for building system-related design documents.

External Contamination Sources

Q: Is the location of makeup air inlets important?

A: Yes. Proper location of air inlets, with respect to exhaust air outlets, is a requirement, according to Section 5.5: *Makeup air inlets and exhaust air outlets shall be located to avoid contamination of the makeup air.*

C: The building designer must position equipment in a way that prevents makeup air contamination; the makeup air inlet location must prevent exhaust air from being reintroduced into the building. This requirement can usually be met in applied systems since makeup air inlets and exhaust fans are in separate units. However, compliance can be more difficult when rooftop unitary equipment must be used, because they typically house both the makeup air inlet and exhaust air outlet. Therefore, designers must determine the likelihood of makeup air contamination and either select equipment designed to avoid this problem or add external ducts or baffles to correct it.

Q: Is the designer required to keep contaminants from outside sources out of the building?

A: No. However, Section 5.5 suggests avoiding such contaminants: *Contaminants from sources such as cooling towers; sanitary vents; vehicular exhaust from parking garages, loading docks, and street traffic should be avoided.* Section 5.12 adds: *Special care should be taken to avoid entrainment of moisture drift from cooling towers into the makeup air and building vents.*

C: Although these are only suggestions, they seem to be corollaries to the requirement for locating the makeup air intake, discussed previously. These are merely additional external contamination sources. If viewed as corollaries, these statements require building designers to account for all possible external contamination sources when locating the system outdoor air intake. In addition, Section 5.12 can be interpreted to include any outdoor equipment with standing water as a potential external contamination source.

Q: Are designers required to keep radon out of buildings?

A: No. Section 5.5 suggests that designers avoid arrangements that foster radon infiltration: *Where soils contain high concentrations of radon, ventilation practices that place crawl spaces, basements, or underground ductwork below atmospheric pressure should be avoided.*

C: Although this section provides only a suggestion, a ventilation system design that actually draws radon into the building is not acceptable to many building owners, occupants, or building codes. For this reason, building designers often interpret this suggestion as a requirement for a good design. HVAC systems should be designed so that subsurface system elements (particularly basements) are maintained at a positive pressure with respect to the atmosphere.

Air System Cleaning

Q: Must the ducts and plenums provide for removing accumulations?

A: Yes. Microorganisms grow in dirt. Section 5.6 requires that duct systems be designed so they can be cleaned: *Ventilating ducts and plenums shall be constructed and maintained to minimize the opportunity for growth and dissemination of microorganisms through the ventilation system.*

C: Building designers commonly extend this requirement to include other air system equipment, such as air handlers and VAV terminal units. It leads to the specification of external insulation, foil-faced insulation, or dual-wall construction and away from internal insulation, fibrous ducts, and duct liners. Designers also allow for adequate access doors and panels for convenient periodic cleaning.

The installer, who actually constructs the air distribution system, is required by the standard to install the air system so that it can be cleaned. A good access door is useless if it is against a brick wall.

It seems, also, that the air system owner is required to maintain (clean) the air system; providing access does not keep the system clean. One major challenge that this requirement presents to the system maintainer is cleaning the ceiling plenum. Many systems use the ceiling plenum as the return air path, but such plenums are often not designed for easy cleaning.

Q: Must access to coils and condensate pans in air handlers or fan-coil units be provided?

A: Yes. Access for cleaning coils and drain pans is a requirement, as specified in Section 5.12: *Provision shall be made for periodic in-situ cleaning of cooling coils and condensate pans. Air-handling and fan coil units shall be easily accessible for inspection and preventive maintenance.*

C: The requirement for in-situ cleaning of cooling coils and condensate pans usually results in equipment specifications calling for adequate access to these components. The equipment supplier must provide equipment that can be cleaned, inspected, and maintained in place. In addition, the equipment's location and installation must ensure adequate access to cooling coils, drain pans, and inspection and maintenance areas.

Local Exhaust

Q: Are local exhaust systems ever needed?

A: Yes. Stationary local contaminant sources require local exhaust systems, as described in Section 5.7: *Contaminants from stationary local sources within the space shall be controlled by collection and removal as close to the source as practicable*. Section 5.8 adds: *Fuel-burning appliances . . . shall be provided with . . . adequate removal of combustion products . . . Combustion system . . . vents shall not be exhausted into attics, crawl spaces, or basements*.

C: Building system designers must provide local exhaust for stationary local air contamination sources, such as copy machines, in compliance with industrial ventilation standards. Fuel-burning appliances are a special local contamination source, and combustion products must be removed via a local exhaust system—never exhausted inside the building, even into unoccupied spaces.

Local Makeup Air

Q: Is local makeup air required?

A: Yes. Air-consuming processes require sufficient air, as provided by Section 5.8: *Fuel-burning appliances . . . shall be provided with sufficient air for combustion. . . . When infiltration supplies all or part of the combustion air, the supply rate shall be demonstrable*.

C: Designers are required to account for the air consumed by indoor fuel-burning appliances. When determining the amount of outdoor air required for a given space, the need for combustion air must be accounted for in design calculations. Special provision for makeup air to appliances may be needed, separate from the ventilation system.

Filtration

Q: Are filters always required?

A: No. If particle removal is needed, the standard suggests using air filters or dust collectors. Section 5.9 states: *When it is necessary to remove particular contaminants, air filters or dust collectors should be used. Dust collectors, not air filters, should be used where dust loading equals or exceeds 10 mg/m^3*.

C: Table 22-1, copied from Table 1 of ASHRAE 62-1989, cites EPA Ambient-Air Quality Standards for Outdoor Air (the acceptable total particulate content for outdoor air). Particle removal is necessary if outdoor or recirculated air exceeds this total particulate concentration.

TABLE 22-1
National Primary Ambient-Air Quality Standards for Outside Air as Set by the U.S. Environmental Protection Agency

	Outdoor Standards						
	Long-Term			Short-Term			
	Concentration		Time	Concentration		Time	ASHRAE
Contaminant	μg/m³	ppm	years	μg/m³	ppm	hours	Table
Sulfur Dioxide	80.0	0.030	1.00	365	0.14	24	1
Total Particulate	50.0		1.00	150		24	1
Carbon Monoxide				40,000	35.00	1	1
Carbon Monoxide				10,000	9.00	8	1
Oxidants (ozone)				235	0.12	1	1
Nitrogen Dioxide	100.0	0.055	1.00				1
Lead	1.6		0.25				1
Asbestos							
Nat'l. Emission Stds.	No visible emissions						C-1
Connecticut				0.001		8	C1
New York	5.0		1.00				C1
Virginia				2.000		24	C1
Formaldehyde							
No Federal Standards							
Connecticut				12.000		8	C1
New York	2.0		1.00				C1
Virginia				12.000		24	C1
Radon	25 mrem/y whole body						C1

If it is necessary to remove particles from the outdoor or recirculated airstream, air filters or dust collectors are recommended but not required. Many building designers interpret this suggestion as a requirement for filtration. Either air filters or dust collectors can be used if dust loading is less than 10 mg/m^3 (low particulate count). If dust loading equals or exceeds 10 mg/m^3 (high particulate count), then the use of dust collectors, not air filters, is suggested.

Q: Is any type of filter or dust collector acceptable?

A: No. If particle removal is needed, proper selection of the filter or collector is required, as specified in Section 5.9: *Air filters and dust collectors shall be selected for the particulate size and loading encountered. Filters shall be tested in accordance with ASHRAE Standard 52-76.*

C: Although the standard does not provide detailed guidelines for filter selection, it does require that the building designer, and presumably the building maintainer, select filters and dust collectors based on the size and quantity of target particles. It also references the test standard.

Q: What about gaseous contaminants?

A: The standard requires the use of sorption methods to control gaseous contaminants in the air, according to Section 5.10: *When compliance with this section does not provide adequate control of gaseous contaminants, methods based on sorption with or without oxidation or other scientifically proven technology shall be used.*

C: If air filtration, as described in Section 5.9, does not clean the outdoor or recirculated air to acceptable levels (Table 22-1), more sophisticated gaseous contaminant removal methods must be used. As a result of this requirement, many building designers specify the use of activated carbon filters in either the outdoor or the supply airstream.

Dehumidification

Q: Is dehumidification required by the standard?

A: No. The standard strongly suggests that high humidity levels be avoided, but does not require dehumidification. Section 5.11: *High humidities can support the growth of pathogenic or allergenic organisms. . . . Relative humidity in habitable spaces preferably should be maintained between 30% and 60% relative humidity . . . to minimize growth of allergenic or pathogenic organisms.*

C: This statement suggests that the humidity level in the occupied space be controlled so that it is below 60 percent relative humidity. As a result, some building designers specify dehumidification control systems to maintain space humidity at this level during occupied hours.

Q: Must relative humidity be limited to 70 percent or less throughout the air system?

A: No. The standard merely observes that fungal contamination can occur at high humidities, as stated in Section 5.12: *If the relative humidity in the occupied spaces and low velocity ducts and plenums exceeds 70%, fungal contamination (for example, mold, mildew, etc.) can occur.*

C: This statement of fact is neither a requirement nor a suggestion, yet some building designers use it to specify that relative humidity of the air anywhere within the building cannot exceed 70 percent. Of course, this is not possible when a cooling coil is used, since the air leaving the cooling coil is often at 55°F and 80 to 95 percent relative humidity. To ensure 70 percent or lower relative humidity in the duct system, some building designers specify that coil air be mixed with warm bypass air before continuing down the duct. Other designers specify induction or fan-powered VAV terminal units to mix saturated primary air and low-humidity plenum air locally. This step, in effect, moves coil bypass from the central coil to each terminal unit.

The standard does not clearly define a low velocity duct. However, air system velocities are usually higher near the fan and lower near the occupied space. The low velocity distinction is probably related to dust cleanliness; that is, high air velocities tend to keep dust and dirt from accumulating. With that in mind, a low air velocity duct is usually the final duct connection or branch to the diffusers.

Ignoring duct system relative humidity, many building designers specify an occupied dehumidification control system to keep the occupied space humidity below 60 percent, thereby preventing fungal contamination. Also, many designers specify an unoccupied dehumidification control system. This is perhaps more important than occupied dehumidification control, since the relative humidity in an unoccupied space rises as the space cools at night. For example, a space at 75°F, 50 percent relative humidity need only cool to approximately 65°F to attain 70 percent relative humidity.

Q: Is humidification required by the standard?

A: No. The standard strongly suggests that low humidity levels be avoided, but does not require humidification. Section 5.11: *Relative humidity in habitable spaces preferably should be maintained between 30% and 60% relative humidity . . . to minimize growth of allergenic or pathogenic organisms*.

C: Some building designers specify humidification control systems to maintain space humidity above 30 percent during occupied hours.

Q: Are steam humidifiers required?

A: No. Steam humidifiers are suggested in Section 5.12 for space humidification: *Steam is preferred as a moisture source for humidifiers*.

C: Though steam is suggested as a moisture source, other moisture sources are acceptable. No statement related to the control of humidification is made, but it is implied that, if moisture is added to the building, a humidity control system is required to modulate the addition.

Q: If evaporative humidifiers are used, must they receive special treatment?

A: No. However, the standard suggests the use of antimicrobial treatments, as pointed out in Section 5.12: *Standing water used in conjunction with water sprays in HVAC air distribution systems should be treated to avoid microbial buildup*.

C: More specifically, the water treatment suggested here is for humidification system water, but some building designers specify water treatment for any standing (or potentially standing) water within the air distribution system. Such treatment may include the use of chemicals (microbiocides) or special coatings. Building designers may specify antimicrobial coatings in drain pans, on cooling coils, and in humidifiers. The EPA has registered the use of an antimicrobial for use in HVAC systems.

Drain Pans

Q: Are drain pans allowed to accumulate water?

A: No. Drain pans must drain completely, according to Section 5.12: *Air handling unit condensate pans shall be designed for self-drainage to preclude the buildup of microbial slime*.

C: Today, building designers specify sloped drain pans to ensure positive draining. Included are drain pans in all HVAC equipment with cooling coils (air handlers, fan coils, unit ventilators, rooftops, etc.). In addition, some designers specify the use of antimicrobial treatments on air handler surfaces to further preclude the buildup of microbial slime. Condensate removal systems are becoming more common and increasingly sophisticated.

Q: Must the drain pans be maintainable?

A: Yes. The standard, in Section 5.12, requires periodic cleaning of coils and drain pans: *Provision shall be made for periodic in-situ cleaning of cooling coils and condensate pans. Air-handling and fan coil units shall be easily accessible for inspection and preventive maintenance.*

C: To allow this access, many building designers specify ducts and equipment with adequate access doors or panels.

Analysis of Section 6.1: Ventilation Rate Procedure

Section 6.1 presents ". . . one way to achieve acceptable indoor air quality," by using space ventilation with outdoor air of specific minimum quality and quantity. This method is sometimes called the "dilution solution." It dictates both the condition and amount of ventilation air necessary to ensure adequate dilution of contaminants in the occupied zone. Specifically, it prescribes:

- The quality of acceptable outdoor air
- The treatment of outside air necessary to make it acceptable
- The minimum volumetric flow rate of outdoor air required to adequately ventilate various occupied space types

Furthermore, Section 6.1 prescribes outdoor airflow rate requirements for systems with multiple spaces served by one air handler, such as VAV systems, and for spaces with intermittent occupancy, such as conference rooms and auditoriums. It also discusses the use of recirculated air for ventilation, and ventilation effectiveness within the occupied zone.

Outdoor Air Quality

Q: Can the building designer simply assume that outdoor air is acceptable for ventilation?

A: No. Designers are required to evaluate the outdoor air that they take in for ventilation, as stated in Section 6.1.1: *This section describes a three-step procedure by which outdoor air shall be evaluated for acceptability.*

C: Because only acceptable outdoor air can be used for ventilation, the outdoor air at a building site must be evaluated to ensure acceptability. The three-step procedure (all three steps are required for compliance) is summarized as follows:

Step 1: Ensure that contaminants in the outdoor air do not exceed the concentrations listed in Table 22-1, which lists air quality standards as set by the EPA.

Building designers have four options for determining whether outdoor air complies with the table. Although the first option appears to involve nothing more than writing a letter, all options actually require a considerable effort.

1. Check government pollution-control agency monitored data for the building location (direct data check).
2. Check monitored data for a location with similar population, geography, weather, and industrial patterns (indirect data check).
3. In lieu of monitored data checks, simply assure that the building location is in a community of less than 20,000 people, with no "substantial" air contamination sources.
4. Monitor the air for three consecutive months.

Step 2: If the outdoor air is ". . . thought to contain any contaminants not listed in the table, . . . guidance on acceptable concentration levels may be obtained by reference to Appendix C of the full standard." The building designer bears the burden of judgment at this step, although it is not clear who is ultimately responsible for "thinking" that unlisted contaminants may be present. It is likely to be the building designer.

Step 3: If Steps 1 and 2 are completed and ". . . there is still a reasonable expectation that the air is unacceptable, sampling shall be conducted in accordance with NIOSH procedures. . . . Finally, acceptable outdoor air quality should be evaluated using the definition for acceptable indoor air quality in Section 3 [of the full standard]." This step places an additional judgment burden on building designers. If he or she has assured that none of the 10 contaminant levels listed by the standard is unacceptable (Steps 1 and 2) but ". . . there is still a reasonable expectation" that the outdoor air is bad (perhaps because of odors or prior experience in the area), Step 3 must be performed. Step 3 requires that air samples be taken and evaluated, apparently for the hundreds of materials listed in the NIOSH *Manual of Analytical Methods*. Step 3 goes on to suggest that the suspected bad outdoor air be further qualified per the definition of unacceptable indoor air quality. In other words, it recommends that building designers carry out a subjective test to demonstrate that ". . . a substantial majority (80 percent or more) of the people exposed do not express dissatisfaction" with the outdoor air.

Outdoor Air Treatment

Q: If the outdoor air is unacceptable, must it be cleaned or filtered?

A: No. The standard suggests that bad outdoor air be treated, but, as Section 6.1.2 is written, it does not require treatment: *If the outdoor air contaminant levels exceed those values given in 6.1.1 [Table 22-1], the air should be treated to control the offending contaminants.*

C: Although this statement refers only to the evaluation of outdoor air based on contaminant levels of Table 22-1, it seems reasonable to apply it to all three evaluation steps required in Section 6.1.1. In other words, if the concentration of any of the identified contaminants (Steps 1 and 2) or any other pollutant (Step 3) exceeds the acceptable level, or if a subjective test fails to accept the air quality, treatment of the outdoor air is suggested.

The building designer, faced with unacceptable outdoor air, is encouraged to treat the outdoor air before it enters the occupied space. Some building designers interpret this suggestion as a requirement for outdoor air treatment (filtering), either in the outdoor air path, before mixing with recirculated air, or in the supply air path. Consequently, air filtration systems (filters, dust collectors and/or air cleaners for particles) and gaseous contaminant removal systems (activated carbon filters for gases and vapors) are often specified.

Q: Is any filter acceptable?

A: Strictly speaking, yes. The standard, in Section 6.1.2, only suggests using an appropriate filter: *Air-cleaning systems suitable for the particle size encountered should be used. For removal of gases and vapors, appropriate air-cleaning systems should be used.*

C: If the total particulate level is too high, using properly selected filters suitable for the particle size is recommended, but not required. Similarly, if the offending contaminants are gaseous, using appropriate air cleaning systems, such as activated carbon filters, is recommended, but not required. These recommendations are often interpreted (and perhaps were intended) as requirements. It seems reasonable to clean the outdoor air before introducing it into the occupied space, especially since the outdoor air contaminant levels also apply to indoor air. As a result, building designers often specify some type of outdoor air treatment.

Ventilation Rates

Q: Can acceptable indoor air quality (conformance to the standard) be achieved via adequate ventilation rate alone, without measuring contaminant levels?

A: Yes. In fact, to conform to the standard using the Ventilation Rate Procedure, the ventilation rates presented by the standard are *required* by Section 6.1.3: *Indoor air quality shall be considered acceptable if the required rates of acceptable outdoor air in Table 2 [summarized here as Table 22-2] are provided for the occupied space.*

C: In effect, when a designer uses the Ventilation Rate Procedure, an occupied zone has acceptable indoor air quality if it is ventilated at the minimum outdoor airflow rate specified in Table 2 of the standard. The table specifies minimum rates of outdoor airflow required for 81 types of occupied-space. In general, these rates were determined by assuming levels of human occupancy and carbon dioxide production, then calculating the dilution rates required to maintain the CO_2 concentration below 1000 ppm.

To use the table, building designers first determine the type of facility and planned use (application) of each space within the facility. The minimum outdoor air requirement for the occupied zone within the space, as found in Table 2 of the standard, is usually expressed as volume flow rate per person. Knowing design occupancy, the actual required minimum outdoor airflow rate for each space is calculated by multiplying the design occupancy by the outdoor air requirement. For some space applications, the required minimum outdoor airflow rate is calculated by multiplying the floor area being served by the outdoor air requirement expressed as volume flow rate per unit area, shown in the last column of the table. In either case, the designer has a defined method for calculating minimum outdoor airflow rate required to comply with the standard.

Q: Do Table 2 values always apply, even when unusual contaminant sources are found in the occupied space?

A: No. This is one of two exceptions to the airflow rates in Table 2. The Standard, in Section 6.1.3, requires that the air quality in spaces with unusual contaminants or sources be controlled by some means other than the outdoor airflow rates prescribed in the table: *Where unusual indoor contaminants or sources are present or anticipated, they shall be controlled at the source or the procedure of [Section] 6.2 shall be followed.*

C: The standard does not define an unusual contaminant or source. However, since the Table 2 rates are based on human occupancy and activities, some

TABLE 22-2
Partial Summary of Table 2 of the Standard

Occupied Space		**Minimum Outdoor Airflow Rate, cfm**	
Facility/ Application	***Space Type***	***per person***	***per sq. ft.***
Commercial (Table 2.1)			
Food/Beverage	Bar, cocktail lounge	30	
	Dining room, cafeteria, fast food	20	
	Kitchen	15	
Hotel, Motel, Dormitory	Baths		35.00
	Bedrooms, living rooms		30.00
	Gambling casinos	30	
	Conference rooms	20	
	Lobbies, assembly rooms, sleep areas	15	
Offices	Office space, conference rooms	20	
	Reception areas	15	
Public Spaces	Elevators		1.00
	Locker and dressing rooms		0.50
	Corridors and utilities		0.05
	Smoking lounges	60	
	Public restrooms	50	
Retail Store, Showroom	Basement and street		0.30
	Upper floors, dressing rooms, malls, arcades		0.20
	Storage rooms, shipping and receiving		0.15
	Warehouse		0.05
Theater	Ticket booths, lobbies	20	
	Auditorium, stages, studios	15	

Continued

TABLE 22-2
Continued

Occupied Space		Minimum Outdoor Airflow Rate, cfm	
Facility/ Application	***Space Type***	***per person***	***per sq. ft.***
Institutional (Table 2.2)			
Education	Corridors		0.10
	Laboratories, training shops	20	
	Classrooms, music rooms, libraries, auditoriums	15	
Hospital, Nursing Home	Autopsy rooms		0.50
	Operating rooms	30	
	Patient rooms	25	
	Medical procedure, recovery, ICU, therapy	15	

building designers conclude that ventilation for spaces with nonhuman sources, such as copy machines, or with unusual human activity, such as exercise machines in an office area, must be determined by means other than Table 2. Once the designer concludes that Table 2 cannot be used as the guideline for acceptable indoor air quality, one of two other paths must be used: either source control or the Indoor Air Quality Procedure (covered in the next section of this chapter).

Source control is not defined by the standard. For some designers, a source control system includes local exhaust airflow with corresponding local makeup airflow. Other designers do not consider local makeup air as part of the source control system; instead, they increase the outdoor airflow rate to provide both the required ventilation airflow and local makeup airflow. In either case, dilution is abandoned in favor of replacement of contaminated air with ventilation air. In other words, outdoor air is used to replace exhausted air rather than to lower the contaminant concentration level in the space through dilution. Contaminants are exhausted locally, at their source, rather than mixed with return air to be exhausted centrally.

Q: Do the Table 2 values always apply, even for areas in industrial facilities not covered by Table 2?

A: No. This is the second of two exceptions to the airflow rates listed in Table 2. The standard requires, in Section 6.1.3, that acceptable air quality in industrial facility areas not mentioned in Table 2 be accomplished as described in a reference source: *For those areas within industrial facilities not covered by Table 2, refer to footnote 15 of Threshold Limit Values and Biological Exposure Indices for 1986-87, American Conference of Governmental Industrial Hygienists.*

C: The standard cannot be used to determine the outdoor airflow required to achieve acceptable indoor air quality in industrial facilities, because the table specifies only commercial, institutional, and residential facilities. Building designers of industrial facilities must use another reference source to find acceptable contaminant concentration levels and, presumably, required dilution rates.

Q: Does the standard effectively prescribe just one ventilation rate for each specific space type by listing estimated maximum occupancy and volume flow rate per person?

A: No. Use of actual occupant density at design is required when calculating the minimum outdoor airflow rate for a space, according to Section 6.1.3: *Where occupant density differs from that in Table 2, use the per occupant ventilation rate for the anticipated occupancy load.*

C: This statement is somewhat confusing, but it seems to mean that the actual value of design occupant density is to be used to calculate the minimum outdoor airflow rate. The building designer determines the design (anticipated) occupancy load for a space by multiplying the design occupant density (persons per 1000 square feet) by the occupiable area of the space. Then the required minimum outdoor airflow rate for the occupied zone is found by multiplying the design occupancy load by the volume flow rate per person from the table. These steps are used regardless of whether the design occupant density matches the estimated maximum occupancy listed in the table.

Q: Does a carbon dioxide concentration below 1000 ppm in a ventilated space mean that acceptable indoor air quality has been achieved?

A: No. The standard, in Section 6.1.3, merely observes that odors are likely to be acceptable if the space is ventilated at a rate that maintains the CO_2 concentration below 1000 ppm: *Comfort (odor) criteria are likely to be satisfied if the ventilation rate is set so that 1000 ppm CO_2 is not exceeded.*

C: This is an observation, not a requirement or suggestion. It informs building designers that odors are likely to be adequately controlled if a space is ventilated as prescribed in Table 2. Outdoor airflow rates presented in Table 2 are calculated to provide sufficient dilution to maintain the CO_2 concentration level below 1000 ppm. The ventilation rate is "set so that 1000 ppm CO_2 is not exceeded" when the outdoor airflow rate to the space is at least as high as the outdoor air requirements listed in the table.

This observation is often misinterpreted to mean that acceptable indoor air quality is achieved by maintaining the CO_2 concentration level in the occupied zone below 1000 ppm. Although a CO_2 level above 1000 ppm indicates unacceptable indoor air quality, the converse is not necessarily true. Many experts claim that there is a weak correlation between low CO_2 levels and acceptable indoor air quality.

Q: If a CO_2 concentration below 1000 ppm is maintained at outdoor airflow rates lower than those listed in Table 2, has acceptable indoor air quality been achieved?

A: No. If the Ventilation Rate Procedure (that is, dilution at the rates presented in Table 2) is not used, other contaminant concentrations may rise, and designers are required to consider this effect, according to Section 6.1.3: *In the event CO_2 is controlled by any method other than dilution, the effects of possible elevation of other contaminants must be considered.*

C: If the CO_2 concentration in a space is maintained below 1000 ppm by any means other than dilution at the rates prescribed by the Ventilation Rate Procedure, other possible contaminants in the space are less diluted, so their concentrations may rise.

Q: Can the Table 2 rates ever be reduced below the values listed?

A: Yes. The standard allows (suggests) temporary reduction in outdoor airflow rates if the outdoor air quality is unacceptable for brief periods, as stated in Section 6.1.2: . . . *the amount of outdoor air may be reduced during periods of high contaminant levels, such as those generated by rush-hour traffic.*

C: If no air cleaning system can adequately clean the outdoor air during times of peak contamination levels, the amount of outdoor air introduced into the building can be reduced during such periods. This suggestion places another judgment burden on building designers, since the amount and duration of reduced outdoor airflow is not specified. It may lead some designers to specify outdoor air quality monitoring.

Analysis of Section 6.2: Indoor Air Quality Procedure

The Ventilation Rate Procedure discussed in this chapter's preceding section presents ". . . one way to achieve acceptable indoor air quality," namely, supplying each occupied space with ventilation air of specific quality and quantity. In contrast, the Indoor Air Quality Procedure (standard's Section 6.2) presents ". . . an alternative performance method . . . for achieving acceptable air quality" by quantitatively describing acceptable indoor air quality. It sets limits on the concentration of known and specifiable contaminants in an effort to achieve acceptable indoor air quality in the occupied zone in a more direct way. More specifically, the IAQ Procedure sets concentration limits for 10 contaminants, prescribes subjective analysis to determine acceptable odor levels, and describes the use of treated recirculated air to reduce the minimum outdoor airflow rates presented in the standard.

Determining Indoor Air Quality

Q: What contaminants are specified by the standard, and what are the acceptable concentration levels of each?

A: Table 22-3 summarizes information given in the standard's Tables 1 and 3, as specified in Section 6.2.1: *Table 1 furnishes information on acceptable contaminant levels in outdoor air [and] . . . also applied indoors for the same exposure times. . . . Table 3 contains limits for four other indoor contaminants.*

C: Tables 1 and 3 of the standard list 11 acceptable indoor concentration levels for 10 contaminants (carbon monoxide is listed for 1- and 8-hour exposures). Indoor concentrations of the contaminants are required to be below the levels shown.

Q: If concentrations in the tables are maintained, is the indoor air quality acceptable?

A: Not necessarily. Section 6.2.1 provides a clear answer: *Tables C-1 and C-3 do not include all known contaminants that may be of concern, and these concentration limits may not, ipso facto, ensure acceptable indoor air quality with respect to other contaminants.*

C: To a certain extent, this observation removes the IAQ Procedure as the performance-based path to acceptable indoor air quality. It implies that, even if the known and specifiable contaminants listed are controlled to the established concentration levels, indoor quality still may not be satisfactory. The building designer has no clear, indisputable definition of acceptable indoor

TABLE 22-3

Indoor Air Quality Standards

	Acceptable Exposure						
	Long-Term			*Short-Term*			
	Concentration		*Time*	*Concentration*		*Time*	
Contaminant	$\mu g/m^3$	*ppm*	*years*	$\mu g/m^3$	*ppm*	*hours*	*Table in standard*
Carbon Dioxide	1.8E6	1000	Cont.				3
Carbon Monoxide (USA)				40,000	35.00	1	1
Carbon Monoxide (USA)				10,000	9.00	8	1
Chlordane (maximum)	5.0	0.0003	Cont.				3
Lead	1.5		0.25				1
Nitrogen Dioxide	100.0	0.055	1.00				1
Oxidants (ozone, USA)				235	0.12	1	1
Ozone	100.0	0.05	Cont.				3
Particulate (total USA)	50.0		1.00	150		24	1
Radon	4 pCi/l		1.00				3
Sulfur Dioxide (USA)	80.0	0.030	1.00	365	0.14	24	1

air quality. Acceptable IAQ is determined by the contaminant levels listed, as well as by unknown levels of unknown contaminants. Therefore, many designers choose to use the Ventilation Rate Procedure in which the requirements are clearer because compliance to those requirements is more easily demonstrated.

Q: Can acceptable indoor air quality be determined solely by sensing contaminant levels?

A: No. The quality of indoor air cannot always be determined by measuring contaminant levels; therefore, subjective evaluation of indoor air quality is required. Section 6.2.1 contains: *To some degree, adequacy of control [of complex contaminant mixtures] must rest upon subjective evaluation.*

C: From the designer's viewpoint, this requirement weakens the IAQ Procedure considerably. For complex contaminant mixtures (usually characterized by odors), the contaminant-related quality of indoor air cannot be measured; it must be determined subjectively, based on the judgment of impartial observers. Building designers are required to design a system that adequately controls contaminants to a level that will be judged acceptable in subjective evaluation by impartial observers, meaning that a positive subjective evaluation is a system design requirement. However, it can occur only after system installation and building occupancy.

Q: If contaminant levels are low enough and a subjective evaluation judges the air quality to be acceptable, is the building design in compliance with the IAQ Procedure?

A: Not necessarily. According to Section 6.2.1: *In the case of some odorless biological aerosols, subjective evaluation is irrelevant. Application of generally acceptable technology, and vigilance regarding adverse influences of reduced ventilation, must therefore suffice.*

C: This requirement further weakens the IAQ Procedure because it provides little concrete direction but adds substantially to the designer's judgment burden with its attendant responsibility. Some contaminants can neither be measured quantitatively nor evaluated subjectively. Building designers are required to use acceptable technology and vigilance when the contaminants cannot be measured or perceived. Again, as a result of this loose definition of acceptable indoor air quality, many building designers choose the more clearly defined Ventilation Rate Procedure.

Subjective IAQ Evaluation

Q: Is a subjective IAQ evaluation required for compliance via the IAQ Procedure?

A: Yes. If odor-causing contaminants cannot be measured objectively, Section 6.2.2 holds that a subjective evaluation is required: *In the absence of objective* *means to assess the acceptability of such [odor-causing] contaminants, the judgment of acceptability must necessarily derive from subjective evaluation of impartial observers.*

C: Many odor-causing contaminants cannot be measured in terms of concentration level or do not have an established harmful concentration level. Therefore, a designer using the IAQ Procedure cannot comply with the standard without a successful subjective evaluation of the completed system.

Q: Is a subjective IAQ evaluation method required?

A: No. Section 6.2.2 refers to Appendix C of the standard, but does not require it: *One method that may be used for measuring subjective response is described in Appendix C.*

C: The suggested method states that the air quality is acceptable ". . . if 80 percent of a panel of at least 20 untrained observers deems the air to be not objectionable under representative conditions of use and occupancy." The observations are intended to detect offensive odors only.

Q: If the system passes the subjective evaluation, is the air quality acceptable?

A: No. Section 6.2.2 adds: *Caution should be used in any subjective evaluation procedure to avoid unacceptable concentrations of other contaminants.*

C: This suggestion warns designers that an acceptable odor level alone does not indicate satisfactory indoor air quality. Unacceptable concentrations of odorless contaminants (including carbon monoxide and radon) can be present in a space, undetected by a subjective evaluation for odors.

Air Cleaning

Q: Can clean, recirculated air be used to reduce the minimum outdoor air rates in Table 2?

A: Yes, but the IAQ Procedure must be used to avoid contaminant accumulation in under-diluted spaces. Sections 6.2.3 and 6.1.3.2 provide details: *Recirculation with air-cleaning systems is also an effective means for controlling contaminants when using the Indoor Air Quality Procedure. Procedure . . . If cleaned, recirculated air is used to reduce the outdoor airflow rate below the values shown in Table 2, the Air Quality Procedure, 6.2, must be used.*

C: To clean air returning from the occupied space, recirculate it as ventilating air, and thereby reduce the minimum outdoor airflow rate listed in Table 2, the designer is required to use the IAQ Procedure—not the Ventilation Rate Procedure. Some contaminants, particularly gaseous ones, may not be adequately cleaned from the recirculated airstream. These contaminants can accumulate in the occupied space, eventually reaching unacceptable concentrations. Therefore, the acceptable indoor air quality must be determined using the IAQ Procedure. To avoid this necessity, many building designers do not specify use of recirculated air-cleaning solely to reduce the minimum outdoor airflow rate.

Q: Is there a method for computing the reduced minimum outdoor airflow rate when recirculated air is used?

A: Yes. The standard includes equations in Appendix E, and suggests their use to calculate reduced outdoor airflow rates. Section 6.2.3: *The allowable contaminant concentration in the occupied zone can be used with the various system models in Appendix E to compute the required outdoor airflow rate.*

C: Using these equations is not a simple matter, partly because several key variables are difficult to determine. For example, the contamination rate (pollution generation rate) for chemicals and furnishings within the space are not readily available; ventilation effectiveness is difficult to determine without testing; and filter efficiency for gaseous contaminants may not be known.

Q: Can recirculated air always be cleaned sufficiently to be used to reduce the outdoor airflow requirement?

A: No. If any contaminant concentrations cannot be maintained to acceptable levels using air cleaning and reduced outdoor airflow, the designer is required to use Table 2 rates. Section 6.2.3 states: . . . *contaminants that are appreciably reduced by the air-cleaning system may be the controlling factor in design and prohibit the reduction of air below that set by the Ventilation Rate Procedure.*

C: Building designers must be aware of the cleansing capability of the air cleaning system. If the level of all contaminants in the occupied space cannot be reduced to acceptable levels by air cleaning, the designer must abandon the IAQ Procedure and use the Ventilation Rate Procedure.

CHAPTER 23

Recordkeeping Forms and Procedures

Keeping records and filing reports are important parts of complying with the requirements of regulators and others. The requirements refer to "occupational injuries and illnesses," but do not differentiate types and causes. Readers of this book will be concerned mainly with injuries and illnesses caused by indoor air quality problems—injuries that can range from an irritated throat to damaged body organs; illnesses that can range from coughs to death. This chapter looks at some of the necessary forms and procedures.

ASHRAE STANDARD 62-1989[1]

Building designers have not completed their responsibilities when they provide acceptable air as described in Chapter 22; they also have to meet the documentation

1. © American Standard Inc. Thanks to the Trane Company, La Crosse, Wisconsin, an American Standard company, for permission to use this material from their *Engineers Newsletters*.

provisions of ASHRAE Standard 62-1989. Following are questions, answers, and comments about documentation procedures.

Q: Is the designer required to document ventilation-related assumptions and calculations?

A: Yes. Clear documentation of the design assumptions is required, according to Sections 6.0 and 6.1.3.1: *Design documentation shall clearly state which assumptions were used in the design so that the limits of the system in removing contaminants can be evaluated by others before the system is operated in a different mode or before new sources are introduced into the space. . . . Design documentation shall specify all significant assumptions about occupants and contaminants.*

C: Building designers must document building design assumptions, calculations, selections, and other information that became part of the final design. This design documentation is in addition to traditional building plans, equipment schedules, and specifications. One of its main purposes is to enable future changes to building operations and usage without unexpected and undesirable indoor air quality consequences.

In addition to the sections just quoted, there are numerous places throughout the standard that require designers to document design assumptions, calculations, and other facts, so the ventilation system can be operated and maintained properly. Thorough documentation allows proper evaluation of the effects of future changes, such as number of occupants and indoor and outdoor contaminant sources.

Q: Does the designer simply record the design, then file the documents away?

A: No. The standard suggests that design documentation be delivered, presumably to the building owner or operator, after system installation. Section 6.3: *Design criteria and assumptions shall be documented and should be made available for operation of the system within a reasonable time after installation.*

C: The building designer must include design documentation as part of the information delivered to the building owner. Normally, transfer of the plans and specifications for the building systems is sufficient, since design assumptions and calculations result in the information summarized in those documents. However, to comply with Section 6.3, additional design-related documentation such as notebooks, spreadsheets, and memos must be included. Furthermore, this design documentation must be delivered within a reasonable period of time after the system is installed, presumably before it is operated.

Q: What information must be documented?

A: Documentation requirements and suggestions appear in many places throughout the standard. A sampling follows:

Section 6.3: *See Sections 4 and 6, as well as 5.2 and 6.1.3 regarding assumptions that should be detailed in the documentation.*

Section 4.2: *Whenever the Ventilation Rate Procedure is used, the design documentation should clearly state that this method was used and that the design will need to be reevaluated if, at a later time, space use changes occur or if unusual contaminants or unusually strong sources of specific contaminants are to be introduced into the space.*

Section 5.2: *The design documentation shall state assumptions that were made in the design with respect to ventilation rates and air distribution.*

Section 6.0: *Design documentation shall clearly state which assumptions were used in the design so that the limits of the system in removing contaminants can be evaluated by others before the system is operated in a different mode or before new sources are introduced into the space.*

Section 6.1.3: *Design documentation shall specify all significant assumptions about occupants and contaminants.*

OSHA RECORDKEEPING REQUIREMENTS

Sections 1904.2, 1904.4, and 1904.5 of the Occupational Safety and Health Act prescribe the records that employers must keep. These records are used for:

- Appropriate enforcement of the act.
- Developing information regarding the causes and prevention of occupational accidents and illnesses.
- Collection, compilation, and analysis of occupational safety and health statistics.

This section of the chapter looks at important parts of the standard and its recordkeeping requirements. In addition to this information, which relates to

OSHA's charge to ensure a healthy environment for every worker, there may be other organizations and government agencies that impose requirements for keeping records.

Employers Who Are Required to Keep Records

All employers in the private sector (OSHA's rules do not apply to governments as employers, but state plans do) with 11 or more employees in the industries shown in Table 23-1 must keep injury and illness records. The industries are determined by their Standard Industrial Classification (SIC).

Another category of employers are those who are normally exempt from the requirements to keep the records, but who might be selected to participate in the mandatory Annual Survey of Occupational Injuries and Illnesses. The exemptions are from recordkeeping only—not from compliance with all applicable safety and health standards. Employers who had no more than 10 employees (full- or part-time) at any time during the previous calendar year are exempt from recordkeeping requirements, but are subject to selection for the mandatory annual survey. Other employers who conduct business primarily in the SICs shown in Table 23-2 are also exempt but subject to selection for the annual survey.

The recordkeeping exemptions apply to all eligible workplaces under the jurisdiction of federal OSHA (that is, states that do not operate under approved state OSHAs). Although state plans are required to meet OSHA standards, there are variations; employers should check with the state agency to see if it has or intends to adopt the exemptions.

Records That OSHA Requires

OSHA Form No. 200, *Log and Summary of Occupational Injuries and Illnesses*, must be maintained by employers not in exempt categories of Table 23-2. The form asks for some brief descriptive information and then provides a simple check-off procedure to maintain a running total of occupational injuries and illnesses for the year. Employers must make the log available to authorized federal and state government officials, employees, and their representatives upon request. OSHA allows employers to use an equivalent form if it provides the same information as Form 200.

OSHA Form No. 101 must also be maintained, to supply supplementary information regarding each injury and illness entered on Form 200. Information requested on this form includes the injured person's name and a description of the

TABLE 23-1
Industries in Which Employers are Required to Keep Records

SIC Group	**Industry**
01–02 and 07–09	Agriculture, forestry, and fishing
13	Oil and gas extraction
15–17	Construction
20–39	Manufacturing
41–42 and 44–49	Transportation, communications, and public utilities
50–51	Wholesale trade
52	Building materials, hardware, garden supply, and mobile home dealers
53	General merchandise stores
54	Food stores
70	Hotels, rooming houses, camps, and other lodging places
75 and 76	Repair services
79	Amusement and recreation services
80	Health services

TABLE 23-2
Employers not Required to Keep Records, but Could be Required to Participate in the Annual Survey of Occupational Injuries and Illnesses

SIC	**Description of Industry**
	Retail Trade
55	Automotive Dealers and Gasoline Service Stations
56	Apparel and Accessory Stores
57	Furniture, Home Furnishings and Equipment Stores
58	Eating and Drinking Places
59	Miscellaneous Retail

Continued

TABLE 23-2
Continued

SIC	*Description of Industry*
	Finance, Insurance and Real Estate
60	Banking
61	Credit Agencies other than Banks
62	Security and Commodity Brokers, and Services
63	Insurance
64	Insurance Agents, Brokers and Services
65	Real Estate
66	Holding and other Investment Offices
	Services
72	Personal Services
73	Business Services
78	Motion Pictures
81	Legal Services
82	Educational Services
83	Social Services
84	Museums, Botanical and Zoological Gardens
86	Membership Organizations
87	Engineering, Accounting, Research, Management, and Related Services
88	Private Households
89	Miscellaneous Services

circumstances of his or her injury or illness. If workers' compensation or other forms provide the same information as Form 101, they may be substituted. Employers must make this information available to authorized government officials upon request.

Required Postings

By no later than February first of each year, employers must post an annual summary of occupational injuries and illnesses for the previous calendar year. It must be posted in a place where notices to employees are normally posted (typically near a time clock), and must remain until at least March first. This posting is required, even if it shows zeros because there were no injuries or illnesses.

Every workplace, regardless of the number of employees, must post OSHA 2203 or the Spanish version, 2200. Commonly called "The OSHA poster," and officially named *Job Safety and Health Protection* poster (or Seguridad en el Trabajo y Protección de la Salud), it informs employees of their rights and responsibilities under the OSH Act. When applicable, the equivalent state poster can be used.

An employer who is cited by OSHA for a violation must post the citation near each place where the violation occurred or in a prominent place where the citation can be easily observed by all affected employees. The citation must remain posted until the violation has been corrected or for three working days, whichever is longer.

Locations of Records, Reports, and Postings

Injury and illness records must be maintained at each workplace. Where workers are on assignments, rather than at a regular workplace, there must be a central location where records are kept. Records, including OSHA 200 and its predecessor forms 100 and 102, must be retained and updated (for example, if complications from an injury develop later) for five years following the calendar year they cover.

Every accident or incident that results in a fatality or the hospitalization of five or more employees must be reported to the nearest OSHA office or the national OSHA hotline (800-321-6742) within 48 hours. This rule applies to all employers, regardless of the number of employees.

The Annual Survey

Each year the Bureau of Labor Statistics (BLS) selects about 280,000 firms to take part in a survey used to calculate job injury and illness rates for various industries nationwide. All employers selected are required by law to participate. Employers who are normally exempt from OSHA recordkeeping are notified of their selection for the survey prior to the calendar year to which the survey relates.

Some of the ways agencies use the survey results are:

- To monitor OSHA's progress
- To assist OSHA in setting standards and evaluating existing standards
- For scheduling inspections
- To evaluate the performance of states and territories that operate their own OSHA-approved safety and health plans

Additional Details

OSHA has tried to anticipate and answer questions that arise from the differences among various organizations, but there will always be questions that must be answered individually. This section looks at some specifics that employers have asked OSHA to clarify.

Data processing records. OSHA Form 200 may be maintained by computer; Form 101 or an acceptable record and the annual summary must be maintained in paper form. Two requirements must be met when maintaining the log by means of data processing:

1. Sufficient information must be available to complete the log within six workdays after receipt of the information that a recordable case has occurred.
2. A copy of the log updated to within 45 calendar days must be present at all times in the establishment.

Availability of log. Questions arise about where to maintain logs and other records, because there are sites that are clearly permanent, workers whose assignments do not take them to a regular location daily, and numerous other workplace arrangements.

Generally, any operation at a given site for more than one year is considered a fixed establishment. When that definition does not apply, the following arrangements are allowed:

- Records may be kept at the field office or mobile base of operations.
- Records may also be kept at an established central location if 1) the address and telephone number of the place where the records are kept is available at the worksite, and 2) there is someone available at the central location during normal business hours to provide information from the records.

Employers who have employees at nonfixed sites must provide them with copies of the summary during the posting period from February first to March first. The requirement is that employees with no fixed worksite receive by direct presentation or by mail a copy of the annual summary.

There are places in the regulations that require certain information be immediately available at designated locations. Available via facsimile or overnight mail does not meet this requirement.

Time limitations. All questions on Forms 101 and 200 must be answered on the forms within six workdays of learning of a recordable injury or illness. If it an employer does not have sufficient information to answer a question within six days, "unknown" may be entered in the block. The employer is required to investigate the case to obtain information to complete the form in its entirety and then update the form.

Citations and Penalties

False statements, misrepresentations, and false certification in records required by OSHA may be punishable by a fine of up to $10,000 or imprisonment for up to six months. Failure to maintain records or file required reports may result in citations and penalties.

PART V

HANDLING IAQ EMERGENCIES

CHAPTER 24

How Planning Can Mitigate Effects of Emergencies

Good planning and thorough precautions reduce the probability of an IAQ emergency, and reduce the harmful effects if it does happen, but no plant can count on a zero probability of such an emergency. Even in this serious topic, with its potential for appalling consequences, precautions reach a point of diminishing returns. This chapter and the four that follow look at emergency-related information from pre-emergency planning to post-emergency examination of what management learned from its own and others' responses and actions.

Indoor air quality disasters, unlike some (but not all) other types of disasters, are more likely to begin in a relatively small area. For example, an earthquake suddenly hits the entire plant, but chlorine from a burst tank initially affects just the surrounding area—and you have a chance to keep it from affecting other areas. Another difference is that some disasters, such as hurricanes, allow time to gather essentials before evacuating; hazardous materials accidents may not allow even a few moments. The shorter the time to act, the more important it is to have a plan. This chapter looks at planning, to smooth the action before, during, and after an emergency.

PREPARING AND IMPLEMENTING AN EMERGENCY PLAN

A plan can prevent an ordinary accident from becoming a major disaster. Information and suggestions in this section are based on the Federal Emergency Management Agency's step-by-step approach to emergency planning, response, and recovery for companies of all sizes. It was sponsored by FEMA in partnership with a public–private alliance that consisted of manufacturers, industry associations, nonprofit organizations, building owners, and others.

The following is in outline form because any attempt to provide an actual plan would yield a document which would be, at best, useless for all except the one company for which it was prepared. An actual plan could inhibit the free flow of ideas necessary for another company's plan. This outline is general, and even includes a step for reviewing and modifying itself; it should be modified and optimized by each company that prepares a plan. One of the first decisions to be made by the team formed in Step 1, unless the decision was made prior to that time, is whether to establish an overall emergency plan at this time, or whether to limit it to indoor air quality emergencies. The last part of this outline is included for those who are preparing an overall emergency plan.

Plan Section 1: Four Steps in the Planning Process

STEP 1—ESTABLISH A PLANNING TEAM

- Form the team.
- Establish authority.
- Issue a mission statement.
- Establish a schedule and budget.

STEP 2—ANALYZE CAPABILITIES AND HAZARDS

Where do you stand right now?

- Review internal plans and policies.
- Meet with outside groups.
- Identify codes and regulations.
- Identify critical products, services and operations.
- Identify internal resources and capabilities.

- Identify external resources.
- Do an insurance review.

Conduct a vulnerability analysis.

- List potential emergencies.
- Estimate probability.
- Assess the potential human impact.
- Assess the potential property impact.
- Assess the potential business impact.
- Assess internal and external resources.
- Add the columns.

STEP 3—DEVELOP THE PLAN

Plan components

- Executive summary
- Emergency management elements
- Emergency response procedures
- Support documents

The Development Process

- Identify challenges and prioritize activities.
- Write the plan.
- Establish a training schedule.
- Continue to coordinate with outside organizations.
- Maintain contact with other corporate offices.
- Review, conduct training, and revise.
- Seek final approval.
- Distribute the plan.

STEP 4—IMPLEMENT THE PLAN

Integrate the plan into company operations and conduct training.

- Planning considerations
- Training activities
- Employee training
- Evaluate and modify the plan

Plan Section 2: Emergency Management Considerations

Direction and control

- Emergency Management Group (EMG)
- Incident Command System (ICS)
- Emergency Operations Center (EOE)
- Planning considerations
- Security
- Coordination of outside response

Communications

- Contingency planning
- Emergency communications
- Family communications
- Notification
- Warning

Life safety

- Evacuation planning
- Evacuation routes and exits
- Assembly areas and accountability
- Shelter
- Training and information
- Family preparedness

Property protection

- Planning considerations
- Protection systems
- Mitigation
- Facility shutdown
- Records preservation

Community outreach

- Involving the community
- Mutual aid agreements

- Community service
- Public information
- Media relations

Recovery and restoration

- Planning considerations
- Continuity of management
- Insurance
- Employee support
- Resuming operations

Administration and logistics

- Administrative actions
- Logistics

Plan Section 3: Hazard-Specific Information

- Hazardous materials incidents
- Fire
- Floods and flash floods
- Hurricanes
- Tornadoes
- Severe winter storms
- Earthquakes
- Technological emergencies

Charting Readiness

After the plan is reviewed and approved, all the effort is wasted if the plan is filed away "until needed." Employees will perform their parts only if they know them well enough to perform them automatically under the pressure of an actual emergency. The FEMA–private industry coalition that designed the preceding emergency management plan included a chart of training drills and exercises, from which Table 24-1 is patterned.

TABLE 24-1

Charting Training Drills and Exercises

	Jan	*Feb*	*Mar*	*Apr*	*May*	*Jun*	*Jul*	*Aug*	*Sep*	*Oct*	*Nov*	*Dec*
Mgmt Orientation & Review												
Employee Orientation & Review												
Contractor Orientation & Review												
Community & Media Orientation & Rev.												
Management Tabletop Exercise												
Response Team Tabletop Exercise												
Walk-through Drill												
Functional Drills												
Evacuation Drills												
Full-scale Exercises & Drills												

It is intended that some items in Table 24-1 be conducted just one time, but they can be repeated if the plan is modified. Other items might be conducted annually, quarterly, or monthly, as required to keep everyone adequately prepared. Proposed dates for each activity can be entered at the tops of cells, and actual dates below them.

OBTAINING FEDERAL GOVERNMENT ASSISTANCE

The Federal Emergency Management Agency provides a wide range of disaster-recovery assistance that few people know about. This section first lists the topics that are searchable on the Internet, and then gives more information about one of the items: Disaster Assistance Employees. Training and other topics are included in Chapter 25.

FEMA on the Internet

FEMA's home page (http://www.fema.gov/) leads to layers of links to information about the agency, such as those in the following list. Although several topics in the list do not appear related to indoor air quality emergencies, they are included because they refer to IAQ or offer suggestions that apply universally to disasters.

Alphabetical Index of FEMA WWW Server

- Archives
- Arson Information Management System (AIMS)
- Business and Industry, Emergency Management Guide for
- Conferences Scheduled
- Director's Speeches
- Disaster Application Center
- Disaster Assistance
- Disaster Assistance Employees
- Disaster Facts by Calendar Year
- Earthquake
- Earthquake Mitigation
- Emergency Management Institute (EMI)
- Emergency Medical Services (EMS)
- Emergency Education NETwork (EENET)
- Emergency Food and Shelter National Board Program
- Emergency Support Functions
- Exercises
- Extreme Heat
- Family Protection Program
- Farmers Home Administration
- Federal Crime Insurance Program

Federal Flood Insurance FAQ
Federal Insurance Administration Directorate
Federal Insurance Companies List
Federal Response Plan
Federal Response Plan Partners
FEMA Radio Network
FEMA Organization
Fire
Flood
Flood Hazard Determination Form (Std.)
Flood Insurance
Flood Insurance Claims
Flood Mitigation
Floods of 1996
FAQ U.S. Fire Administration USFA Frequently Asked Questions
Guide to Recovery Programs
Global Emergency Management System (GEMS)
Help after a Disaster
Hurricane
I Want to Know
Immediate Help
Independent Study Program
Individual and Family Protection Program
Individual Assistance
Information
Information Systems Directorate
Separate Internet Mailing Service—News Releases and Situation Reports
Landslides and Mudflows
Learning Resource Center
Library
Listserve Services
Map Service Center
Matching Mitigation Funds
Mission Statement
Mitigation
Mitigation Directorate
Myths and Facts About the NFIP
National Emergency Training Center
National Fallen Firefighters Memorial
National Fire Academy (NFA)
National Fire Data Center (NFDC)
National Fire Incident Reporting System (NFIRS)
National Flood Insurance Program Community Status Book
National Flood Insurance Program, Myth and Fact
News Desk
Oklahoma City Disaster Archives
Operations Support Directorate
Partners
Photo Library
Planning
Preparedness
Preparedness, Training and Exercises Directorate
Preparing for a Disaster
Presidential Disaster Declaration
Prevention
Public Fire Education
Publications
Publications, U.S. Fire Administration
Questions
Recovery
Recovery Channel
Recovery Times
Regional and Area Offices

Response
Response and Recovery Directorate
Risk Reduction
Upcoming Seminars and Symposiums
Small Business Administration
Thunderstorms
Tornadoes
Training
Tropical Storm Watch
Tsunamis
Urban Search and Rescue
United States Fire Administration
Vision Statement
Volcanoes
Weather Link
What's New
What You Need to Know about Federal Disaster Assistance
What We Do
Who We Are
Wildland Fires
Winter Driving

Disaster Assistance Employees

FEMA has an unusual arrangement with other federal agencies. When large-scale disasters strike, FEMA augments its regular full-time work force with Disaster Assistance Employees (DAEs), also called reservists. They are federal employees hired for a two-year appointment to work on a disaster for a few days to a few weeks, then return to their regular jobs. A DAE may work with FEMA two or three times a year, depending on assignments, availability, frequency of disasters, and scale of disasters. Following is a sampling of the assistance DAEs can provide:

- Interviewing disaster victims
- Inspecting damaged facilities
- Handling telephone calls
- Handling administrative matters
- Installing and operating computer networks
- Data entry and records management

Because the reservist program changes frequently, it is best to check the FEMA Internet home page for the latest information.

GETTING ASSISTANCE FROM THE PRIVATE SECTOR

The main business of some companies is responding to emergency situations; some of them specialize in spills and other accidents that could be IAQ disasters.

Specializing in this field takes them to various locations and gains them familiarity with regulations and requirements of all levels of government. They know of cleanup contractors and the quality of work each does. This section suggests services you should evaluate when selecting a company to assist with an emergency.[1]

- *Type of Accident Handled.* Some companies transport many kinds of hazardous materials, and others use just one kind in laboratory work. A contractor who has ideal qualifications for companies in the transport category might be unsatisfactory for companies using one toxic substance in a laboratory.
- *Type of Material Handled.* Just because a spill contractor has done excellent work in cleaning radioactive spills does not mean he has the experience or equipment to handle a nitric acid spill.
- *Response Time.* A contractor with excellent recommendations will be of little help to a manufacturer the contractor cannot reach in a short time.
- *Notification and Reporting.* It is important that the contractor be familiar with local, state, and federal requirements for immediate reporting of accidents. Many authorities are strict in the follow-up reports they demand.
- *Cleanup Contractor.* If the spill contractor is to select, monitor, and regulate the cleanup contractor, it is important to hire a spill contractor who knows the work of available cleanup contractors.
- *Costs.* The spill contractor should have experience in avoiding or minimizing costs such as fines and lawsuits.

1. Information provided by *The Spill Center,* Acton, Massachusetts.

CHAPTER 25

Preparation and Training for Emergencies

One part of preparing the plant and the staff for indoor air quality emergencies is the plan that was covered in Chapter 24. The present chapter continues with a look at how to collect information about emergency preparations, and then suggests ways to train staff to handle their emergency assignments.

DISASTER INFORMATION SOURCES

Chapter 24 explained how to obtain a wide variety of related federal government information on the Internet. Many of those locations also provide addresses you can use to send for additional information, e-mail addresses for faster requests and responses, and telephone numbers for immediate discussion of specific problems.

Another medium for receiving information, especially updates, news releases, daybook advisories, and Editor's Tip Sheets, is the fax machine. The Federal Emergency Management Agency maintains a 24-hour automated information system, which is not restricted to indoor air quality problems. Its documents are stored in five sections, each with its own index, as follows:

> *Section I Disaster Information*—Contains the latest information on current disaster activities nationwide, including a list of contacts at the Disaster Field Offices. Historical disaster profiles (cost and damage figures) and annual disaster activity reports are also available. Document Number 1000.
>
> *Section II News Desk*—Contains news releases for the current year, daybook advisories, FEMA Radio Network Advisories, and Editor's Tip Sheets. Document Number 2000.
>
> *Section III Emergency Preparedness Information*—Background information and fact sheets on what to do before, during, and after a disaster. This section includes a list of publications available on similar topics. Document Number 3000.
>
> *Section IV FEMA*—About the agency and what it does. FEMA summary, informal citations for FEMA authorities, director and staff biographies, agency mission statement and organization chart, speech files, and issue statements. Organizational charts of FEMA directorates, regional offices, and descriptions of programs are also included in this section. Document Number 4000.
>
> *Section V Miscellaneous Issues, Topics, and Policy Information*—Policy papers, white papers, and other documents related to FEMA and its mission. Document Number 5000.

You can receive any of these documents free on a fax machine by dialing (202) 646-3362 [(202) 646-FEMA] on a touch-tone telephone. Voice prompts lead you to make your choices by pressing numbers, and then you follow other voice prompts to complete your request. The prompts tell you how to receive the complete document index (approximately 20 pages), a specific section's index (updated weekly), or a specific document.

EMERGENCY MANAGEMENT INDEPENDENT STUDY

FEMA's Emergency Management Institute (EMI) Independent Study (IS) Program consists of 12 self-paced courses designed for people who have emergency management responsibilities as well as for other employees and the general public. There are no prerequisites or enrollment fees for these courses. Each Independent Study Course, usually completed in 10 to 12 hours, includes lessons with practice exercises and a final examination. Those who score 75 percent or higher are issued a certificate of completion by EMI. All materials that are on the Internet are available to anyone who can access them, but official enrollment in the courses and scoring of final examinations is limited to United States residents.

Several other courses are also available, designed for specific emergency management functions. Emergency Management Professionals can obtain more information and enroll in these audience-specific courses by contacting the Independent Study Office at EMI (see the section "More Information from FEMA" at the end of this chapter).

Individual Enrollment Procedures

An applicant can obtain and submit FEMA Form L-173, or write a letter of application. The letter must include name, social security number, mailing address, and course number and name. Only one course per student per request will be accepted.

Group Enrollment Procedures

If five or more participants will take a course, they must first appoint a group coordinator, and then submit L-173 forms as a group. Alternatively, they can submit a student roster with each student's name, social security number, address, and telephone number.

Course Descriptions

The following five courses are available to all applicants.

> *IS-1 Emergency Program Manager: An Orientation to the Position.* This study course provides an introduction to Comprehensive Emergency Management (CEM) and the Integrated Emergency Management System (IEMS). Included is an in-depth look at the four phases of comprehensive emergency management: mitigation, preparedness, response, and recovery. The text is accompanied by illustrations, diagrams, and figures. In most units, there are worksheets, exercises, and tasks to complete.

IS-2 Emergency Preparedness, USA. This study course contains information about natural and technological hazards and national security issues. Participants are led through the development of personal emergency preparedness plans and are encouraged to become involved in the local emergency preparedness network. The text is accompanied by illustrations, maps, charts, and diagrams.

IS-3 Radiological Emergency Management. This study course contains information on a variety of radiological topics, including:

- Fundamental principles of radiation
- Nuclear threat and protective measures
- Nuclear power plants
- Radiological transportation accidents
- Other radiological hazards

IS-3 is available in computer-based training or textbook format. To take the course in computer-based training requires access to an IBM-compatible personal computer with 640K of memory and a 5 1/4″ diskette drive.

IS-5 Hazardous Materials: A Citizen's Orientation. This study course provides a general introduction to hazardous materials, and can serve as a foundation for more specific research. The course has five units and has been designed with the objective of helping individuals to:

- Recognize the dangers posed by hazardous materials.
- Identify places where hazardous materials are likely to be encountered.
- Understand when a hazard may exist.
- Contact the appropriate persons or agencies to give or receive specific hazardous materials information.
- Identify procedures to minimize personal and community exposure to hazardous materials.

IS-7 A Citizen's Guide to Disaster Assistance. This study course provides a basic understanding of the roles and responsibilities of the local community, state, and federal governments in providing disaster assistance. It is appropriate for both the general public and those involved in emergency management who need a broad introduction to disaster assistance.

National Training Center Programs

FEMA's National Emergency Training Center in Emmitsburg, Maryland, is home of the Emergency Management Institute and the National Fire Academy. There, emergency managers, firefighters, and others can take courses in many areas of emergency management. Typical courses include emergency planning, exercise design and evaluation, disaster management, hazardous materials response, and fire service management.

FEMA courses are also given by many states. Special seminars and workshops are offered via satellite as part of FEMA's Emergency Education Network, called EENET. A course of special interest to engineers, architects, and building code officials interested in IAQ consequences is the Multihazard Building Design Summer Institute.

More Information From FEMA

FEMA's home page on the Internet (http://www.fema.gov/) is a typical World Wide Web page, with underlined hyperlinks to other pages and other links. From here you can locate a form for submitting comments online.

For comments or additional information, the telephone number is (202) 646-4600. You can write FEMA's Office of Emergency Information and Public Affairs at 500 C Street S.W., Room 820, Washington, D.C. 20472-0001.

CHAPTER 26

Considering Emergencies in Designing Building Systems

When a new plant or a major renovation is in the planning stages, it would be prudent to include provisions for possible indoor air quality emergencies. Even if it were possible (it isn't) to reduce the probability of an internal accident to zero, there are emergencies such as railroad and highway accidents that are beyond the plant's control but can affect the plant through its external environment.

The arrangement of this chapter is to look at a complete centrally-controlled building system as might be found in a large modern facility. Although the installation described would be desirable for every plant, it would be unnecessary and not feasible for most and even laughable for many. Therefore, it is suggested that readers study every item and assign an "interest rating" to each. A reader who uses a scale of one to five could design an interest rating system as follows:

1. We need this item.
2. Will try to get approval for this item.
3. We should make provision for adding this item later.
4. This item would be nice to have, but it's hard to justify the cost.
5. We have no need for this item.

The rest of this chapter looks at items to be considered for inclusion in a safe building design. They are in no special order, because "order of importance" would change from reader to reader, depending on hundreds of variables related to the plants where they work.

MONITORING CONTINUOUSLY FOR POLLUTANTS

Automatic alarms can warn workers when pollutant concentrations have exceeded preselected levels. Whatever trigger levels are selected, workers should be trained to understand the alarm's meaning. For example, the alarm might be triggered at a point which means one of the following:

- The level is dangerous; everyone evacuate immediately.
- The level is approaching the hazardous concentration; turn on exhaust fans.
- Immediately stop a certain type of production.
- Continue working; a designated individual will ensure that the supervisor is aware of the alarm.

The monitoring system could be designed with more than one trigger level. For example, a horn might tell workers to turn exhaust fans on and, if air quality continued deteriorating, a bell could tell workers to evacuate. To prevent the pollution from migrating to adjoining areas, the monitoring system could close doors between the areas. Of course, this arrangement would be acceptable only if workers in the polluted area had another means of leaving that area.

Another way to protect unaffected areas is to tie the automatic monitoring system to the HVAC system and have it close off return air intakes in the affected rooms. This can be one of the most important features of a system because, without it, the effects of an accident could be quickly distributed throughout the plant.

If the monitoring and HVAC systems cannot be arranged for automatic shutoff of return air in the affected rooms, one or more employees can be assigned to close them manually when the alarm sounds. The team writing the emergency management plan should give careful thought to such an assignment if the alarm signals immediate evacuation; the consequences of assigning one or more selected employees to be the last ones out can be far-reaching.

CENTRAL MONITORING

A central monitoring system that is basically a frame in which modules can be plugged allows flexibility for selecting critical areas to monitor. If a potentially dangerous chemical is no longer being used in a certain area, but another area will be the critical one for a while, this flexible system could be ideal.

There is no definition of *complete central monitoring*, because there seems to be no end to the items that can be monitored and transmitted to the central location, usually the security office. In addition to chemical detectors for IAQ, temperature, smoke, sound, and video are some of the more common items to monitor. The importance of video monitoring is that it can back up or verify signals from other monitors. For example, if an air contaminant leaks slowly into an area, workers may not realize how it is affecting them—workers can pass out from some contaminants without knowing they are in danger. Security personnel can observe such workers in time to take action.

There are several choices of readouts in the central monitoring station:

- Each area being monitored can have a complete set of readouts in the central location. Monitoring personnel scan through them regularly. It is good to have a flashing signal that will attract attention when a reading at some location is out of range.
- One set of readouts can be switched from area to area. Monitoring personnel watch the set while the system switches them through the areas one at a time.
- When there is one set that switches from area to area, the switching can be automatic and programmed, or it can be switched manually. If automatic, it should be possible for a person to override it and switch manually, or hold it at a certain area.
- Security personnel at the central location have additional control if there are loudspeakers and microphones that allow two-way, hands-free communication with each area.
- Any alarms that are activated in an area should also sound in the central location.
- It is sometimes advisable to allow alarms to be set off manually from the central location.

USING YOUR VENTILATION SYSTEM EFFECTIVELY

Usually integral with the HVAC system, the ventilation system has a vital part in any indoor air quality concern, especially during and after an emergency. An essential measure for keeping contaminants in one area from polluting the entire plant is to be able to close the return air intake(s) in any polluted area. The choices are:

- Automatic closure initiated by monitors
- Manual closure in the area
- Remote closure controlled automatically from the central location
- Remote closure controlled manually

If return air intakes are closed, supply air continuing to enter the area will create a positive pressure that forces air from that area to adjacent areas. Although that may not be as damaging as distributing polluted air to the entire plant, neither is it consistent with the objective of keeping the polluted air contained. Therefore, it is important that, when return air intakes are closed, room air must be exhausted directly outside. Open windows will relieve the positive pressure; forced ventilation is better, and can be designed to create a negative pressure which will ensure that any air exchange with adjacent areas will be into the polluted area.

Providing for closing the outside air intake adds an additional safety factor. However, few plants monitor outside air, so this closure is almost always under manual control. The most serious worry is usually from widespread pollution due to transportation accidents (highway and railroad spills), but other events such as smoggy days give sufficient reason for not bringing in outside air.

COMMUNICATING THE EMERGENCY MANAGEMENT PLAN

It may seem obvious, but it is easy to leave a major gap in the emergency management plan if this simple list is not written down. Prepare a list of those who must be notified in the event of an IAQ emergency. In most plants, the following list identifies such individuals by groups:

- Workers in the affected area, who must protect themselves

- Emergency management team, whose members have assignments such as controlling ventilation
- Communication facilitator
- Plant emergency response team, usually within the fire-fighting department
- Outside emergency response team, usually within the municipal fire department
- Medical personnel. If not included in the response teams, then usually called by the emergency coordinator if needed.

After preparing the list by groups, you can determine how to communicate the emergency to the members of each group. Some of the forms of communication to consider are:

- Alarms, mentioned earlier in this chapter. Audible alarms might have to be supplemented with visual alarms in special situations which include handicapped workers and noisy environments (whether workers wear hearing protection or not).
- Alerts automatically sent to the main security office and others.
- Telephone tree.
- Telephone automatic dialer. There are systems on the market that can be programmed to dial a list of numbers. When someone answers, the dialer gives a voice message describing the emergency and then it listens for a prearranged touch-tone response. These systems can be set into action automatically by monitoring sensors or they can require manual initiation.
- Pagers. Many plant engineers and other key personnel can be reached at any time this way.
- Fire and police departments and ambulance services. Some local regulations prohibit automatic notification by monitoring equipment. Some prefer being called by a person because it reduces the concern about false alarms, and because it makes two-way questioning of the nature of the emergency possible.

THE STATISTICAL RELIABILITY OF WARNINGS

When comparing the equipment provided by different manufacturers, it is helpful to understand what they mean when they use statisticians' phrases such as "Type I error less than 1 percent" and "Type II error less than 3 percent." A Type

I error occurs when the system misses something that should be detected; for IAQ equipment, an air contaminant is the item that should be detected. Manufacturers also refer to this error as an alpha (α) error. For example, suppose an instrument triggers an alarm if carbon monoxide concentration exceeds 35 parts per million, and the instrument specifications include, "Type I error less than 1 percent." The specification means that, if the concentration exceeds 35 ppm 100 times, the instrument can trigger the alarm 99 times but fail to detect one of the excesses.

The opposite is a Type II error, also called a beta (β) error, which is the probability of a false detection, or false alarm. For example, the same instrument might include "Type II error less than 2 percent." This means that, if the instrument takes 100 readings, and the concentration is actually less than 35 ppm every time, the instrument could trigger the alarm twice.

If it is more important that an instrument initiate an alarm when it should, and less important that there are more frequent false alarms, some emergency managers take the "brute force" approach of using redundant instruments. Setting up two monitors that have the same Type I error specification brings up a classical example of binomial probability, which follows the formula:

$$P(r) = \frac{n!}{r!(n-r)!} p^r q^{n-r}$$

Where: p = probability of each instrument reading correctly
q = probability of each instrument reading incorrectly
n = number of readings
r = number of correct readings

If two instruments, each with a 0.99 probability of reading correctly, are connected so that an alarm will sound if either or both instruments sense excess carbon monoxide, the variables have the following values: $p = 0.99$; $q = 0.01$; $n = 100$; $r = 1$ when either instrument triggers the alarm; $r = 2$ when both instruments trigger the alarm. Since the alarm sounds when either or both instruments trigger it, the probability of having the alarm sound correctly is equal to the probability of having one instrument trigger it plus the probability of having two instruments trigger it. Statistically it is written:

$$P(1 \text{ or } 2) = P(1) + P(2)$$

and it is calculated by substitution as follows:

$$P(1 \text{ or } 2) = \frac{2!}{2!(2-1)!}\, 0.99^{1} 0.01^{2-1} + \frac{2!}{2!(2-2)!}\, 0.99^{2} 0.01^{2-2}$$
$$= 0.9999$$

Using redundant instruments has brought the reliability closer to a perfect 1.0—from 0.99 to 0.9999.

CHAPTER 27

Using Respirators in an Emergency

Chapter 7 gave an overall look at respirator usage; the current chapter covers the special use of respirators to provide a safe way for workers to leave an area in an emergency. If there is a reasonable possibility of an indoor air accident, escape devices should be stored where workers can get them quickly, and workers should be trained in their use.

USING RESPIRATORS TO ESCAPE EMERGENCIES

Respirators designed to allow a person working in a normally safe environment sufficient time to escape a respiratory hazard that occurs suddenly are called escape devices or escape apparatus. As with other respirators, described in Chapter 7, escape devices can be examined in two categories: air-purifying respirators and self-contained breathing apparatus.

Air-Purifying Respirators

An air-purifying respirator takes in air from the room, removes contaminants by sorbent and/or filter media, and provides breathable air to the person wearing it. These respirators cannot be used in an oxygen-deficient atmosphere because they do not create oxygen. If the surrounding atmosphere is deficient in oxygen, the wearer will not receive satisfactory air to breathe. Types of air-purifying escape respirators include the escape gas mask (canister) respirator, the gas mask (canister) respirator, and the filter self-rescuer.

The escape gas mask consists of a half-mask or a mouthpiece respirator. The half-mask filters contaminants from the air, and can also be used to escape from low concentrations of organic vapor or acid gas. The mouthpiece respirator can be used for short periods of time to escape from low concentrations of organic vapor or acid gas.

An escape gas mask respirator equipped with a full facepiece can be used also to escape from Immediately Dangerous to Life and Health (IDLH) conditions but not from an oxygen-deficient atmosphere.

The filter self-rescue unit is a mouthpiece device that is designed specifically to protect against less than one percent carbon monoxide.

Self-Contained Breathing Apparatus (SCBA)

The main feature of SCBA is that it includes tanks of air and therefore provides air to users for escape from oxygen-deficient environments. They are commonly used with full facepieces or hoods and, depending on the supply of air, are usually rated as 3- to 60-minute units. Self-contained self-rescuer (SCSR) devices have been approved by MSHA/NIOSH specifically for escape from mines, but may also have application in other similar environments. They are mouthpiece respirators that provide a source of oxygen-enriched air for up to 60 minutes.

Selecting Escape Apparatus

When selecting escape apparatus, give careful consideration to potential eye irritation. This potential is important for determining whether a gas mask or SCBA equipped with full facepiece should be selected rather than a device equipped with a half-mask or mouthpiece.

Most gas masks or escape gas masks can be used in situations involving gas(es), vapor(s), or particulates. The selection for escaping from particulate-contaminated environments must be an air-purifying element that will provide

TABLE 27-1
Selection Options for Escape Respirators

Escape Conditions	*Type of Respirator*
Short distance to exit, no obstacles, no oxygen-deficiency	Any escape gas mask[a] (canister respirator) or gas mask[b] (canister respirator)
	Any escape self-contained breathing apparatus having a suitable service life[c]
	Any acceptable device for entry into emergency situations
Long distance to exit or obstacles along the way. No oxygen deficiency.	Any gas mask[b]
	Any escape self-contained breathing apparatus having a suitable service life[c]
	Any self-contained self-rescuer having a suitable service life
Potential oxygen deficiency	Any escape self-contained breathing apparatus having a suitable service life[c]
	Any self-contained self-rescuer having a suitable service life

[a] An escape gas mask is a respirator designed for use only during escape from IDLH or non-IDLH atmospheres. It may consist of a half-mask facepiece or mouthpiece, appropriate air-purifying element for the contaminant, and associated connections. Maximum use concentrations for these types of respirators are designated by the manufacturer.

[b] A gas mask consists of a full facepiece and either chin-style or front- or back-mounted canisters with associated connections. Maximum use concentrations for canister air-purifying elements are listed in Table 27-2.

[c] Escape self-contained breathing apparatus can have rated service lives of 3 to 60 minutes. All acceptable devices for entry into emergency situations can also be used.

TABLE 27-2

NIOSH Recommended Maximum Use Concentrations, in ppm, for Gas and Vapor Air-Purifying Elements

Type of Gas or Vapor	Classification of Gas and Vapor Air-Purifying Elements		
	Cartridge(s)	Chin-style canister	Front- or back-mounted canister
Organic vapors	1,000[a]	5,000[b]	20,000[b]
Acid gases			
Sulfur dioxide (SO_2)	50	100	100
Chlorine (Cl_2)	10	25	25
Hydrochloric (HCl)	50	100	100
Ammonia (NH_3)	300	500	500
Methyl Amine (CH_3NH_2)	100	—	—
Carbon Monoxide (CO)	NA	NA	1,500

[a] Maximum use concentration will be 1,000 ppm or the immediately dangerous to life or health value for the specific organic vapor, whichever is lower.

[b] Maximum use concentration for "entry into" will be limited to the value listed or to the immediately dangerous to life and health value for the specific organic vapor, whichever is lower.

protection from the given type of particulate. Use the information in Table 27-1 to select escape apparatus appropriate for your plant's layout and other conditions.

STORING RESPIRATORS

Proper storage is an important factor for respirators that are intended for emergency, rather than regular, use. Serious illness, even death, can result if the respirators prove to be unreliable when that accident (which everyone hoped would never occur) finally happens. OSHA requires that stored respirators be protected against dust, sunlight, heat, extreme cold, excessive moisture, and damaging chemicals.

It is strongly recommended that freshly cleaned respirators be placed in heat-sealed or reusable plastic bags until reissue. They should be stored in a clean, dry location away from direct sunlight. On shelves, they should not be in multiple layers and facepieces and exhalation valves should be in approximately a normal position to prevent the rubber or plastic from becoming permanently distorted.

Air-purifying respirators kept ready for nonroutine or emergency use should be stored in a cabinet in individual compartments. Each compartment should be clearly labeled with the user's name or other identification. The storage cabinet should be unlocked and readily accessible. (It is tempting to store material in front of it because, "no one uses it.") All workers should know of its location; safe evacuation may depend on how quickly workers can get to the emergency respirators.

Self-contained breathing apparatus for emergency evacuation of a given area should be stored in a nearby area that is more protected from potential emergencies. The reason for this storage arrangement is that even thoroughly trained workers take 30 to 60 seconds to put these devices on. In a highly contaminated atmosphere such as might be created by a massive release of a toxic material, a worker could be severely injured or incapacitated in that length of time. Generally, the best procedure after an accident is to escape as quickly as possible to an uncontaminated area where the SCBA is, put the equipment on, and then—if necessary—re-enter the contaminated area. For any particular plant, the emergency manager should evaluate the potential hazard, taking into account the area's physical configuration, before making a decision as to the best storage location for SCBA.

CHAPTER 28

Conducting Your Post-Emergency Analysis

Included in the Emergency Management Plan should be a provision for reviewing everything that led up to the emergency, how it was handled, what procedures might be changed, and how the plan should be modified. The review offers an opportunity to examine company policies in general; for example, one question asks how workers with handicaps fared during the emergency.

This chapter gives suggestions in the form of statements and questions. Some of them may not apply to your plant and some may not apply to a given emergency, but they will lead to other questions you can ask. Each reader can add questions that apply uniquely to his or her plant and/or to a recent emergency.

While it is important to determine what caused the emergency—and to learn how to avoid one in the future—pointing fingers, finding fault, or placing blame is not the purpose of the post-emergency analysis. This analysis should be kept to a search for facts; if desired, a separate investigation can look at other issues.

ITEMS RELATING TO EVENTS BEFORE THE EMERGENCY

Did all conditions conform to OSHA requirements?

Did a defect in equipment, processes, materials, or procedure initiate the emergency?

Was this a sudden event, or was there a slow buildup?

If there was a slow buildup, would a lower trigger level have been better?

What sequence of events led to the emergency?

Do employees take emergency training seriously?

Have there been false alarms recently?

Were identified pollutant pathways found to be correct?

Was available information about the contaminants correct?

Now that it has happened, can you look back and recall subtle signs that could have been warnings?

ITEMS RELATING TO THE EMERGENCY

What alarms were actuated manually?

What alarms were actuated automatically?

Did automatic equipment (alarms, ventilation controls, etc.) work properly?

Did employees understand the emergency's seriousness from the beginning?

ITEMS RELATING TO COMMUNICATIONS DURING THE EMERGENCY

How well did communications function?

How effective was the automatic telephone dialer in assembling the response team?

ITEMS RELATING TO ACTIVITIES DURING THE EMERGENCY

How well did the evacuation plan function?

Did personal equipment (respirators and others) work properly?

After the problem first began, did anything exacerbate it?

What contaminant(s) were involved?

What was the highest contaminant concentration?

If recording equipment was operating, describe the time-concentration pattern.

Was proper first aid given to those who needed it?

Were handicapped workers or visitors in more danger than others?

If there were visitors or other outsiders in the plant, did someone take control of them?

If consultants were involved, was their performance satisfactory?

If outside services (fire departments and others) were involved, were they satisfactory?

AFTER THE EMERGENCY

If there was an internal spill, was the correct cleanup equipment on hand?

How long did the cleanup take?

Were there lingering odors after the area was declared safe?

How were the company's employee relations affected?

How were the company's public relations affected?

Is a dollar figure available for the accident's direct cost?

QUESTIONS RELATING TO BETTER PREPAREDNESS

How could the emergency have been avoided?

What additional training would have helped?

Should inspection procedures be changed?

Would the results have been different if there had been different drills (more frequent, more intense, different focus, etc.)?

Should there be any changes in the monitoring equipment?

Would tags and instructional labels reduce the chance of a similar emergency?

Would posters and signs reduce the chance of a similar emergency?

Would more be known now if someone had been assigned to record events and times during the emergency?

Is there material to put in a file which might be called, "Items to consider when designing our new building."?

APPENDICES

APPENDIX A

Case Studies

Indoor air quality problems can be very real and dangerous—even deadly in some situations. At other times the problem can be so elusive that it is difficult or impossible to determine if an IAQ problem really exists. Investigators like to quickly identify "what's in the air," write a report that includes recommendations for cleaning the air, and know they have solved the problem. Less satisfying is to work for months, find nothing definite, and write a report saying some changes were made and occupants' symptoms have stopped (but, if there was an actual problem, will it return?).

This appendix gives examples of both types of indoor air quality situations. In the first case study, investigators found and fixed several instances of poor practices; and the second demonstrates a situation in which exhaustive investigation revealed nothing substantial. The fact that both case studies relate to government buildings means nothing more than that the documents were readily available; it neither means that government buildings have more IAQ problems nor that governments as employers are quicker to react to and solve reported problems.

Both reports are considerably condensed for this appendix. Background information, theories, histories, and other information in the reports have been omitted, but complete identification of the reports is in the references at the end of this appendix for readers who want to examine the cases thoroughly.

SOCIAL SECURITY ADMINISTRATION, PETERSBURG, VIRGINIA

A request from the American Federation of Governmental Employees (AFGE) to NIOSH set the Health Hazard Evaluation (HHE) in motion. The request stated that for two years employees and the public had complained of high relative humidity, breathing problems, burning eyes, and lack of air-conditioning at the SSA office in Petersburg, Virginia.

Results of Investigation

Inspection of the air handling unit (AHU) revealed many problems with preventive maintenance of the HVAC. Visual inspection of the acoustic lining of the mixed air plenum suggested the existence of biological growth and an accumulation of dry scaly debris in the condensate pan. Laboratory analysis of bulk material samples collected from the condensate pan and acoustic liner confirmed amplification of biological material in the building's AHU. In addition, the mechanical linkage for the pneumatic return air damper was disconnected, which may have prevented proper operation of the AHU.

The filtration system consists of a fiberglass roll filter, a material that is less than 20 percent efficient based on ASHRAE dust spot criteria. The purpose of these filters is to keep lint and dust from accumulating on the system's heating and cooling coils.

Of 121 air supply slots in the area occupied by the SSA, 113 were accessible, and airflow measurements were taken at all 113. In open spaces and private offices the flows ranged from 18 to 89 cfm. At some locations with similar usage and occupant density, the flow rate varied widely (e.g., 27 cfm to 81 cfm). This wide range of flow rates suggests a seriously unbalanced system, and no test and balance report could be found. An unbalanced system can cause poor air distribution and a lack of temperature control. It could be responsible for complaints of thermal discomfort, and can affect the perceived comfort of individuals in the occupied space. The unbalance can explain why some occupants complained of too much airflow while others complained of too little. Investigators did not attempt to improve the system balance or even adjust airflow from the diffusers because those changes should be made only by trained HVAC professionals.

Table A-1 shows the ranges of carbon dioxide, temperature, and relative humidity measurements at six locations in the offices. CO_2 concentrations exceeded the ASHRAE recommendation of 1000 ppm during the midday sample period at two locations, and temperatures exceeded the ASHRAE thermal comfort range of 64°F to 74°F for winter months. The measurements were made in January, and outdoor CO_2 concentration, at 350 ppm, was within the normal outdoor range. Relative humidity values in Table A-1 are within the acceptable range (defined as at least 90 percent of individuals feeling thermally comfortable) for persons clothed in typical winter clothing and performing mainly sedentary tasks.

Analytical results of bulk samples collected from the acoustic lining of the AHU mixed air plenum and the condensate pan are given in Table A-2. All the fungal taxa identified are normal constituents of the environment, but the concentrations observed indicate the presence of flourishing fungal cultures. The predominance of yeast colonies is characteristic of an atmosphere high in moisture. Although yeasts have not been documented to cause immunologic problems, their existence and the quantity present indicate an environment favorable to the growth of microorganisms. Gram negative (Gram –) bacteria were also found in the bulk material sample collected from the condensate pan, which again indicates the presence of stagnant water at some point in time. Gram negative bacteria can produce endotoxins, which have been documented to produce hypersensitivity pneumonitis in some exposed individuals. Although there were no established criteria regarding "acceptable" concentrations of fungi and/or bacteria in ventilation system interiors, the concentrations observed indicate that there was a microbial reservoir, and amplification in that reservoir was occurring.

TABLE A-1

CO_2, Temperature, and Relative Humidity Measurements

Time Reference	*Ranges*		
	CO_2 in ppm	*Temperature in °F*	*Relative Humidity in %*
7:30 a.m.	575 - 650	76 - 77	35 - 38
11:25 a.m.	825 - 1125	76 - 78	28 - 30
2:35 p.m.	625 - 750	77 - 78	26 - 30

TABLE A-2
Microbiological Results of Bulk Samples

Sample Location	*Fungi*		*Bacteria*	
	CFU/gm	*Taxa Rank*	*CFU/gm*	*Taxa Rank*
Air Plenum	3,400,000	Pen>Clad	ND	Alc>Cor
Condensate Pan	520,000	Yea>>Pen	9,300,000	Xan>Ps>TA

Notes: Yea = Yeast
Pen = Penicillium
Clad = Cladosporium
Alc = Alcaligines faecalis
Cor = Corynebacterium pdeudodiphtheriticum
Xan = Xanthomonas campestris (Gram –)
Ps = Pseudomonas avenae (Gram –)
TA = Thermoactinomyces
ND = Non-detectable
CFU = Colony forming unit

Employees were questioned about their perceptions of various environmental conditions on their floors. Informal discussions with employees revealed that environmental concerns about the building included a lack of air movement, detecting cigarette smoke, thermal discomfort, and odors. Reports of employees being too hot or too cold did not appear to be related to one specific work area, maybe due to individual locations in the work area having markedly different conditions than others. It was possibly due to proximity to the thermal sensor or ventilation ducts (supply and return), or because of improper balance of the HVAC system.

Conclusions

Significant deficiencies that may be related to both symptoms and comfort complaints were noted in the offices' indoor environment. Carbon dioxide levels in excess of ASHRAE recommendations were observed at some locations in the building, indicating an insufficient amount of fresh air supply (dilution ventilation) to the building at certain times. Visual inspection of the HVAC unit revealed inadequate maintenance, as evidenced by the disconnected mechanical linkage for the pneumatic return air damper. In addition, visual inspection of the interior of the AHU mixed air plenum revealed the presence of microbiological growth, and subsequent analysis of the bulk material samples collected confirmed the presence of fungi and bacteria.

The air filtration system is a low efficiency fiberglass roll filter. Airflow measurements suggest an unbalanced HVAC system. An unbalanced system can result in poor air distribution and can affect the perceived comfort of certain individuals in the occupied spaces.

Recommendations

The following recommendations were discussed at the closing conference. Subsequent analysis of the bulk material samples support these recommendations:

1. A qualified HVAC firm should be contracted to conduct a mechanical system audit to verify that the system is adequately sized and designed for the current application. The amount of outside air delivered to the space should be determined and the system adjusted to deliver a minimum of 20 cfm per person during periods of normal occupancy. If the system is not capable of delivering an adequate amount of outside air, it may be undersized and need to be reconfigured. The entire HVAC system should also be tested and balanced.
2. The SSA space should be placed under slight positive pressure relative to the outdoor environment and adjacent spaces (SSA does not occupy the entire building) to minimize chances of air entering from other spaces. To achieve this pressure balance, the AHU should provide an outside air intake volume which is at least 10 percent larger than the exhaust air volumes from the area served.
3. The thermal sensor configuration should be evaluated to determine if additional sensors or temperature averaging sensors would result in increased thermal comfort.
4. A preventive maintenance plan should institute regular inspections of facilities so that equipment can be maintained in a way that does not promote growth of microbial contamination.
5. The AHU should be inspected and cleaned on a monthly basis. A record of all cleaning should be kept and any potential problems corrected.
6. The condenser coil condensate pan should be cleaned and disinfected with a biocide. When choosing a biocide, it is important to consider that they are designed to kill living cells, and can pose health effects if used incorrectly or in improper dilutions. The existence of spores is possible when there are molds, so the biocide should also contain a sporicide. Because the

potential exists for bioaerosol exposure to individuals involved in the cleanup process, personnel should use respirators with high efficiency particulate air (HEPA) filters.

7. The disconnected mechanical pneumatic linkage for the return air damper should be repaired and tested to determine if it is in proper working order.
8. The fiberglass roll filter system should be upgraded to the maximum efficiency possible without affecting the HVAC system performance.
9. Communication between management and employees should be increased to facilitate the exchange of concerns about environmental conditions at the building. Employees should be made aware of the problems, and of the decisions made by management to address those problems.

OREGON EMPLOYMENT BUILDING

This case study is an example of frustration. Employees complained of burning eyes, nausea, and even of plants in the office dying. The building owner investigated, measured, hired consultants, involved prestigious laboratories, and made modifications. However, nothing was identified that could have caused the complaints, which gradually tapered off and stopped. Information for this section is from documents that were distributed to employees, posted, and/or used for presentations (see References, p. 418).

Symptoms

Some complaints were discussed among coworkers; some were told informally and orally to supervisors; some were brought up in meetings, regardless of the agenda or topic; and some were described in official complaint filings. The following is a summary of symptoms listed in the complaint file:

- Blistering or rashes
- Blotches
- Headache
- Itching
- Nausea
- Respiratory complaints: cough, dry throat, scratchy throat, throat irritation

- Scaling
- Vomiting

Breadth of Investigation

Starting with the first complaint, the division tried to identify an indoor air contaminant. As a state government agency, it drew upon resources of the state to make measurements and take precautions. As the complaints continued but measurements and investigations revealed nothing, the division began calling in and hiring consultants. By the investigation's end, the list of outside consultants, advisors, and agencies was long, and included:

- Bonneville Power Administration, Portland, Oregon
- Department of Administrative Services, Salem, Oregon
- Health Division, DHR, Portland, Oregon
- Indoor Air Quality Consultants
- Lawrence Berkeley Laboratory, Berkeley, California
- National Institute for Occupational Safety & Health, Denver, Colorado
- Northwest Engineering, Tigard, Oregon
- Oregon Health Sciences University, Portland, Oregon
- OR-OSHA, Salem, Oregon
- Rogers Engineering, Eugene, Oregon
- SAIF Corporation, Salem, Oregon
- Ventilation Consultants, Bellevue, Washington
- Waterlab, Salem, Oregon
- Workers Compensation, Salem, Oregon

Tests Conducted

The division itself, various state agencies, outside consultants, and others conducted tests. They tested everything about the HVAC system's operation. They tested for contaminants most likely to be found in offices, then for contaminants less likely to be found, and finally for contaminants that could not possibly be present. Following is a summary of the testing:

- Airflow—in, out, local, total
- Air makeup
- Air temperature—at a variety of times and locations

- Air ventilation and exchange rates
- Asbestos
- Bacteria
- Carbon dioxide
- Carbon monoxide
- Carbon tetrachloride
- Chlorine
- Contaminates—air exhaust to air intake
- Contaminates—from around the building
- Dusts
- Fiberglass
- Fibers
- Formaldehyde
- Fungal spores
- Light emissions
- Nitrogen dioxide
- Ozone
- Radon
- Relative humidity
- Sulfur dioxide
- Vinyl chloride

Building Modifications

The investigation revealed some items that could be improved. Although none of the deficiencies could have caused the complaints, the following modifications were made to the building:

- Arranged HVAC so that pulling 100% outside air can be selected.
- Cafeteria stove vent converted to automatic "on."
- Decks were sealed on 2nd and 3rd floors.
- Exhaust fans were added to cafeteria and other areas.
- Filters—95 percent were converted to electronic assembly plant type.
- HVAC was fully balanced and electronic controls were installed.
- Installed venting over high speed paper handlers.
- Insulation was changed and/or installed in plenums.
- Installed an air filtration system in the mail center.
- Meters were installed to isolate critical computer.
- Plenums were cleaned throughout building.

- Reflector screens were installed on windows where needed.
- Roof was replaced.
- Skylights were replaced.
- Smoking was eliminated in the building.
- Supply fans were modified to allow better air throw.

Possible "Non-Building" Complaint Causes

The experts agreed that, although some deficiencies were found, and some modifications were made, there had not been any problem within the building that could have caused the symptoms and complaints (see the next section on "Conclusions"). Following are some possible explanations that consultants and management provided:

- Individual allergies that can be diagnosed only through extensive allergy testing by a personal physician.
- Cold, influenza, fatigue, or injury symptoms that are mistakenly attributed to office air quality conditions.
- Psychosocial symptoms that relate to stress or work relationships rather than to chemical, physical, or biological agents; or which aggravate symptoms that might otherwise go unnoticed.
- Common symptoms that occur at a similar rate in all groups and are not ordinarily reported because they are not serious enough for members of small groups to discuss with each other.

Conclusions

Interim reports, the final report, and other documentation presented a record of the investigation and the outcome. The following quotations express the consensus of investigators and analysts regarding overall conclusions.

> ". . . we looked at materials used by the housekeeping, custodial maintenance, and grounds keeping personnel. This included soaps, pesticides, carpet cleaner and static treatment, detergents and shampoos. No misuses were noted, nor were the reported symptoms associated with these materials."

> ". . . we looked at the chemical makeup of the papers used in the office and of chemicals used in photocopiers. Our evaluation and tests run on dust and paper samples showed no dangerous products or handling abuses, nor was there any association found between workers and products that would account for the complaints."

"[Engineering studies] were made of the HVAC system . . . to see if there were deficiencies in design or maintenance that could cause air quality deterioration. None of the studies identified a clear causative problem . . ."

"Ambient air tests taken inside the building for a wide variety of contaminants have continually shown satisfactory air quality, and carbon dioxide readings taken at desk levels have consistently shown adequate fresh air circulation."

"Medical evaluations of affected persons have not produced a common diagnosis nor identified a causative agent. There have been diagnoses indicating that some of the symptoms could be allergic responses, but no specific allergen has been identified."

"In all cases tests have shown the measured substances to have been well within recommended levels for indoor environments. Temperature and humidity levels have shown the indoor air to be within recommended comfort zones and the measured carbon dioxide levels have consistently indicated that there is ample fresh air supplied by the air conditioning system."

"We believe that we have evaluated every available aspect of the problem. We have looked at the possible indoor air contaminants, the design and operation of the HVAC system, and the symptoms of all affected persons who have reported to us. We have also consulted with the other agencies involved in these complaints and sought the council of private physicians who have seen affected employees.

"We have not found a particular agent, condition, or factor that is linked to the complaints or that suggests a common cause. We have not found evidence either that symptoms occur more frequently among employees of the Employment Division than occur among any other large group of people. Nor have we found any evidence that air quality problems have caused any of the illnesses or reports of illness."

References

NIOSH Health Hazard Evaluation Report, by U.S. Department of Health and Human Services, *HETA 92-0374-2402*, Social Security Administration, Petersburg, Virginia.

Air Quality Review—Central Office, one-page summary, March 8, 1994.

State of Oregon, Employment Div., memo to all employees, Nov. 21, 1985. No. BSM:GB:SUP 3.

Business Services, Employment Dept., *Central Office Air Quality Review*, Mar. 8, 1994.

APPENDIX B

Exposure Limits for Air Contaminants

Table Z-1 of Federal Regulation 1910.1000 is the basis for establishing acceptable exposure limits for workers. Many regulatory and advisory bodies conduct their own investigations, do their own analyses, and, in various ways, establish sets of exposure limits that are sometimes tighter and sometimes looser than Table Z-1. The safest policy for exployers is to keep the working environment within the tightest limits; keeping the level of a hazardous substance lower than necessary can never be more harmful.

Table B-1 includes all the information from Table Z-1, plus its extensive footnotes. References are made in a few places to other publications, sources, and tables, where readers can find additional details about specific substances. However, Table B-1 will serve most purposes because it gives exposure limits for substances whose identifications range from esoteric through such common names as "grain dust." All substances in the table can affect workers via the respiratory tract. In addition, the skin is a significant route of exposure for substances that have an X in the Skin Designation column.

Table B-1
Limits for Air Contaminants—Table Z-1 of Federal Regulation 1910.1000

Substance	CAS No. (c)	*ppm (a)(1)*	*mg/m(3) (b)(1)*	*Skin designation*
Acetaldehyde	75-07-0	200	360	
Acetic acid	64-19-7	10	25	
Acetic anhydride	108-24-7	5	20	
Acetone	67-64-1	1000	2400	
Acetonitrile	75-05-8	40	70	
2-Acetylaminofluorene; see 1910.1014	53-96-3			
Acetylene dichloride; see 1,2-Dichloroethylene				
Acetylene tetrabromide	79-27-6	1	14	
Acrolein	107-02-8	0.1	0.25	
Acrylamide	79-06-1		0.3	X
Acrylonitrile; see 1910.1045	107-13-1			
Aldrin	309-00-2		0.25	X
Allyl alcohol	107-18-6	2	5	X
Allyl chloride	107-05-1	1	3	
Allyl glycidyl ether (AGE)	106-92-3	(C)10	(C)45	
Allyl propyl disulfide	2179-59-1	2	12	
alpha-Alumina	1344-28-1			
Total dust			15	
Respirable fraction			5	
Aluminum Metal (as Al)	7429-90-5			
Total dust			15	

TABLE B-1
Continued

Substance	CAS No. (c)	*ppm (a)(1)*	*mg/m(3) (b)(1)*	*Skin designation*
Respirable fraction			5	
4-Aminodiphenyl; see 1910.1011	92-67-1			
2-Aminoethanol; see Ethanolamine				
2-Aminopyridine	504-29-0	0.5	2	
Ammonia	7664-41-7	50	35	
Ammonium sulfamate	7773-06-0			
Total dust			15	
Respirable fraction			5	
n-Amyl acetate	628-63-7	100	525	
sec-Amyl acetate	626-38-0	125	650	
Aniline and homologs	62-53-3	5	19	X
Anisidine (o-,p-isomers)	29191-52-4		0.5	X
Antimony and compounds (as Sb)	7440-36-0		0.5	
ANTU (alpha Naphthylthiourea)	86-88-4		0.3	
Arsenic, inorganic compounds (as As); see 1910.1018	7440-38-2			
Arsenic, organic compounds (as As)	7440-38-2		0.5	
Arsine	7784-42-1	0.05	0.2	
Asbestos; see 1910.1001	(4)			
Azinphos-methyl	86-50-0		0.2	X

Continued

TABLE B-1
Continued

Substance	CAS No. (c)	*ppm (a)(1)*	*mg/m(3) (b)(1)*	*Skin designation*
Barium, soluble compounds (as Ba)	7440-39-3		0.5	
Barium sulfate	7727-43-7			
Total dust			15	
Respirable fraction			5	
Benomyl	17804-35-2			
Total dust			15	
Respirable fraction			5	
Benzene; See 1910.1028	71-43-2			
See Table Z-2 for the limits applicable in the operations or sectors excluded in 1910.1028(d)				
Benzidine; see 1910.1010	92-87-5			
p-Benzoquinone; see Quinone				
Benzo(a)pyrene; see Coal tar pitch volatiles				
Benzoyl peroxide	94-36-0		5	
Benzyl chloride	100-44-7	1	5	
Beryllium and beryllium compounds (as Be)	7440-41-7		(2)	
Biphenyl; see Diphenyl				
Bismuth telluride,				
Undoped	1304-82-1			
Total dust			15	

TABLE B-1
Continued

Substance	*CAS No. (c)*	*ppm (a)(1)*	*mg/m(3) (b)(1)*	*Skin designation*
Respirable fraction			5	
Boron oxide	1303-86-2			
Total dust			15	
Boron trifluoride	7637-07-2	(C)1	(C)3	
Bromine	7726-95-6	0.1	0.7	
Bromoform	75-25-2	0.5	5	X
Butadiene (1,3-Butadiene)	106-99-0	1000	2200	
Butanethiol; see Butyl mercaptan				
2-Butanone (Methyl ethyl ketone)	78-93-3	200	590	
2-Butoxyethanol	111-76-2	50	240	X
n-Butyl-acetate	123-86-4	150	710	
sec-Butyl acetate	105-46-4	200	950	
tert-Butyl-acetate	540-88-5	200	950	
n-Butyl alcohol	71-36-3	100	300	
sec-Butyl alcohol	78-92-2	150	450	
tert-Butyl alcohol	75-65-0	100	300	
Butylamine	109-73-9	(C)5	(C)15	X
tert-Butyl chromate (as CrO(3))	1189-85-1		(C)0.1	X
n-Butyl glycidyl ether (BGE)	2426-08-6	50	270	
Butyl mercaptan	109-79-5	10	35	
p-tert-Butyltoluene	98-51-1	10	60	

Continued

Table B-1
Continued

Substance	*CAS No. (c)*	*ppm (a)(1)*	*mg/m(3) (b)(1)*	*Skin designation*
Cadmium (as Cd); see 1910.1027	7440-43-9			
Calcium Carbonate	1317-65-3			
Total dust			15	
Respirable fraction			5	
Calcium hydroxide	1305-62-0			
Total dust			15	
Respirable fraction			5	
Calcium oxide	1305-78-8		5	
Calcium silicate	1344-95-2			
Total dust			15	
Respirable fraction			5	
Calcium sulfate	7778-18-9			
Total dust			15	
Respirable fraction			5	
Camphor, synthetic	76-22-2		2	
Carbaryl (Sevin)	63-25-2		5	
Carbon black	1333-86-4		3.5	
Carbon dioxide	124-38-9	5000	9000	
Carbon disulfide	75-15-0		(2)	
Carbon monoxide	630-08-0	50	55	
Carbon tetrachloride	56-23-5		(2)	
Cellulose	9004-34-6			
Total dust			15	
Respirable fraction			5	

TABLE B-1
Continued

Substance	*CAS No. (c)*	*ppm (a)(1)*	*mg/m(3) (b)(1)*	*Skin designation*
Chlordane	57-74-9		0.5	X
Chlorinated camphene	8001-35-2		0.5	X
Chlorinated diphenyl oxide	55720-99-5		0.5	
Chlorine	7782-50-5	(C)1	(C)3	
Chlorine dioxide	10049-04-4	0.1	0.3	
Chlorine trifluoride	7790-91-2	(C)0.1	(C)0.4	
Chloroacetaldehyde	107-20-0	(C)1	(C)3	
a-Chloroacetophenone (Phenacyl chloride)	532-27-4	0.05	0.3	
Chlorobenzene	108-90-7	75	350	
o-Chlorobenzylidene malononitrile	2698-41-1	0.05	0.4	
Chlorobromomethane	74-97-5	200	1050	
2-Chloro-1,3-butadiene; See beta-Chloroprene				
Chlorodiphenyl (42% Chlorine)(PCB)	53469-21-9		1	X
Chlorodiphenyl (54% Chlorine)(PCB)	11097-69-1		0.5	X
1-Chloro-2, 3-epoxypropane; See Epichlorohydrin				
2-Chloroethanol; See Ethylene chlorohydrin				
Chloroethylene; See Vinyl chloride				

Continued

TABLE B-1
Continued

Substance	CAS No. (c)	ppm (a)(1)	mg/m(3) (b)(1)	Skin designation
Chloroform (Trichloro-methane)	67-66-3	(C)50	(C)240	
bis(Chloromethyl) ether; see 1910.1008	542-88-1			
Chloromethyl methyl ether; see 1910.1006	107-30-2			
1-Chloro-1-nitropropane	600-25-9	20	100	
Chloropicrin	76-06-2	0.1	0.7	
beta-Chloroprene	126-99-8	25	90	X
2-Chloro-6 (trichloro-methyl) pyridine	1929-82-4			
Total dust			15	
Respirable fraction			5	
Chromic acid and chromates (as CrO(3))	(4)		(2)	
Chromium (II) compounds (as Cr)	7440-47-3		0.5	
Chromium (III) compounds (as Cr)	7440-47-3		0.5	
Chromium metal and insol. salts (as Cr).	7440-47-3		1	
Chrysene; see Coal tar pitch volatiles				
Clopidol	2971-90-6			
Total dust			15	
Respirable fraction			5	

Table B-1
Continued

Substance	*CAS No. (c)*	*ppm (a)(1)*	*mg/m(3) (b)(1)*	*Skin designation*
Coal dust (less than 5% SiO(2)), respirable fraction			(3)	
Coal dust (greater than or equal to 5% SiO(2)), respirable fraction			(3)	
Coal tar pitch volatiles (benzene soluble fraction), anthracene, BaP, phenanthrene, acridine, chrysene, pyrene	65966-93-2		0.2	
Cobalt metal, dust, and fume (as Co)	7440-48-4		0.1	
Coke oven emissions; see 1910.1029				
Copper	7440-50-8			
Fume (as Cu)			0.1	
Dusts and mists (as Cu)			1	
Cotton dust (e), see 1910.1043			1	
Crag herbicide (Sesone)	136-78-7			
Total dust			15	
Respirable fraction			5	
Cresol, all isomers	1319-77-3	5	22	X
Crotonaldehyde	123-73-9	2	6	
	4170-30-3			
Cumene	98-82-8	50	245	X
Cyanides (as CN)	(4)		5	X

Continued

TABLE B-1
Continued

Substance	*CAS No. (c)*	*ppm (a)(1)*	*mg/m(3) (b)(1)*	*Skin designation*
Cyclohexane	110-82-7	300	1050	
Cyclohexanol	108-93-0	50	200	
Cyclohexanone	108-94-1	50	200	
Cyclohexene	110-83-8	300	1015	
Cyclopentadiene	542-92-7	75	200	
2,4-D (Dichlorophen-oxyacetic acid)	94-75-7		10	
Decaborane	17702-41-9	0.05	0.3	X
Demeton (Systox)	8065-48-3		0.1	X
Diacetone alcohol (4-Hydroxy-4-methyl-2-pentanone)	123-42-2	50	240	
1,2-Diaminoethane; see Ethylenediamine				
Diazomethane	334-88-3	0.2	0.4	
Diborane	19287-45-7	0.1	0.1	
1,2-Dibromo-3-chloro-propane (DBCP); see 1910.1044	96-12-8			
1,2-Dibromoethane; see Ethylene dibromide				
Dibutyl phosphate	107-66-4	1	5	
Dibutyl phthalate	84-74-2		5	
o-Dichlorobenzene	95-50-1	(C)50	(C)300	
p-Dichlorobenzene	106-46-7	75	450	
3,3'-Dichlorobenzidine; see 1910.1007	91-94-1			

TABLE B-1
Continued

Substance	*CAS No. (c)*	*ppm (a)(1)*	*mg/m(3) (b)(1)*	*Skin designation*
Dichlorodifluoromethane	75-71-8	1000	4950	
1,3-Dichloro-5,5-dimethyl hydantoin	118-52-5		0.2	
Dichlorodiphenyltrichloroethane (DDT)	50-29-3		1	X
1,1-Dichloroethane	75-34-3	100	400	
1,2-Dichloroethane; see Ethylene dichloride				
1,2-Dichloroethylene	540-59-0	200	790	
Dichloroethyl ether	111-44-4	(C)15	(C)90	X
Dichloromethane; see Methylene chloride				
Dichloromonofluoromethane	75-43-4	1000	4200	
1,1-Dichloro-1-nitroethane	594-72-9	(C)10	(C)60	
1,2-Dichloropropane; see Propylene dichloride				
Dichlorotetrafluoroethane	76-14-2	1000	7000	
Dichlorvos (DDVP)	62-73-7		1	X
Dicyclopentadienyl iron	102-54-5			
Total dust			15	
Respirable fraction			5	
Dieldrin	60-57-1		0.25	X
Diethylamine	109-89-7	25	75	

Continued

Table B-1
Continued

Substance	*CAS No. (c)*	*ppm (a)(1)*	*mg/m(3) (b)(1)*	*Skin designation*
2-Diethylaminoethanol	100-37-8	10	50	X
Diethyl ether; see Ethyl ether				
Difluorodibromomethane	75-61-6	100	860	
Diglycidyl ether (DGE)	2238-07-5	(C)0.5	(C)2.8	
Dihydroxybenzene; see Hydroquinone				
Diisobutyl ketone	108-83-8	50	290	
Diisopropylamine	108-18-9	5	20	X
4-Dimethylaminoazo-benzene; see 1910.1015	60-11-7			
Dimethoxymethane; see Methylal				
Dimethyl acetamide	127-19-5	10	35	X
Dimethylamine	124-40-3	10	18	
Dimethylaminobenzene; see Xylidine				
Dimethylaniline (N,N-Dimethylaniline)	121-69-7	5	25	X
Dimethylbenzene; see Xylene				
Dimethyl-1,2-dibromo-2,2-dichloroethyl phosphate	300-76-5		3	
Dimethylformamide	68-12-2	10	30	X
2,6-Dimethyl-4-heptanone; see Diisobutyl ketone				

TABLE B-1
Continued

Substance	CAS No. (c)	*ppm (a)(1)*	*mg/m(3) (b)(1)*	*Skin designation*
1,1-Dimethylhydrazine	57-14-7	0.5	1	X
Dimethylphthalate	131-11-3		5	
Dimethyl sulfate	77-78-1	1	5	X
Dinitrobenzene				
(all isomers)			1	X
(ortho)	528-29-0			
(meta)	99-65-0			
(para)	100-25-4			
Dinitro-o-cresol	534-52-1		0.2	X
Dinitrotoluene	25321-14-6		1.5	X
Dioxane (Diethylene dioxide)	123-91-1	100	360	X
Diphenyl (Biphenyl)	92-52-4	0.2	1	
Diphenylmethane diisocyanate; see Methylene bisphenyl isocyanate				
Dipropylene glycol methyl ether	34590-94-8	100	600	X
Di-sec octyl phthalate (Di-(2-ethylhexyl) phthalate)	117-81-7		5	
Emery	12415-34-8			
Total dust			15	
Respirable fraction			5	
Endrin	72-20-8		0.1	X
Epichlorohydrin	106-89-8	5	19	X

Continued

Table B-1
Continued

Substance	CAS No. (c)	*ppm (a)(1)*	*mg/m(3) (b)(1)*	*Skin designation*
EPN	2104-64-5		0.5	X
1,2-Epoxypropane; see Propylene oxide				
2,3-Epoxy-1-propanol; see Glycidol				
Ethanethiol; see Ethyl mercaptan				
Ethanolamine	141-43-5	3	6	
2-Ethoxyethanol (Cellosolve)	110-80-5	200	740	X
2-Ethoxyethyl acetate (Cellosolve acetate)	111-15-9	100	540	X
Ethyl acetate	141-78-6	400	1400	
Ethyl acrylate	140-88-5	25	100	X
Ethyl alcohol (Ethanol)	64-17-5	1000	1900	
Ethylamine	75-04-7	10	18	
Ethyl amyl ketone (5-Methyl-3-heptanone)	541-85-5	25	130	
Ethyl benzene	100-41-4	100	435	
Ethyl bromide	74-96-4	200	890	
Ethyl butyl ketone (3-Heptanone)	106-35-4	50	230	
Ethyl chloride	75-00-3	1000	2600	
Ethyl ether	60-29-7	400	1200	
Ethyl formate	109-94-4	100	300	
Ethyl mercaptan	75-08-1	(C)10	(C)25	

TABLE B-1
Continued

Substance	*CAS No. (c)*	*ppm (a)(1)*	*mg/m(3) (b)(1)*	*Skin designation*
Ethyl silicate	78-10-4	100	850	
Ethylene chlorohydrin	107-07-3	5	16	X
Ethylenediamine	107-15-3	10	25	
Ethylene dibromide	106-93-4		(2)	
Ethylene dichloride (1,2-Dichloroethane)	107-06-2		(2)	
Ethylene glycol dinitrate	628-96-6	(C)0.2	(C)1	X
Ethylene glycol methyl acetate; see Methyl cellosolve acetate				
Ethyleneimine; see 1910.1012	151-56-4			
Ethylene oxide; see 1910.1047	75-21-8			
Ethylidene chloride; see 1,1-Dichlorethane				
N-Ethylmorpholine	100-74-3	20	94	X
Ferbam	14484-64-1			
Total dust			15	
Ferrovanadium dust	12604-58-9		1	
Fluorides (as F)	(4)		2.5	
Fluorine	7782-41-4	0.1	0.2	
Fluorotrichloromethane (Trichlorofluoro-methane)	75-69-4	1000	5600	
Formaldehyde; see 1910.1048	50-00-0			

Continued

TABLE B-1
Continued

Substance	CAS No. (c)	*ppm (a)(1)*	*mg/m(3) (b)(1)*	*Skin designation*
Formic acid	64-18-6	5	9	
Furfural	98-01-1	5	20	X
Furfuryl alcohol	98-00-0	50	200	
Grain dust (oat, wheat barley)			10	
Glycerin (mist)	56-81-5			
Total dust			15	
Respirable fraction			5	
Glycidol	556-52-5	50	150	
Glycol monoethyl ether; see 2-Ethoxyethanol				
Graphite, natural respirable dust	7782-42-5		(3)	
Graphite, synthetic				
Total dust			15	
Respirable Fraction			5	
Guthion; see Azinphos methyl				
Gypsum	13397-24-5			
Total dust			15	
Respirable fraction			5	
Hafnium	7440-58-6		0.5	
Heptachlor	76-44-8		0.5	X
Heptane (n-Heptane)	142-82-5	500	2000	
Hexachloroethane	67-72-1	1	10	X
Hexachloronaphthalene	1335-87-1		0.2	X

TABLE B-1
Continued

Substance	CAS No. (*c*)	*ppm (a)(1)*	*mg/m(3) (b)(1)*	*Skin designation*
n-Hexane	110-54-3	500	1800	
2-Hexanone (Methyl n-butyl ketone)	591-78-6	100	410	
Hexone (Methyl isobutyl ketone)	108-10-1	100	410	
sec-Hexyl acetate	108-84-9	50	300	
Hydrazine	302-01-2	1	1.3	X
Hydrogen bromide	10035-10-6	3	10	
Hydrogen chloride	7647-01-0	(C)5	(C)7	
Hydrogen cyanide	74-90-8	10	11	X
Hydrogen fluoride (as F)	7664-39-3		(2)	
Hydrogen peroxide	7722-84-1	1	1.4	
Hydrogen selenide (as Se)	7783-07-5	0.05	0.2	
Hydrogen sulfide	7783-06-4		(2)	
Hydroquinone	123-31-9		2	
Iodine	7553-56-2	(C)0.1	(C)1	
Iron oxide fume	1309-37-1		10	
Isomyl acetate	123-92-2	100	525	
Isomyl alcohol (primary and secondary)	123-51-3	100	360	
Isobutyl acetate	110-19-0	150	700	
Isobutyl alcohol	78-83-1	100	300	
Isophorone	78-59-1	25	140	
Isopropyl acetate	108-21-4	250	950	
Isopropyl alcohol	67-63-0	400	980	

Continued

TABLE B-1
Continued

Substance	*CAS No. (c)*	*ppm (a)(1)*	*mg/m(3) (b)(1)*	*Skin designation*
Isopropylamine	75-31-0	5	12	
Isopropyl ether	108-20-3	500	2100	
Isopropyl glycidyl ether (IGE)	4016-14-2	50	240	
Kaolin	1332-58-7			
Total dust			15	
Respirable fraction			5	
Ketene	463-51-4	0.5	0.9	
Lead inorganic (as Pb); see 1910.1025	7439-92-1			
Limestone	1317-65-3			
Total dust			15	
Respirable fraction			5	
Lindane	58-89-9		0.5	X
Lithium hydride	7580-67-8		0.025	
L.P.G. (Liquified petroleum gas)	68476-85-7	1000	1800	
Magnesite	546-93-0			
Total dust			15	
Respirable fraction			5	
Magnesium oxide fume	1309-48-4			
Total Particulate			15	
Malathion	121-75-5			
Total dust			15	X
Maleic anhydride	108-31-6	0.25	1	

TABLE B-1
Continued

Substance	*CAS No. (c)*	*ppm (a)(1)*	*mg/m(3) (b)(1)*	*Skin designation*
Manganese compounds (as Mn)	7439-96-5		(C)5	
Manganese fume (as Mn)	7439-96-5		(C)5	
Marble	1317-65-3			
Total dust			15	
Respirable fraction			5	
Mercury (aryl and inorganic) (as Hg)	7439-97-6		(2)	
Mercury (organo) alkyl compounds (as Hg)	7439-97-6		(2)	
Mercury (vapor) (as Hg)	7439-97-6		(2)	
Mesityl oxide	141-79-7	25	100	
Methanethiol; see Methyl mercaptan				
Methoxychlor	72-43-5			
Total dust			15	
2-Methoxyethanol; (Methyl cellosolve)	109-86-4	25	80	X
2-Methoxyethyl acetate (Methyl cellosolve acetate)	110-49-6	25	120	X
Methyl acetate	79-20-9	200	610	
Methyl acetylene (Propyne)	74-99-7	1000	1650	
Methyl acetylene propadiene mixture (MAPP)		1000	1800	
Methyl acrylate	96-33-3	10	35	X

Continued

TABLE B-1
Continued

Substance	CAS No. (c)	*ppm (a)(1)*	*mg/m(3) (b)(1)*	*Skin designation*
Methylal (Dimethoxy-methane)	109-87-5	1000	3100	
Methyl alcohol	67-56-1	200	260	
Methylamine	74-89-5	10	12	
Methyl amyl alcohol; see Methyl Isobutyl carbinol				
Methyl n-amyl ketone	110-43-0	100	465	
Methyl bromide	74-83-9	(C)20	(C)80	X
Methyl butyl ketone; see 2-Hexanone				
Methyl cellosolve; see 2-Methoxyethanol				
Methyl cellosolve acetate; see 2-Methoxyethyl acetate				
Methyl chloride	74-87-3		(2)	
Methyl chloroform (1,1,1-Trichloroethane)	71-55-6	350	1900	
Methylcyclohexane	108-87-2	500	2000	
Methylcyclohexanol	25639-42-3	100	470	
o-Methylcyclohexanone	583-60-8	100	460	X
Methylene chloride	75-09-2		(2)	
Methyl ethyl ketone (MEK); see 2-Butanone				
Methyl formate	107-31-3	100	250	

Table B-1
Continued

Substance	*CAS No. (c)*	*ppm (a)(1)*	*mg/m(3) (b)(1)*	*Skin designation*
Methyl hydrazine (Monomethyl hydrazine)	60-34-4	(C)0.2	(C)0.35	X
Methyl iodide	74-88-4	5	28	X
Methyl isoamyl ketone	110-12-3	100	475	
Methyl isobutyl carbinol	108-11-2	25	100	X
Methyl isobutyl ketone; see Hexone				
Methyl isocyanate	624-83-9	0.02	0.05	X
Methyl mercaptan	74-93-1	(C)10	(C)20	
Methyl methacrylate	80-62-6	100	410	
Methyl propyl ketone; see 2-Pentanone				
alpha-Methyl styrene	98-83-9	(C)100	(C)480	
Methylene bisphenyl isocyanate (MDI)	101-68-8	(C)0.02	(C)0.2	
Mica; see Silicates				
Molybdenum (as Mo)	7439-98-7			
Soluble compounds			5	
Insoluble Compounds				
Total dust			15	
Monomethyl aniline	100-61-8	2	9	X
Monomethyl hydrazine; see Methyl hydrazine				
Morpholine	110-91-8	20	70	X
Naphtha (Coal tar)	8030-30-6	100	400	

Continued

Table B-1
Continued

Substance	*CAS No. (c)*	*ppm (a)(1)*	*mg/m(3) (b)(1)*	*Skin designation*
Naphthalene	91-20-3	10	50	
alpha-Naphthylamine; see 1910.1004	134-32-7			
beta-Naphthylamine; see 1910.1009	91-59-8			
Nickel carbonyl (as Ni)	13463-39-3	0.001	0.007	
Nickel, metal and insoluble compounds (as Ni)	7440-02-0		1	
Nickel, soluble compounds (as Ni)	7440-02-0		1	
Nicotine	54-11-5		0.5	X
Nitric acid	7697-37-2	2	5	
Nitric oxide	10102-43-9	25	30	
p-Nitroaniline	100-01-6	1	6	X
Nitrobenzene	98-95-3	1	5	X
p-Nitrochlorobenzene	100-00-5		1	X
4-Nitrodiphenyl; see 1910.1003	92-93-3			
Nitroethane	79-24-3	100	310	
Nitrogen dioxide	10102-44-0	(C)5	(C)9	
Nitrogen trifluoride	7783-54-2	10	29	
Nitroglycerin	55-63-0	(C)0.2	(C)2	X
Nitromethane	75-52-5	100	250	
1-Nitropropane	108-03-2	25	90	
2-Nitropropane	79-46-9	25	90	

Table B-1
Continued

Substance	CAS No. (c)	*ppm (a)(1)*	*mg/m(3) (b)(1)*	*Skin designation*
N-Nitrosodimethylamine; see 1910.1016				
Nitrotoluene (all isomers)		5	30	X
o-isomer	88-72-2			
m-isomer	99-08-1			
p-isomer	99-99-0			
Nitrotrichloromethane; see Chloropicrin				
Octachloronaphthalene	2234-13-1		0.1	X
Octane	111-65-9	500	2350	
Oil mist, mineral	8012-95-1		5	
Osmium tetroxide (as Os)	20816-12-0		0.002	
Oxalic acid	144-62-7		1	
Oxygen difluoride	7783-41-7	0.05	0.1	
Ozone	10028-15-6	0.1	0.2	
Paraquat, respirable dust	4685-14-7		0.5	X
	1910-42-5			
	2074-50-2			
Parathion	56-38-2		0.1	X
Particulates not otherwise regulated (PNOR)(f)				
Total dust			15	
Respirable fraction			5	
PCB; see Chlorodiphenyl (42% and 54% chlorine)				

Continued

Table B-1
Continued

Substance	CAS No. (c)	*ppm (a)(1)*	*mg/m(3) (b)(1)*	*Skin designation*
Pentaborane	19624-22-7	0.005	0.01	
Pentachloronaphthalene	1321-64-8		0.5	X
Pentachlorophenol	87-86-5		0.5	X
Pentaerythritol	115-77-5			
Total dust			15	
Respirable fraction			5	
Pentane	109-66-0	1000	2950	
2-Pentanone (Methyl propyl ketone)	107-87-9	200	700	
Perchloroethylene (Tetrachloroethylene)	127-18-4		(2)	
Perchloromethyl mercaptan	594-42-3	0.1	0.8	
Perchloryl fluoride	7616-94-6	3	13.5	
Petroleum distillates (Naphtha) (Rubber Solvent)		500	2000	
Phenol	108-95-2	5	19	X
p-Phenylene diamine	106-50-3		0.1	X
Phenyl ether, vapor	101-84-8	1	7	
Phenyl ether-biphenyl mixture, vapor		1	7	
Phenylethylene; see Styrene				
Phenyl glycidyl ether (PGE)	122-60-1	10	60	
Phenylhydrazine	100-63-0	5	22	X

TABLE B-1
Continued

Substance	CAS No. *(c)*	*ppm (a)(1)*	*mg/m(3) (b)(1)*	*Skin designation*
Phosdrin (Mevinphos)	7786-34-7		0.1	X
Phosgene (Carbonyl chloride)	75-44-5	0.1	0.4	
Phosphine	7803-51-2	0.3	0.4	
Phosphoric acid	7664-38-2		1	
Phosphorus (yellow)	7723-14-0		0.1	
Phosphorus pentachloride	10026-13-8		1	
Phosphorus pentasulfide	1314-80-3		1	
Phosphorus trichloride	7719-12-2	0.5	3	
Phthalic anhydride	85-44-9	2	12	
Picloram	1918-02-1			
Total dust			15	
Respirable fraction			5	
Picric acid	88-89-1		0.1	X
Pindone (2-Pivalyl-1,3-indandione)	83-26-1		0.1	
Plaster of paris	26499-65-0			
Total dust			15	
Respirable fraction			5	
Platinum (as Pt)	7440-06-4			
Metal				
Soluble Salts			0.002	
Portland cement	65997-15-1			
Total dust			15	
Respirable fraction			5	

Continued

Table B-1
Continued

Substance	CAS No. (c)	*ppm (a)(1)*	*mg/m(3) (b)(1)*	*Skin designation*
Propane	74-98-6	1000	1800	
beta-Propriolactone; see 1910.1013	57-57-8			
n-Propyl acetate	109-60-4	200	840	
n-Propyl alcohol	71-23-8	200	500	
n-Propyl nitrate	627-13-4	25	110	
Propylene dichloride	78-87-5	75	350	
Propylene imine	75-55-8	2	5	X
Propylene oxide	75-56-9	100	240	
Propyne; see Methyl acetylene				
Pyrethrum	8003-34-7		5	
Pyridine	110-86-1	5	15	
Quinone	106-51-4	0.1	0.4	
RDX: see Cyclonite				
Rhodium (as Rh), metal fume and insoluble compounds	7440-16-6		0.1	
Rhodium (as Rh), soluble compounds	7440-16-6		0.001	
Ronnel	299-84-3		15	
Rotenone	83-79-4		5	
Rouge				
Total dust			15	
Respirable fraction			5	

TABLE B-1
Continued

Substance	CAS No. (*c*)	*ppm (a)(1)*	*mg/m(3) (b)(1)*	*Skin designation*
Selenium compounds (as Se)	7782-49-2		0.2	
Selenium hexafluoride (as Se)	7783-79-1	0.05	0.4	
Silica, amorphous, precipitated and gel	112926-00-8		(3)	
Silica, amorpous, diatomaceous earth, containing less than 1% crystalline silica	61790-53-2		(3)	
Silica, crystalline cristobalite, respirable dust	14464-46-1		(3)	
Silica, crystalline quartz, respirable dust	14808-60-7		(3)	
Silica, crystalline tripoli (as quartz), respirable dust	1317-95-9		(3)	
Silica, crystalline tridymite, respirable dust	15468-32-3		(3)	
Silica, fused, respirable dust	60676-86-0		(3)	
Silicates (less than 1% crystalline silica) Mica (respirable dust)	12001-26-2		(3)	
Soapstone, total dust			(3)	
Soapstone, respirable dust			(3)	

Continued

TABLE B-1
Continued

Substance	CAS No. (c)	*ppm (a)(1)*	*mg/m(3) (b)(1)*	*Skin designation*
Talc (containing asbestos): use asbestos limit: see 29 CFR 1910.1001			(3)	
Talc (containing no asbestos), respirable dust	14807-96-6		(3)	
Tremolite, asbestiform; see 1910.1001				
Silicon	7440-21-3			
Total dust			15	
Respirable fraction			5	
Silicon carbide	409-21-2			
Total dust			15	
Respirable fraction			5	
Silver, metal and soluble compounds (as Ag)	7440-22-4		0.01	
Soapstone; see Silicates				
Sodium fluoroacetate	62-74-8		0.05	X
Sodium hydroxide	1310-73-2		2	
Starch	9005-25-8			
Total dust			15	
Respirable fraction			5	
Stibine	7803-52-3	0.1	0.5	
Stoddard solvent	8052-41-3	500	2900	
Strychnine	57-24-9		0.15	
Styrene	100-42-5		(2)	

Table B-1
Continued

Substance	CAS No. (c)	*ppm (a)(1)*	*mg/m(3) (b)(1)*	*Skin designation*
Sucrose	57-50-1			
Total dust			15	
Respirable fraction			5	
Sulfur dioxide	7446-09-5	5	13	
Sulfur hexafluoride	2551-62-4	1000	6000	
Sulfuric acid	7664-93-9		1	
Sulfur monochloride	10025-67-9	1	6	
Sulfur pentafluoride	5714-22-7	0.025	0.25	
Sulfuryl fluoride	2699-79-8	5	20	
Systox; see Demeton				
2,4,5-T (2,4,5-trichloro-phenoxyacetic acid)	93-76-5		10	
Talc; see Silicates				
Tantalum, metal and oxide dust	7440-25-7		5	
TEDP (Sulfotep)	3689-24-5		0.2	X
Tellurium and compounds (as Te)	13494-80-9		0.1	
Tellurium hexafluoride (as Te)	7783-80-4	0.02	0.2	
Temephos	3383-96-8			
Total dust			15	
Respirable fraction			5	
TEPP (Tetraethyl pyrophosphaate)	107-49-3		0.05	X
Terphenylis	26140-60-3	(C)1	(C)9	

Continued

TABLE B-1
Continued

Substance	CAS No. (c)	*ppm (a)(1)*	*mg/m(3) (b)(1)*	*Skin designation*
1,1,1,2-Tetrachloro-2,2-difluoroethane	76-11-9	500	4170	
1,1,2,2-Tetrachloro-1,2-difluoroethane	76-12-0	500	4170	
1,1,2,2-Tetrachloroethane	79-34-5	5	35	X
Tetrachoroethylene; see Perchloroethylene				
Tetrachloromethane; see Carbon tetrachloride				
Tetrachloronaphthalene	1335-88-2		2	X
Tetraethyl lead (as Pb)	78-00-2		0.075	X
Tetrahydrofuran	109-99-9	200	590	
Tetramethyl lead, (as Pb)	75-74-1		0.075	X
Tetramethyl succinonitrile	3333-52-6	0.5	3	X
Tetranitromethane	509-14-8	1	8	
Tetryl (2,4,6-Trinitrophenylmethylnitramine)	479-45-8		1.5	X
Thallium, soluble compounds (as Tl)	7440-28-0		0.1	X
4,4'-Thiobis(6-tert, Butyl-m-cresol)	96-69-5			
Total dust			15	
Respirable fraction			5	
Thiram	137-26-8		5	
Tin, inorganic compounds (except oxides) (as Sn)	7440-31-5		2	

TABLE B-1
Continued

Substance	CAS No. (c)	*ppm (a)(1)*	*mg/m(3) (b)(1)*	*Skin designation*
Tin, organic compounds (as Sn)	7440-31-5		0.1	
Titanium dioxide	13463-67-7			
Total dust			15	
Toluene	108-88-3		(2)	
Toluene-2,4-diisocyanate (TDI)	584-84-9	(C)0.02	(C)0.14	
o-Toluidine	95-53-4	5	22	X
Toxaphene; see Chlorinated camphene				
Tremolite; see Silicates				
Tributyl phosphate	126-73-8		5	
1,1,1-Trichloroethane; see Methyl chloroform				
1,1,2-Trichloroethane	79-00-5	10	45	X
Trichloroethylene	79-01-6		(2)	
Trichloromethane; see Chloroform				
Trichloronaphthalene	1321-65-9		5	X
1,2,3-Trichloropropane	96-18-4	50	300	
1,1,2-Trichloro-1,2,2-trifluoroethane	76-13-1	1000	7600	
Triethylamine	121-44-8	25	100	
Trifluorobromomethane	75-63-8	1000	6100	
2,4,6-Trinitrophenol; see Picric acid				

Continued

TABLE B-1
Continued

Substance	*CAS No. (c)*	*ppm (a)(1)*	*mg/m(3) (b)(1)*	*Skin designation*
2,4,6-Trinitrophenyl-methyl nitramine; see Tetryl				
2,4,6-Trinitrotoluene (TNT)	118-96-7		1.5	X
Triorthocresyl phosphate	78-30-8		0.1	
Triphenyl phosphate	115-86-6		3	
Turpentine	8006-64-2	100	560	
Uranium (as U)	7440-61-1			
Soluble compounds			0.05	
Insoluble compounds			0.25	
Vanadium	1314-62-1			
Respirable dust (as V(2)O(5))			(C)0.5	
Fume (as V(2)O(5))			(C)0.1	
Vegetable oil mist				
Total dust			15	
Respirable fraction			5	
Vinyl benzene; see Styrene				
Vinyl chloride; see 1910.1017	75-01-4			
Vinyl cyanide; see Acrylonitrile				
Vinyl toluene	25013-15-4	100	480	
Warfarin	81-81-2		0.1	

TABLE B-1
Continued

Substance	CAS No. (c)	*ppm (a)(1)*	*mg/m(3) (b)(1)*	*Skin designation*
Xylenes (o-, m-, p-isomers)	1330-20-7	100	435	
Xylidine	1300-73-8	5	25	X
Yttrium	7440-65-5		1	
Zinc chloride fume	7646-85-7		1	
Zinc oxide fume	1314-13-2		5	
Zinc oxide	1314-13-2			
Total dust			15	
Respirable fraction			5	
Zinc stearate	557-05-1			
Total dust			15	
Respirable fraction			5	
Zirconium compounds (as Zr)	7440-67-7		5	

Footnote(1) The PELs are 8-hour TWAs unless otherwise noted; a (C) designation denotes a ceiling limit. They are to be determined from breathing-zone air samples.

Footnote(a) Parts of vapor or gas per million parts of contaminated air by volume at 25 degrees C and 760 torr.

Footnote(b) Milligrams of substance per cubic meter of air. When entry is in this column only, the value is exact; when listed with a ppm entry, it is approximate.

Footnote(c) The Chemical Abstracts Service (CAS) number is for information only. Enforcement is based on the substance name. For an entry covering more than one metal compound measured as the metal, the CAS number for the metal is given—not CAS numbers for the individual compounds.

Footnote(d) The final benzene standard in 1910.1028 applies to all occupational exposures to benzene except in some circumstances the distribution and sale of fuels, sealed containers and pipelines, coke production, oil and gas drilling and production, natural gas processing, and the percentage exclusion for liquid mixtures; for the excepted subsegments, the benzene limits in Table Z-2 apply. See 1910.1028 for specific circumstances.

Footnote(e) This 8-hour TWA applies to respirable dust as measured by a vertical elutriator cotton dust sampler or equivalent instrument. The time-weighted average applies to the cotton waste processing operations of waste recycling (sorting, blending, cleaning and willowing) and garnetting. See also 1910.1043 for cotton dust limits applicable to other sectors.

Footnote(f) All inert or nuisance dusts, whether mineral, inorganic, or organic, not listed specifically by substance name are covered by the Particulates Not Otherwise Regulated (PNOR) limit which is the same as the inert or nuisance dust limit of Table Z-3.

Footnote(2) See Table Z-2.

Footnote(3) See Table Z-3.

Footnote(4) Varies with compound.

APPENDIX C

Medical Reference Chart

The medical reference chart presents a condensation of information about chemicals that harm workers via inhalation. Other information that applies to all or most items in the table are:

- There should be a preplacement medical examination specified for each item in the table.
- There should be an annual medical examination for workers working with every item except benzene, O-Toluidine, and Toluene. Examinations for benzene and toluene should be bi-annual, and for O-Toluidine they should be quarterly.
- Required training in exposure education should be specified for every item in the list.

Table C-1
OSHA Medical Reference Table

Chemical or Condition	Status	Regulation or Criteria Document
Acetylene	Pending	NIOSH 76-195
Acrylamide	Pending	NIOSH 77-112
Acrylonitrile	Approved	1910.1045
Alkanes	Pending	NIOSH 77-151
Allyl Chloride	Pending	NIOSH 76-204
Ammonia	Pending	NIOSH 74-136
Anesthetic Gases	Pending	NIOSH 77-140
Antimony	Pending	NIOSH 78-216
Asbestos	Approved	1910/1926.1001
Asphalt Fumes	Pending	NIOSH 78-106
Benzedine Congenger Dyes	Pending	NIOSH 83-105
Benzene	Approved	1910.1028
Benzoyl Peroxide	Pending	NIOSH 76-166
Benzyl Chloride	Pending	NIOSH 78-182
Beryllium	Pending	NIOSH 72-10268
Boron Trifluoride	Pending	NIOSH 77-122
1,3-Butadiene	Approved	1910.1030
Cadmium	Approved	1910.1027
Carbaryl	Pending	NIOSH 77-107
Carbon Black	Pending	NIOSH 78-204
Carbon Dioxide	Pending	NIOSH 76-194
Carbon Disulfide	Pending	NIOSH 77-156
Carbon Monoxide	Pending	NIOSH 73-11000
Carbon Tetrachloride	Pending	NIOSH 76-133
Carcinogens	Approved	1910.1002, 1016

Table C-1
Continued

Chemical or Condition	*Status*	*Regulation or Criteria Document*
Chlorine	Pending	NIOSH 76-170
Chloroform	Pending	NIOSH 75-114
Chloroprene	Pending	NIOSH 77-210
Chromic Acid	Pending	NIOSH 73-11021
Chromium (VI)	Pending	NIOSH 76-129
Coal Gasification	Pending	NIOSH 78-191
Coal Liquification	Pending	NIOSH 80-122
Cobalt	Pending	NIOSH 82-107
Coke Oven Emissions	Approved	1910.1029
Compressed Air Atmosphere	Approved	1926.803 (B)
Confined Space	Approved	1910.146
Cotton Dust	Approved	1910.1043
Crystalline Silica	Pending	NIOSH 75-120
Cyanide	Pending	NIOSH 77-108
DDT	Pending	NIOSH 78-200
Dibromochloropropane	Approved	1910.1044
Dinitro-Ortho-Cresol	Pending	NIOSH 78-131
Dinitrotoluenes	Pending	NIOSH 85-109
Dioxane	Pending	NIOSH 76-226
Epichlorohydrin	Pending	NIOSH 76-206
Ethylene Dibromide	Approved	1910.1048
Ethylene Dichloride	Pending	NIOSH 76-139
Ethylene Oxide	Approved	1910.1047
Ethylene Thiourea	Pending	NIOSH 78-144
Fibrous Glass	Pending	NIOSH 77-152

Continued

TABLE C-1
Continued

Chemical or Condition	Status	Regulation or Criteria Document
Fire Brigade	Approved	1910.156
Fluoro Polymers, byproducts of	Pending	NIOSH 77-193
Formaldehyde	Approved	1910.1048
Furturyl Alcohol	Pending	NIOSH 79-133
Glycidyl Ethers	Approved	1910-1000
Hazardous Chemicals in Lab.	Approved	1910.1450
Hazardous Waste	Approved	1910.120
Hot Environments	Pending	NIOSH 80-132
Hydrazines	Pending	NIOSH 78-172
Hydrogen Fluoride	Pending	NIOSH 76-143
Hydrogen Sulfide	Pending	NIOSH 77-158
Hydroquinone	Pending	NIOSH 78-155
Inorganic Arsenic	Approved	1910.1018
Inorganic Fluoride	Pending	NIOSH 76-103
Inorganic Lead	Approved	1910.1025
Inorganic Mercury	Pending	NIOSH 73-11024
Isopropyl Alcohol	Pending	NIOSH 76-142
Kepone	Pending	NIOSH MEMO 1-76
Ketones	Pending	NIOSH 78-173
Malathion	Pending	NIOSH 77-166
Methylenedianilines	Approved	1910.1000
Methyl Alcohol	Pending	NIOSH 76-148
Methyl Parathion	Pending	NIOSH 77-106
Methylene Chloride	Approved	1910.1000
Monohalomethanes	Pending	NIOSH 84-117

TABLE C-1
Continued

Chemical or Condition	Status	Regulation or Criteria Document
Nickel	Pending	NIOSH 77-164
Nickel Carbonyl	Pending	NIOSH 77-184
Nitric Acid	Pending	NIOSH 76-141
Nitriles	Pending	NIOSH 78-212
Nitrogen Oxides	Pending	NIOSH 76-149
Nitroglycerine	Pending	NIOSH 78-167
Organotin Compounds	Pending	NIOSH 77-115
O-Toluidine	Pending	NIOSH 78-179
Phenol	Pending	NIOSH 76-196
Phosgene	Pending	NIOSH 76-137
Polychlorinated Biphenols	Pending	NIOSH 77-225
Precast Concrete Products	Pending	NIOSH 84-103
Print/Allied Coating Products	Pending	NIOSH 84-115
Propylene Oxide	Pending	CIB51-89
Rendering Processes	Pending	NIOSH 81-133
Sodium Hydroxide	Pending	NIOSH 76-105
Sulfur Dioxide	Pending	NIOSH 74-111
Sulphuric Acid	Pending	NIOSH 74-128
TB	Pending	OSHA 5/92
TCDO-Dioxin	Pending	NIOSH 84-104
1,1,2,2,-Tetrachloroethane	Pending	NIOSH 77-121
Tetrachloro-ethylene	Pending	NIOSH 76-185
Thiols	Pending	NIOSH 78-213
Toluene	Pending	NIOSH 73-11023
Toluene Diisocynanate	Pending	NIOSH 73-11022

Continued

Table C-1
Continued

Chemical or Condition	*Status*	*Regulation or Criteria Document*
1,1,1,-Trichloroethane	Pending	NIOSH 76-184
Trichloroethylene	Pending	NIOSH 73-11025
Tungsten Carbide	Pending	NIOSH 77-227
Vanadium	Pending	NIOSH 77-222
Vinyl Acetate	Pending	NIOSH 78-205
Vinyl Chloride	Approved	1910.1017
Vinyl Halides	Pending	NIOSH 79-102
Wood Preservative Chemicals	Pending	NIOSH 83-106
Xylene	Pending	NIOSH C-75-168
Zinc Oxide	Pending	NIOSH 76-104

APPENDIX D

Glossary

This appendix presents a list of terms, acronyms, abbreviations, and names that have application in the field of indoor air quality. Most of the entries are used and explained throughout the book, but are repeated here to give readers one place they can turn to for meanings. Entries not used in this book are included as a resource for readers who are using other references.

Absorption Holding a removed gas or vapor by penetration, usually deeply into molecular spaces of the sorbent. See also *adsorption* and *chemisorption*.

ACGIH American Conference of Governmental Industrial Hygienists.

Adsorption Similar to absorption (q.v.), but adsorption retains the contaminant molecule on the surface of the sorbent granule by physical attraction. See also *chemisorption*.

Antimicrobial Refers to several agents that kill microbial growth. See *disinfectant*, *sanitizer*, and *sterilizer*.

AOEC Association of Occupational and Environmental Clinics.

ASHRAE American Society of Heating, Refrigerating, and Air-Conditioning Engineers.

ATSDR Agency for Toxic Substances and Disease Registry.

Bactericide A chemical that kills bacteria.

Biological contaminants Also called "microbiologicals" and "microbials," these are agents derived from or that are living organisms (e.g., viruses, bacteria, fungi, and mammal and bird antigens) that can be inhaled and can cause many types of health effects including allergic reactions, respiratory disorders, hypersensitivity diseases, and infectious diseases.

Breathing zone Area of a room in which occupants draw breath as they work. Investigators study working arrangements in a room (sitting, walking around, working on ladders, etc.) to define a breathing zone. It is usually defined by specified distances from floor, ceiling, and each wall.

BRI Building-related illness (q.v.).

Building envelope Elements of the building that enclose the internal space, including all external building materials, windows, and walls.

Building related illness (BRI) Diagnosable illness whose symptoms can be identified and whose cause can be directly attributed to airborne building pollutants (examples are Legionnaire's disease and hypersensitivity pneumonitis).

CAS Chemical Abstracts Service (of the American Chemical Society). The service assigns a numeric designation that uniquely identifies a specific chemical compound regardless of the name or naming system used.

Ceiling plenum Space below the flooring above the suspended ceiling. It accommodates the mechanical and electrical equipment and is used as part of the air distribution system. The space should be kept under negative pressure.

CET Corrected effective temperature.

CFR Code of Federal Regulations.

Chemisorption Similar to adsorption (q.v.), but chemisorption retains the contaminant molecule on the surface of the sorbent granule by chemical forces. See also *absorption*.

CHRIS Chemical Hazards Response Information System. It provides information on handling and disposing of toxic substances.

CL Ceiling limit.

Commissioning Start-up of a building that includes testing and adjusting HVAC, electrical, plumbing, and other systems to ensure proper functioning and adherence to design specifications. Also includes instruction of building representatives in using the building and its systems.

Conditioned air Air that has been heated, cooled, humidified, or dehumidified to maintain an interior space within specified conditions. Sometimes referred to as "tempered air" to differentiate it from the popular use of "conditioned air" to mean cooled air.

Constant air volume system Air handling system that provides a constant airflow while varying the temperature to meet heating and cooling needs.

CPC Chemical protective clothing.

CWXSP Coal Workers X-Ray Surveillance Program.

Damper Control that varies airflow through an outlet, inlet, or duct. A damper position may be immovable, manually adjustable, or part of an automated control system.

DHHS Department of Health and Human Services.

Diffuser Component of the air distribution system, usually where a duct opens to an area, that determines how supply air entering the area will diffuse and circulate. See also *grille*.

Disinfectant One of three groups of antimicrobials registered by the EPA for public health use, it destroys or irreversibly inactivates infectious or other undesirable organisms, but not necessarily their spores. EPA registers three types of disinfectant products, based upon submitted efficacy data: limited, general or broad spectrum, and hospital disinfectant.

DOL Department of Labor.

DRDS Division of Respiratory Disease Studies.

Dust When referring to a particulate filtering respirator, dust is a solid, mechanically produced particle.

ET Effective temperature.

ETS Environmental tobacco smoke.

Fume When referring to a particulate filtering respirator, fume is a solid condensation particulate, usually of a vaporized metal.

Grille A protection, usually located where return air enters a duct. See also *diffuser*.

Helmet When referring to a respirator, a helmet is a rigid protective headgear incorporated into the design.

HEPA High efficiency particulate arrestance, referring to filters.

HHE Health hazard evaluation.

Hood When referring to a respirator, a hood is a light flexible device covering only the head and neck; or head, neck, and shoulders.

HSDB Hazardous Substances Data Bank, is a part of the National Library of Medicine System. It contains health and safety profiles for over 4100 chemicals. Included are 144 data elements in 10 categories including use information, substance identification, environmental fate, standards, personal protective equipment, and physical and chemical properties.

HVAC Heating, ventilating, and air conditioning.

Hypersensitivity diseases Diseases characterized by allergic responses to animal antigens. The hypersensitivity diseases most clearly associated with indoor air quality are asthma, rhinitis, and hypersensitivity pneumonitis.

Hysteresis The result of a system's condition that causes its output to depend on the direction in which its input changes. In a mechanical system the condition is usually slack that must be taken up after an input change of direction before there will be a change in the output.

IARC International Agency for Research on Cancer is a research organization authorized by the World Health Organization in 1965. It publishes and updates a series of monographs on a substantial number of toxic chemicals and substances in which the carcinogenic risk of these chemicals is evaluated.

IEQ Indoor environmental quality.

IMIS Integrated Management Information System is a database developed by OSHA in 1979 with sampling information on more than 100,000 substances. This database contains exposure measurements obtained by OSHA compliance officers during thousands of health inspections.

Indicator compounds Chemical compounds, such as carbon dioxide, whose presence at known concentrations may be used to estimate certain building conditions.

LC_{50} Lethal concentration for 50 percent of exposed laboratory animals.

LD_{50} Lethal dose for 50 percent of exposed laboratory animals.

Makeup air Air brought into a building from outdoors through the ventilation system to replace air that has been exhausted.

MCS Multiple chemical sensitivity (q.v.).

Microbiologicals See *biological contaminants*.

Mist When referring to a particulate filtering respirator, a mist is a liquid condensation particle.

MSDS Material Safety Data Sheet is a compilation of data and information on an individual hazardous chemical. OSHA's Hazard Communication Standard requires manufacturers and importers to prepare MSDSs. They contain data on chemical identification; current exposure limits; chemical reactivity; fire and explosion limits; and information on health hazards, emergency procedures, spill, leak, and disposal procedures; and any needed special protection or precautions.

MSHA Mine Safety and Health Administration.

Multiple chemical sensitivity Refers to a condition in which a person is considered to be sensitive to a number of chemicals at very low concentrations.

There are a number of views about the existence, potential causes, and possible remedial actions regarding this phenomenon.

NCHS National Center for Health Statistics.

NHIS National Health Interview Survey.

NIOSH National Institute for Occupational Safety and Health.

NOES National Occupational Exposure Survey is a database completed in 1982 by NIOSH. It contains a sample of the number of persons exposed by substance and industry from approximately 4500 businesses in 98 geographic areas in the U.S. These surveys provide national estimates of potential exposure to workplace hazards, by industry and occupational group.

NOHSM National Occupational Health Survey of Mining.

NTP National Toxicology Program.

OCIS OSHA Computerized Information System is a comprehensive database that contains information and data on standards interpretation, chemical information, hazardous waste activity, citations, a health hazard evaluation index, training materials, and other information.

ODTS Organic dust toxic syndrome.

OSHA Occupational Safety and Health Administration.

PAPR Powered air-purifying respirator.

PEL Permissible Exposure Limit is a limit developed by OSHA to indicate the maximum airborne concentration of a contaminant to which an employee maybe exposed over the duration specified by the type of PEL assigned to that contaminant.

PFT Pulmonary function test.

Plenum Air compartment connected to the duct system.

ppb Parts per billion.

ppm Parts per million.

RD_{50} Concentration eliciting a 50 percent decrease in respiratory rate.

Re-entrainment Air exhausted from a building that is immediately brought back through the air intake or other openings in the building envelope.

REL Recommended Exposure Limit, issued by NIOSH to aid in controlling hazards in the workplace. These limits are generally expressed as 8- or 10-hour TWAs for a 40-hour workweek and/or ceiling levels with time limits ranging from instantaneous to 120 minutes.

REL Recommended exposure limit (recommendation by NIOSH).

RTECS Registry of Toxic Effects of Chemical Substances.

Sanitizer One of three groups of antimicrobials registered by EPA for public health uses, it reduces but does not necessarily eliminate all the microorganisms on a treated surface. To be a registered sanitizer, test results for a product must show a reduction of at least 99.9 percent in the number of each test microorganism over the parallel control.

SBS Sick building syndrome.

SENSOR Sentinel Event Notification System for Occupational Risks.

SHE Sentinel health event.

Sick building syndrome A term used when building occupants experience acute health and/or comfort effects that appear to be linked to time spent in the building, but where no specific illness or cause can be identified. The complaints may be localized to a particular room or zone, or spread throughout the building.

Soil gases Gases that enter a building from the surrounding ground, such as radon, volatile organic compounds, and pesticides.

Sporicide A chemical that kills spores.

Stack effect Airflow produced by convection as warm air rises, creating a positive pressure at the top of a building and a negative pressure at the bottom. The stack effect can override the mechanical system and disrupt intended ventilation and circulation patterns.

Static pressure Equilibrium condition that exists when the amount of air supplied to a space equals the amount exhausted.

STEL Short term exposure limit.

Sterilizer One of three groups of antimicrobials registered by EPA for public health uses, it destroys or eliminates all forms of bacteria, fungi, viruses, and their spores. Because spores are considered the most difficult form of a microorganism to destroy, EPA labels "sporicide" as synonymous with "sterilizer."

TLV Threshold limit value (guideline recommended by ACGIH).

Torr A unit of pressure that supports a column of mercury one millimeter high.

Tracer gases Compounds, such as sulfur hexafluoride, which are used to identify suspected pollutant pathways and to quantify ventilation rates. Tracer gases may be detected qualitatively by their odor or quantitatively by air monitoring equipment.

TVOC Total volatile organic compound.

TWA Time weighted average.

Variable air volume system Air handling system that conditions the air to a constant temperature and varies the outside airflow that mixes with it to control the room temperature.

VAV Variable air volume system (q.v.).

Ventilation air Total air; a combination of outside air brought into the system, and recirculated air. The expression is sometimes used to refer only to outside air brought into the system.

VOC Volatile organic compound (q.v.).

Volatile organic compound (VOC) A compound that evaporates from organic chemicals. In the indoor air quality field, VOCs are usually evaporated from housekeeping, maintenance, and building products, either being used or in storage. In sufficient quantities, VOCs can cause eye, nose, and throat irritations; headaches; dizziness; visual disorders; and memory impairment. Some are known to cause cancer in animals, some are known to cause cancer in humans, and some are suspected of causing cancer in humans.

WBGT Wet bulb globe thermometer.

WGT Wet globe temperature.

WISH Workers' Institute for Safety and Health.

APPENDIX E

Agencies and Associations

For readers who want to dig further, get specialized information, find the latest changes in regulations, or obtain services, this section gives names, addresses, and telephone numbers. Government agencies, industry associations, consumer organizations, and others are included.

GOVERNMENT AGENCIES

Environmental Protection Agency

This agency conducts a non-regulatory indoor air quality program that emphasizes research, information dissemination, technical guidance, and training. It issues regulations and carries out other activities that affect indoor air quality under the laws for pesticides, toxic substances, and drinking water.

Public Information Center (PM-211B)
401 M Street, SW
Washington, DC 20460
(202) 260-2080
Distributes indoor air quality publications.

TSCA Hotline Service
(202) 554-1404
Provides information on asbestos and other toxic substances.

Occupational Safety and Health Administration (OSHA)

Promulgates safety and health standards, facilitates training and consultation, and enforces regulations to ensure that workers are provided healthful working conditions.

U.S. Department of Labor
Occupational Safety and Health Administration
3rd and Constitution Avenues, NW
Washington, D.C. 20210

Most publications and other information are available from the ten regional offices. States and territories identified with an asterisk in the following list operate their own OSHA-approved job safety and health programs (Connecticut and New York plans cover public employees only).

Region I (CT*, MA, ME, NH, RI, VT*)
133 Portland Street
1st Floor
Boston, MA 02114
(617) 565-7164

Region II (NJ, NY*, PR*, VI*)
201 Varick Street - Room 670
New York, NY 10014
(212) 377-2378

Region III (DC, DE, MD*, PA, VA*, WV)
Gateway Building - Suite 2100
3535 Market Street
Philadelphia, PA 19104
(215) 596-1201

Region IV (AL, FL, GA, KY*, MS, NC*, SC*, TN*)
1375 Peachtree Street, NE - Suite 587
Atlanta, GA 30367
(404) 347-3573

Region V (IL, IN*, MI*, MN*, OH, WI)
230 South Dearborn Street - Room 3244
Chicago, IL 60604
(312) 353-2220

Region VI (AR, LA, NM*, OK, TX)
525 Griffin Street - Room 602
Dallas, TX 75202
(214) 767-4731

Region VII (IA*, KS, MO, NE)
911 Walnut Street - Room 406
Kansas City, MO 64106
(816) 426-5861

Region VIII (CO, MT, ND, SD, UT*, WY*)
Federal Building - Room 1576
1961 Stout Street
Denver, CO 80294
(303) 844-3061

Region IX (American Samoa, AZ8, CA*, Guam, HI*, Trust Territories of the Pacific)
71 Stevenson Street - 4th Floor
San Francisco, CA 94105
(415) 744-6670

Region X (AK*, ID, OR*, WA*)
1111 Third Avenue - Suite 715
Seattle, WA 98101-3212
(206) 553-5930

OTHER FEDERAL AGENCIES

Bonneville Power Administration
PO Box 3621-RMRD
Portland, OR 97208
(503) 230-5475
Provides radon-resistant construction techniques, source control, and removal technology for indoor air pollutants.

Consumer Product Safety Commission
5401 Westbard Avenue
Bethesda, MD 20207
(800) 638-CPSC
Reviews complaints regarding the safety of consumer products and takes action to ensure product safety.

General Services Administration
18th and F Streets, NW
Washington, DC 20405
(202) 501-1464
Writes indoor air quality policy for federal buildings. Provides proactive indoor air quality building assessments. Assesses complaints and provides remedial action.

U.S. Department of Energy
Office of Conservation and Renewable Energy
1000 Independence Avenue, SW, CE-43
Washington, DC 20585
Quantifies the relationship among reduced infiltration, adequate ventilation, and acceptable indoor air quality.

U.S. Department of Health and Human Services
Office on Smoking and Health
National Center for Chronic Disease Prevention and Health
Promotion Centers for Disease Control
1600 Clifton Road, NE
Mail Stop K50
Atlanta, GA 30333
(404) 488-5705
Disseminates information about the health effects of passive smoking and strategies for eliminating exposure to environmental tobacco smoke.

Tennessee Valley Authority
Occupational Hygiene Department
328 Multipurpose Building
Muscle Shoals, AL 35660
(205) 386-2314
Provides building surveys and assessments associated with employee indoor air quality complaints.

PRIVATE SECTOR CONTACTS

Building Management Associations

Association of Physical Plant Administrators of Universities and Colleges
1446 Duke Street
Alexandria, VA 22314-3492
(703) 684-1446

Building Owners and Managers Association International
1201 New York Avenue, NW, Suite 300
Washington, DC 20005
(202) 408-2684

Institute of Real Estate Management
430 North Michigan Avenue
Chicago, IL 60611
(312) 661-1930

International Council of Shopping Centers
1199 North Fairfax Street, Suite 204
Alexandria, VA 22314
(703) 549-7404

International Facilities Management Association
Summit Tower, Suite 1710
11 Greenway Plaza
Houston, TX 77046
(713) 623-4361

National Association of Industrial and Office Parks
1215 Jefferson Davis Highway, Suite 100
Washington, DC 20005
(703) 979-3400

Professional and Standard Setting Organizations

Air and Waste Management Association
P.O. Box 2861
Pittsburgh, PA 15230
(412) 232-3444

Air Conditioning and Refrigeration Institute
1501 Wilson Blvd., Suite 600
Arlington, VA 22209
(703) 524-8800

American Conference of Governmental Industrial Hygienists
6500 Glenway Avenue, Building D-7
Cincinnati, OH 45211
(513) 661-7881

American Industrial Hygiene Association
P.O. Box 8390
345 White Pond Drive
Akron, OH 44320
(216) 873-2442

American Society for Testing and Materials
1916 Race Street
Philadelphia, PA 19103
(215) 299-5571

American Society of Heating, Refrigerating, and Air-Conditioning Engineers
1791 Tullie Circle, Ne
Atlanta, GA 30329
(404) 636-8400

National Conference of States on Building Codes and Standards, Inc.
505 Huntmar Park Drive, Suite 210
Herndon, VA 22070
(703) 437-0100

Product Manufacturers

Adhesive and Sealant Council
1627 K Street, NW, Suite 1000
Washington, DC 20006-1707
(202) 452-1500

Asbestos Information Association
1745 Jefferson Davis Highway, Room 509
Arlington, VA 22202
(703) 979-1150

Business Council on Indoor Air Quality
1225 19th Street, Suite 300
Washington, DC 20036
(202) 775-5887

Carpet and Rug Institute
310 Holiday Avenue
Dalton, GA 30720
(404) 278-3176

Chemical Specialties Manufacturers Association
1913 I Street, NW
Washington, DC 20006
(202) 872-8110

Electric Power Research Institute
PO Box 10412
Palo Alto, CA 94303
(415) 855-2902

Formaldehyde Institute, Inc.
1330 Connecticut Avenue, NW
Washington, DC 20036
(202) 822-6757

Foundation of Wall and Ceiling Industries
1600 Cameron Street
Alexandria, VA 22314-2705
(703) 548-0374

Gas Research Institute
8600 West Bryn Mawr Avenue
Chicago, IL 60631
(312) 399-8304

National Paint and Coatings Association
1500 Rhode Island Avenue, NW
Washington, DC 20005
(202) 462-6272

Thermal Insulation Manufacturers Association Technical Services
Air Handling Committee
1420 King Street
Alexandria, VA 22314
(703) 684-0474

Building Service Associations

Air-Conditioning and Refrigeration Institute
1501 Wilson Boulevard, 6th Floor
Arlington, VA 22209
(703) 524-8800

Air-Conditioning Contractors of America
1513 16th Street, NW
Washington, DC 20036
(202) 483-9370

American Consulting Engineers Council
1015 15th Street, NW, Suite 802
Washington, DC 20005
(202) 347-7474

Associated Air Balance Council
1518 K Street, NW
Washington, DC 20005
(202) 737-0202

Association of Energy Engineers
4025 Pleasantdale Road, Suite 420
Atlanta, GA 30340
(404) 447-5083

Association of Specialists in Cleaning and Restoration International
10830 Annapolis Junction Road, Suite 312
Annapolis Junction, MD 20701
(301) 604-4411

National Air Duct Cleaners Association
1518 K Street, NW, Suite 503
Washington, DC 20005
(202) 737-2926

National Association of Power Engineers
3436 Haines Way, Suite 101
Falls Church, VA 22041
(703) 845-7055

National Energy Management Institute
601 North Fairfax Street, Suite 160
Alexandria, VA 22314
(703) 739-7100

National Environmental Balancing Bureau
1385 Piccard Drive
Rockville, MD 20850
(301) 977-3698

National Pest Control Association
8100 Oak Street
Dunn Loring, VA 22027
(703) 573-8330

Sheet Metal and Air Conditioning Contractors National Association
4201 LaFayette Center Drive
Chantilly, VA 22021
(703) 803-2980

ENVIRONMENTAL/HEALTH/CONSUMER ORGANIZATIONS

American Academy of Allergy and Immunology
611 East Wells Street
Milwaukee, WI 53202
(414) 272-6071

American Lung Association
or your local lung association
1740 Broadway
New York, NY 10019

Consumer Federation of America
1424 16th Street, NW, Suite 604
Washington, DC 20036

National Center for Environmental Health Strategies
1100 Rural Avenue
Voorhees, NJ 08043
(609) 429-5358

National Environmental Health Association
720 South Colorado Blvd.
South Tower, Suite 970
Denver, CO 80222
(303) 756-9090

National Foundation for the Chemically Hypersensitive
P.O. Box 9
Wrightsville Beach, NC 28480
(517) 697-3989

Occupational Health Foundation
1126 16th Street, NW
Washington, DC 20036
(202) 842-7840

APPENDIX F

Forms and Checklists

Step-by-step investigation implies orderly, systematic moving through the entire process. An aid that encourages this orderly approach is a supply of previously-prepared forms, carefully designed to obtain the necessary information. This section presents a series of forms, from the public domain, that you can use directly or with modifications to meet your specific needs.

Table F-1 is a summary of the forms and checklists in this section. Unless a page count is shown, the form is one page long. Some of these forms are referred to in Chapters 10 and 15.

TABLE F-1
Summary of Forms in Appendix F

Figure	*Title*	*Description*
F-1	IAQ Management Checklist (4 pages)	Designed to help keep track of the elements of the IAQ profile and IAQ management plan
F-2	Pollutant Pathway Record for IAQ Profiles	Emphasis is on positive and negative pressure areas that should be maintained
F-3	Zone/Room Record	For recording information on a room-by-room basis for topics of room use, ventilation, and occupant population
F-4	Ventilation Worksheet	For use in conjunction with Fig F-3 when calculating quantities of outdoor air that are being supplied to individual zones or rooms
F-5	Incident Log	Systematic documentation of each IAQ complaint or problem and how it was handled
F-6	Occupant Interview (2 pages)	To be filled in by interviewer. Allows room for additional comments.
F-7	Occupant Diary	Encourages recording of observations as they occur
F-8	Log of Activities and System Operation	For recording activities and equipment operating schedules as they occur
F-9	HVAC Checklist, Short Form (4 pages)	To gather information for investigating an IAQ problem, or for periodic inspections. Duplicate pages 2–4 for each large air handling unit.
F-10	Pollutant and Source Inventory (6 pages plus a general purpose sheet)	To be used as a general checklist of potential indoor and outdoor pollutant sources
F-11	Hypothesis Form (2 pages)	For summarizing what has been learned during the investigation, and collecting the investigator's thoughts

IAQ Management Checklist

Building Name: ______________________ Date: ____________

Address: ______________________

Completed by (name/title): ______________________

Use this checklist to make sure that you have included all necessary elements in your IAQ profile and IAQ management plan. *Sections 4 and 5* discuss the development of the IAQ profile and IAQ management plan.

Item	Date begun or completed (as applicable)	Responsible person (name, telephone)	Location ("NA" if the item is not applicable to this building)
IAQ PROFILE			
Collect and Review Existing Records			
HVAC design data, operating instructions, and manuals			
HVAC maintenance and calibration records, testing and balancing reports			
Inventory of locations where occupancy, equipment, or building use has changed			
Inventory of complaint locations			
Conduct a Walkthrough Inspection of the Building			
List of responsible staff and/or contractors, evidence of training, and job descriptions			
Identification of areas where positive or negative pressure should be maintained			
Record of locations that need monitoring or correction			
Collect Detailed Information			
Inventory of HVAC system components needing repair, adjustment, or replacement			
Record of control settings and operating schedules			

IAQ Management Checklist

Item	Date begun or completed (as applicable)	Responsible person (name, telephone)	Location ("NA" if the item is not applicable to this building)
Plan showing airflow directions or pressure differentials in significant areas			
Inventory of significant pollutant sources and their locations			
MSDSs for supplies and hazardous substances that are stored or used in the building			
Zone/Room Record			
IAQ MANAGEMENT PLAN			
Select IAQ Manager			
Review IAQ Profile			
Assign Staff Responsibilities/ Train Staff			
Facilities Operation and Maintenance			
▪ confirm that equipment operating schedules are appropriate			
▪ confirm appropriate pressure relationships between building usage areas			
▪ compare ventilation quantities to design, codes, and ASHRAE 62-1989			
▪ schedule equipment inspections per preventive maintenance plan or recommended maintenance schedule			
▪ modify and use HVAC Checklist(s); update as equipment is added, removed, or replaced			
▪ schedule maintenance activities to avoid creating IAQ problems			

IAQ Management Checklist

Item	Date begun or completed (as applicable)	Responsible person (name, telephone)	Location ("NA" if the item is not applicable to this building)
■ review MSDSs for supplies; request additional information as needed			
■ consider using alarms or other devices to signal need for HVAC maintenance (e.g., clogged filters)			
Housekeeping			
■ evaluate cleaning schedules and procedures; modify if necessary			
■ review MSDSs for products in use; buy different products if necessary			
■ confirm proper use and storage of materials			
■ review trash disposal procedures; modify if necessary			
Shipping and Receiving			
■ review loading dock procedures (*Note:* If air intake is located nearby, take precautions to prevent intake of exhaust fumes.)			
■ check pressure relationships around loading dock			
Pest Control			
■ consider adopting IPM methods			
■ obtain and review MSDSs; review handling and storage			
■ review pest control schedules and procedures			
■ review ventilation used during pesticide application			

IAQ Management Checklist

Item	Date begun or completed (as applicable)	Responsible person (name, telephone)	Location ("NA" if the item is not applicable to this building)
Occupant Relations			
■ establish health and safety committee or joint tenant/ management IAQ task force			
■ review procedures for responding to complaints; modify if necessary			
■ review lease provisions; modify if necessary			
Renovation, Redecorating, Remodeling			
■ discuss IAQ concerns with architects, engineers, contractors, and other professionals			
■ obtain MSDSs; use materials and procedures that minimize IAQ problems			
■ schedule work to minimize IAQ problems			
■ arrange ventilation to isolate work areas			
■ use installation procedures that minimize emissions from new furnishings			
Smoking			
■ eliminate smoking in the building			
■ if smoking areas are designated, provide adequate ventilation and maintain under negative pressure			
■ work with occupants to develop appropriate non-smoking policies, including implementation of smoking cessation programs			

Pollutant Pathway Record For IAQ Profiles

This form should be used in combination with a floor plan such as a fire evacuation plan.

Building Name: ______________________ File Number: ______________

Address: ____

Completed by: ______________________ Title: ______________ Date: ______________

Sections 2, 4 and 6 discuss pollutant pathways and driving forces.

Building areas that contain contaminant sources (e.g., bathrooms, food preparation areas, smoking lounges, print rooms, and art rooms) should be maintained under negative pressure relative to surrounding areas. Building areas that need to be protected from the infiltration of contaminants (e.g., hallways in multi-family dwellings, computer rooms, and lobbies) should be maintained under positive pressure relative to the outdoors and relative to surrounding areas.

List the building areas in which pressure relationships should be controlled. As you inspect the building, put a Y or N in the "Needs Attention" column to show whether the desired air pressure relationship is present. Mark the floor plan with arrows, plus signs (+) and minus signs (-) to show the airflow patterns you observe using chemical smoke or a micromanometer.

Building areas that appear isolated from each other may be connected by airflow passages such as air distribution zones, utility tunnels or chases, party walls, spaces above suspended ceilings (whether or not those spaces are serving as air plenums), elevator shafts, and crawlspaces. If you are aware of pathways connecting the room to identified pollutant sources (e.g., items of equipment, chemical storage areas, bathrooms), it may be helpful to record them in the "Comments" column, on the floor plan, or both.

Building Area (zone, room)	Use	Intended Pressure		Needs Attention? (Y/N)	Comments
		Positive (+)	Negative (-)		

Zone/Room Record

Building Name: ______________________ File Number: ______________ Date: __________

Address: ______________________ Completed by: ______________ Title: __________

This form is to be used differently depending on whether the goal is to *prevent* or to *diagnose* IAQ problems. During the development of a profile, this form should be used to record more general information about the entire building; during an investigation, the form should be used to record more detailed information about the complaint area and areas surrounding the complaint area or connected to it by pathways.

Use the last three columns when underventilation is suspected. Use the **Ventilation Worksheet** and *Appendix A* to estimate outdoor air quantities. Compare results to the design specifications, applicable building codes, or ventilation guidelines such as ASHRAE 62-1989. (See *Appendix A* for some outdoor air quantities required by ASHRAE 62-1989.) *Note:* For VAV systems, minimum outdoor air under reduced flow conditions must be considered.

PROFILE AND DIAGNOSIS INFORMATION					DIAGNOSIS INFORMATION ONLY		
Building Area (Zone/Room)	**Use****	**Source of Outdoor Air***	**Mechanical Exhaust? (Write "No" or estimate cfm airflow)**	**Comments**	**Peak Number of Occupants or Sq. Ft. Floor Area****	**Total Air Supplied (in cfm)*****	**Outdoor Air Supplied per Person or per 150 Sq. Ft. Area (in cfm)******

* Sources might include air handling unit (e.g., AHU-4), operable windows, transfer from corridors
** Underline the information in this column if current use or number of occupants is different from design specifications
*** Mark the information with a **P** if it comes from the mechanical plans or an **M** if it comes from the actual measurements, such as recent test and balance reports.
**** ASHRAE 62-1989 gives ventilation guidance per 150 sq. ft.

Ventilation Worksheet

Building Name: ______________________ File Number: ______________

Address: ____

Completed by (name): ______________________ Date: ______________

This worksheet is designed for use with the **Zone/Room Record.** *Appendix A* provides guidance on methods of estimating the amount of ventilation (outdoor) air being introduced by a particular air handling unit. *Appendix B* discusses the ventilation recommendations of ASHRAE Standard 62-1989, which was developed for the purpose of preventing indoor air quality problems. Formulas are given below for calculating outdoor air quantities using thermal or CO_2 information.

The equation for calculating outdoor air quantities **using thermal measurements** is:

$$\text{Outdoor air (in percent)} = \frac{T_{\text{return air}} - T_{\text{mixed air}}}{T_{\text{return air}} - T_{\text{outdoor air}}} \times 100$$

Where: T = temperature in degrees Fahrenheit

The equation for calculating outdoor quantities **using carbon dioxide measurements** is:

$$\text{Outdoor air (in percent)} = \frac{C_s - C_r}{C_0 - C_r} \times 100$$

Where: C_s = ppm of carbon dioxide in the supply air (if measured in a room), or
C_s = ppm of carbon dioxide in the mixed air (if measured at an air handler)
C_r = ppm of carbon dioxide in the return air
C_o = ppm of carbon dioxide in the outdoor air

Use the table below to estimate the ventilation rate in any room or zone. *Note:* ASHRAE 62-1989 generally states ventilation (outdoor air) requirements on an occupancy basis; for a few types of spaces, however, requirements are given on a floor area basis. Therefore, this table provides a process of calculating ventilation (outdoor air) on either an occupancy or floor area basis.

Zone/Room	Percent of Outdoor Air A	Total Air Supplied to Zone/Room (cfm) B	Peak Occupancy (number of people) or Floor Area (square feet) C	$D = \frac{B}{C}$ Total Air Supplied Per Person (or per square foot area) D	E = (Ax100) x D Outdoor air Supplied Per Person (or per square foot area) E

Incident Log

Building Name: ______________________ Dates (from): __________ (to): __________

Address: ______________________ Completed by (name): ______________________

File Number	Date	Problem Location	Investigation Record (check the forms that were used)									Outcome/Comments (use more than one line if needed)	Log Entry By (initials)
			Complaint Form	Occupant Interview	Occupant Diary	Log of Activities	Zone/Room Record	HVAC Checklist	Pollutant Pathway	Source Inventory	Hypothesis Form		

Occupant Interview

Building Name: ________________________ File Number: ____________

Address: ____

Occupant Name: ____________________ Work Location: ____________________

Completed by: ____________________ Title: ______________ Date: __________

Section 4 discusses collecting and interpreting information from occupants.

SYMPTOM PATTERNS

What kind of symptoms or discomfort are you experiencing?

Are you aware of other people with similar symptoms or concerns? Yes ______ No ______

If so, what are their names and locations? ____

____

Do you have any health conditions that may make you particularly susceptible to environmental problems?

- ❑ contact lenses
- ❑ allergies
- ❑ chronic cardiovascular disease
- ❑ chronic respiratory disease
- ❑ chronic neurological problems
- ❑ undergoing chemotherapy or radiation therapy
- ❑ immune system suppressed by disease or other causes

TIMING PATTERNS

When did your symptoms start?

When are they generally worst?

Do they go away? If so, when?

Have you noticed any other events (such as weather events, temperature or humidity changes, or activities in the building) that tend to occur around the same time as your symptoms?

Occupant Interview

SPATIAL PATTERNS

Where are you when you experience symptoms or discomfort?

Where do you spend most of your time in the building?

ADDITIONAL INFORMATION

Do you have any observations about building conditions that might need attention or might help explain your symptoms (e.g., temperature, humidity, drafts, stagnant air, odors)?

Have you sought medical attention for your symptoms?

Do you have any other comments?

Occupant Diary

Occupant Name: ______________________ Title: ______________________ Phone: ________________

Location: ____ File Number :________________

On the form below, please record each occasion when you experience a symptom of ill-health or discomfort that you think may be linked to an environmental condition in this building.

It is important that you record the time and date and your location within the building as accurately as possible, because that will help to identify conditions (e.g., equipment operation) that may be associated with your problem. Also, please try to describe the severity of your symptoms (e.g., mild, severe) and their duration (the length of time that they persist). Any other observations that you think may help in identifying the cause of the problem should be noted in the "Comments" column. Feel free to attach additional pages or use more than one line for each event if you need more room to record your observations.

Section 6 discusses collecting and interpreting occupant information.

Time/Date	Location	Symptom	Severity/Duration	Comments

Log of Activities and System Operation

Building Name: ______________ Address: ______________ File Number : ______________

Completed by: ______________ Title: ______________ Phone: ______________

On the form below, please record your observations of the HVAC system operation, maintenance activities, and any other information that you think might be helpful in identifying the cause of IAQ complaints in this building. Please report any other observations (e.g., weather, other associated events) that you think may be important as well.

Feel free to attach additional pages or use more than one line for each event.

Equipment and activities of particular interest:

Air Handler(s): ______________

Exhaust Fan(s): ______________

Other Equipment or Activities: ______________

Date/Time	Day of Week	Equipment Item/Activity	Observations/Comments

HVAC Checklist - Short Form

Building Name: ______________________ Address: ______________________

Completed by: ______________________ Date: __________ File Number: __________

Sections 2, 4 and 6 and Appendix B discuss the relationships between the HVAC system and indoor air quality.

MECHANICAL ROOM

- Clean and dry? ______________ Stored refuse or chemicals? ______________
- Describe items in need of attention ______________________

MAJOR MECHANICAL EQUIPMENT

- Preventive maintenance (PM) plan in use? ______________________

Control System

- Type ______________________
- System operation ______________________
- Date of last calibration ______________________

Boiler

- Rated Btu input __________ Condition ______________________
- Combustion air: is there at least one square inch free area per 2,000 Btu input? ______________
- Fuel or combustion odors ______________________

Cooling Tower

- Clean? no leaks or overflow? ______________ Slime or algae growth? ______________
- Eliminator performance ______________________
- Biocide treatment working? (list type of biocide) ______________________
- Spill containment plan implemented? ______________ Dirt separator working? ______________

Chillers

- Refrigerant leaks? ______________________
- Evidence of condensation problems? ______________________
- Waste oil and refrigerant properly stored and disposed of? ______________________

HVAC Checklist - Short Form

Building Name: ______________________ Address: ______________________

Completed by: ______________________ Date: __________ File Number: __________

AIR HANDLING UNIT

- Unit identification ______________________ Area served ______________________

Outdoor Air Intake, Mixing Plenum, and Dampers

- Outdoor air intake location ______________________
- Nearby contaminant sources? (describe) ______________________
- Bird screen in place and unobstructed? ______________________
- Design total cfm __________ outdoor air (O.A.) cfm __________ date last tested and balanced __________
- Minimum % O.A. (damper setting) __________ Minimum cfm O.A. $\frac{\text{(total cfm x minimum \% O.A.)}}{100}$ = __________
- Current O.A. damper setting (date, time, and HVAC operating mode) ______________________
- Damper control sequence (describe) ______________________
- Condition of dampers and controls (note date) ______________________

Fans

- Control sequence ______________________
- Condition (note date) ______________________
- Indicated temperatures supply air ________ mixed air ________ return air ________ outdoor air ________
- Actual temperatures supply air ________ mixed air ________ return air ________ outdoor air ________

Coils

- Heating fluid discharge temperature ________ ΔT ______ cooling fluid discharge temperature ________ ΔT ______
- Controls (describe) ______________________
- Condition (note date) ______________________

Humidifier

- Type ______________________ If biocide is used, note type ______________________
- Condition (no overflow, drains trapped, all nozzles working?) ______________________
- No slime, visible growth, or mineral deposits? ______________________

HVAC Checklist - Short Form

Building Name: ______________________ Address: ______________________

Completed by: ______________________ Date: __________ File Number: __________

DISTRIBUTION SYSTEM

Zone/ Room	System Type	Supply Air		Return Air		Power Exhaust		
		ducted/ unducted	cfm	ducted/ unducted	cfm	cfm	control	serves (e.g. toilet)

Condition of distribution system and terminal equipment (note locations of problems)

- Adequate access for maintenance? ______________________
- Ducts and coils clean and obstructed? ______________________
- Air paths unobstructed? supply ______ return ______ transfer ______ exhaust ______ make-up ______
- Note locations of blocked air paths, diffusers, or grilles ______________________
- Any unintentional openings into plenums? ______________________
- Controls operating properly? ______________________
- Air volume correct? ______________________
- Drain pans clean? Any visible growth or odors? ______________________

Filters

Location	Type/Rating	Size	Date Last Changed	Condition (give date)

HVAC Checklist - Short Form

Building Name: ______________________ Address: ______________________

Completed by: ______________________ Date: __________ File Number: __________

OCCUPIED SPACE

Thermostat types ______________________

Zone/ Room	Thermostat Location	What Does Thermostat Control? (e.g., radiator, AHU-3)	Setpoints		Measured Temperature	Day/ Time
			Summer	Winter		

Humidistat/Dehumidistat types ______________________

Zone/ Room	Humidistat/ Dehumidistat Location	What Does It Control?	Setpoints (%RH)	Measured Temperature	Day/ Time

- Potential problems (note location) ______________________
- Thermal comfort or air circulation problems (drafts, obstructed airflow, stagnant air, overcrowding, poor thermostat location)

____

____

- Malfunctioning equipment ______________________
- Major sources of odors or contaminants (e.g., poor sanitation, incompatible uses of space)

____

Pollutant and Source Inventory

Building Name: ______________________ Address: ______________________

Completed by: ______________________ Date: __________ File Number: ______________

Using the list of potential source categories below, record any indications of contamination or suspected pollutants that may require further investigation or treatment. Sources of contamination may be constant or intermittent or may be linked to single, unrepeated events. For intermittent sources, try to indicate the time of peak activity or contaminant production, including correlations with weather (e.g., wind direction).

Sections 2, 4 and 6 discuss pollutant sources. Appendix A provides guidance on common measurements.

Source Category	Checked	Needs Attention	Location	Comments
SOURCES OUTSIDE BUILDING				
Contaminated Outdoor Air				
Pollen, dust				
Industrial contaminants				
General vehicular contaminants				
Emissions from Nearby Sources				
Vehicle exhaust (parking areas, loading docks, roads)				
Dumpsters				
Re-entrained exhaust				
Debris near outside air intake				
Soil Gas				
Radon				
Leaking underground tanks				
Sewage smells				
Pesticides				

Pollutant and Source Inventory

Building Name: ______________________ Address: ______________________

Completed by: ______________________ Date: __________ File Number: __________

Using the list of potential source categories below, record any indications of contamination or suspected pollutants that may require further investigation or treatment. Sources of contamination may be constant or intermittent or may be linked to single, unrepeated events. For intermittent sources, try to indicate the time of peak activity or contaminant production, including correlations with weather (e.g., wind direction).

Source Category	Checked	Needs Attention	Location	Comments
Moisture or Standing Water				
Rooftop				
Crawlspace				
EQUIPMENT				
HVAC System Equipment				
Combustion gases				
Dust, dirt, or microbial growth in ducts				
Microbial growth in drip pans, chillers, humidifiers				
Leaks of treated boiler water				
Non HVAC System Equipment				
Office Equipment				
Supplies for Equipment				
Laboratory Equipment				

Pollutant and Source Inventory

Building Name: ______________________ Address: ______________________

Completed by: ______________________ Date: __________ File Number: __________

Using the list of potential source categories below, record any indications of contamination or suspected pollutants that may require further investigation or treatment. Sources of contamination may be constant or intermittent or may be linked to single, unrepeated events. For intermittent sources, try to indicate the time of peak activity or contaminant production, including correlations with weather (e.g., wind direction).

Source Category	Checked	Needs Attention	Location	Comments
HUMAN ACTIVITIES				
Personal Activities				
Smoking				
Cosmetics (odors)				
Housekeeping Activities				
Cleaning materials				
Cleaning procedures (e.g., dust from sweeping, vacuuming)				
Stored supplies				
Stored refuse				
Maintenance Activities				
Use of materials with volatile compounds (e.g., paint, caulk, adhesives)				
Stored supplies with volatile compounds				
Use of pesticides				

Pollutant and Source Inventory

Building Name: ______________________ Address: ______________________

Completed by: ______________________ Date: ____________ File Number: ______________

Using the list of potential source categories below, record any indications of contamination or suspected pollutants that may require further investigation or treatment. Sources of contamination may be constant or intermittent or may be linked to single, unrepeated events. For intermittent sources, try to indicate the time of peak activity or contaminant production, including correlations with weather (e.g., wind direction).

Source Category	Checked	Needs Attention	Location	Comments
BUILDING COMPONENTS FURNISHINGS				
Locations Associated with Dust or Fibers				
Dust-catching area (e.g., open shelving)				
Deteriorated furnishings				
Asbestos-containing materials				
Unsanitary Conditions/Water Damage				
Microbial growth in or on soiled or water-damaged furnishings				

Pollutant and Source Inventory

Building Name: ______________________ Address: ______________________

Completed by: ______________________ Date: ____________ File Number: ______________

Using the list of potential source categories below, record any indications of contamination or suspected pollutants that may require further investigation or treatment. Sources of contamination may be constant or intermittent or may be linked to single, unrepeated events. For intermittent sources, try to indicate the time of peak activity or contaminant production, including correlations with weather (e.g., wind direction).

Source Category	Checked	Needs Attention	Location	Comments
Chemicals Released From Building Components or Furnishings				
Volatile compounds				
OTHER SOURCES				
Accidental Events				
Spills (e.g., water, chemicals, beverages)				
Water leaks or flooding				
Fire damage				

Pollutant and Source Inventory

Building Name: ______________________ Address: ______________________

Completed by: ______________________ Date: __________ File Number: ______________

Using the list of potential source categories below, record any indications of contamination or suspected pollutants that may require further investigation or treatment. Sources of contamination may be constant or intermittent or may be linked to single, unrepeated events. For intermittent sources, try to indicate the time of peak activity or contaminant production, including correlations with weather (e.g., wind direction).

Source Category	Checked	Needs Attention	Location	Comments
Special Use/Mixed Use Areas				
Smoking lounges				
Food preparation areas				
Underground or attached parking garages				
Laboratories				
Print shops, art rooms				
Exercise rooms				
Beauty salons				
Redecorating/Repair/Remodeling				
Emissions from new furnishings				
Dust, fibers from demolition				
Odors, volatile compounds				

Hypothesis Form

Building Name: ______________________ File Number: ______________

Address: ______________________

Completed by: ______________________

Complaint Area (may be revised as the investigation progresses):

____

____

____

____

Complaints (e.g., summarize patterns of timing, location, number of people affected):

____

____

HVAC: Does the ventilation system appear to provide adequate outdoor air, efficiently distributed to meet occupant needs in the complaint area? If not, what problems do you see?

____

____

Is there any apparent pattern connecting the location and timing of complaints with the HVAC system layout, condition or operating schedule?

____

____

Pathways: What pathways and driving forces connect the complaint area to locations of potential sources?

____

Are the flows opposite to those intended in the design? ______________________

Sources: What potential sources have been identified in the complaint area or in locations associated with the complaint area (connected by pathways)?

____

Is the pattern of complaints consistent with any of these sources? ______________________

Hypothesis Form

Hypothesis: Using the information you have gathered, what is your best explanation for the problem?

Hypothesis testing: How can this hypothesis be tested?

If measurements have been taken, are the measurement results consistent with this hypothesis?

Results of Hypothesis Testing:

Additional Information Needed:

Index

A

E

F

G

H

I

K

L

M

N

O

P

R

S

T

V

W

X

Z